Responsible Editor : Chen Wenfang

Published by Science Press
16 Donghuangchenggen North Street
Beijing 100717, China

Distribution right throughout the world, excluding
China, Hong Kong and Macao, is granted to
CRC Press, Inc.
2000 Corporate Blvd., N. W.
Boca Raton, FL 33431
U. S. A.

Printed in Hong Kong

Library of Congress Cataloging-in-Publication Data
Zhou, Lixing
Theory and numerical modeling of turbulent gas-particle flows and
combustion / by Zhou Lixing
239p. 20.2×12.5cm.
Includes bibliographical references and index.
ISBN 0-8493-7721-8
1. Turbulence—Mathematical models. 2. Combustion—Mathematical
models. 3. Laminar flow—Mathematical models. I. Title.
TA357.5.T87Z46 1993 92-38051
620.1' 064-dc20 CIP
ISBN 7-03-003493-7/TB·110, Science Press, China
ISBN 0-8493-7721-8, CRC Press, Inc., USA

THEORY AND NUMERICAL MODELING OF TURBULENT GAS-PARTICLE FLOWS AND COMBUSTION

Zhou Lixing

Department of Engineering Mechanics, Tsinghua University

SCIENCE PRESS
Beijing, New York

CRC PRESS, Inc.
Florida, U. S. A.

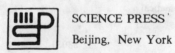

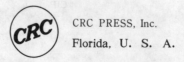

CONTENTS

PART Ⅱ MODELING OF TURBULENT GAS-PARTICLE FLOWS AND COMBUSTION

PREFACE

This volume was written as a textbook for master and doctoral graduate students on the basis of the teaching experience of the author in his course "Combustion Theory" for nearly 30 years and his course "Numerical Modeling of Turbulent Gas-particle Flows and Combustion" for more than ten years given in the Department of Engineering Mechanics, Tsinghua University, and also the lectures given at more than ten universities of West Germany, the United States and the Soviet Union during his visit to these countries.

The main goal of this book is to give the readers, first, a clear idea of the theoretical treatment of basic combustion phenomena in laminar flows by classical analytical solutions, and second, the numerical modeling theory, methods and results of turbulent single-phase flows and gas-particle flows and combustion. It was intended to combine the classical theory, such as ignition and flame propagation with the continuum mechanics, basic conservation equations of reacting fluids and boundary-layer theory, the classical analytic studies with recently developed numerical studies. Although many existing books deal with combustion theory and numerical modeling of combustion processes, but none of them serve the purpose of combining these two fields together and focussing attention on such complex a problem as turbulent gas-particle flows and combustion which takes place in many practical engineering facilities, such as fluid machines, heat exchangers, combustors and furnaces.

The first part of this book gives fundamentals of combustion and gas-particle flows. including basic conservation equations of multicomponent reacting fluids, ignition and extinction, laminar premixed combustion, droplet and particle combustion, combustion in laminar boundary layers and fundamentals of gas-particle flows. In this part the physical phenomena are described by rather simple mathematics. Whenever possible, the analytical results are compared with experimental results. However, it is not intended to include the detailed experimental data and experimental methods.

The second part of this book describes numerical modeling of turbulent gas-particle flows and combustion. Here the emphasis is the phenomena directly related to fluid flow, so only the modeling of single-phase turbulent flows, turbulent two-phase flows (gas-particle flows) and turbulent single-phase and two-phase combustion are discussed. Other problems, such as the modeling of radiative heat transfer and pollutant formation, general solution procedure of fluid flow, can be found in many existing books. In this part, not only the the model formulation and equations and their solution procedure are presented, but also the modeling results, the application and assessment of these models are discussed.

This book is written to be applicable for the faculty and students, research associates and assistants majoring in mechanical engineering (thermal science and fluid mechanics), astronautical and aeronautical, metallurgical, chemical and nuclear engineering and other related fields.

Many important research results obtained in the Two-phase Flow and Combustion Laboratory, Department of Engineering Mechanics, Tsinghua University, are included in this book.

Special thanks to Professors H. Ning, G. J. Zhou and C. K. Wu for their valuable comments and suggestions. Thanks also to my colleagues and students, Prof. W. Y. Lin, Prof. R. X. Li, Dr. X. Q. Huang, Dr. J. Zhang and Dr. T. Hong for their contribution in numerical modeling of turbulent gas-particle flows and combustion under the author's direction, which enriched the contents of this book.

The author will much appreciate, if any reader would give his/her own comments or suggestions.

Lixing Zhou

Tsinghua Yuan, Beijing

August, 1992

INTRODUCTION

1. TURBULENT GAS-PARTICLE FLOWS

Most of fluid flows in engineering facilities, such as flows in hydraulic channels, gas pipes, heat exchangers, fluid machines, chemical reactors, combustors and furnaces, are turbulent flows due to the size of the geometric system, the velocity range and the presence of various barriers or expansions leading to flow separation, and frequently they are complex turbulent flows, such as recirculating flows, swirling flows, buoyant flows. Moreover, there are seldom pure liquid or gas flows. Most of practical fluid flows are laden with particles, droplets or gas bubbles.

Gas containing a large amount of solid particles or liquid droplets is called gas-particle flows, or two-phase flows, or multiphase flows, or suspension flows. The word "Two-phase" or "Three-phase" in its thermodynamic sense denotes the gas-solid, gas-liquid, liquid-solid, or gas-liquid-solid system, however, in the dynamic sense, particles or droplets of different sizes, having different velocities and different temperatures could be different phases. Hence we use the terminology "Multiphase". There are a variety of gas-particle flows in natural environment and engineering, such as cloud and fog (rain droplets), dusty air, sandy rivers, pneumatic/hydraulic conveying, dust separation and collection, spray coating, drying and cooling, spray/pulverized-coal combustion, plasma chemistry, fluidized bed, flows in gun barrels, solid-rocket exaust, steam-droplet flows in turbines and gas-fibre flows.

In many cases we are dealing with turbulent gas-particle flows with heat and mass transfer, chemical reactions (exothermic or endothermic), and even electrostatic effects (for gas flows laden with fine particles through metal or plastic pipes) or magnetic effects (for gas flows in plasma torch and MHD generators). The object to be discussed in this book is turbulent gas-particle flows with heat and mass transfer and chemical reactions.

2. COMBUSTION

The word "Combustion" denotes chemical reactions with high heat release and light radiation. These reactions, in the first place are oxidation of solid fuel (nonmetals or metals), liquid fuel or gaseous fuel, but chloridation, fluoridation, nitridation and dissociation reactions can also be considered as combustion, if they have high heating effect. As a matter of fact, combustion is a complex process of gas flow or gas-particle flow with heat and mass transfer and chemical reactions. Conventionally, combustion may be categorized into diffusion combustion and premixed combustion according to unmixed or premixed fuel and oxidizer prior to combustion. There are two limiting cases of reacting flows depending on so-called First Damköhler number D_I which is defined as the ratio of flow time (residence time) to reaction time. If D_I is much smaller than unity, or reaction is very slow and negligible compared with the flow process, the reacting flow will be a frozen flow. If D_I is much greater than unity, or reaction is very fast and near to the equilibrium state, then the reacting flow will be an equilibrium flow. For most of combusting flows D_I is of the order of unity. There are two limiting cases of combustion depending on the so-called Second Damköhler number D_{II} which is defined as the ratio of diffusion time to reaction time. If D_{II} is much greater than unity, or diffusion is very slow compared with reaction, there will be a diffusion-controlled combustion. If D_{II} is much smaller than unity, or reaction is very slow compared with diffusion, there will be a kinetics-controlled combustion. For example, the gaseous diffusion flame is certainly diffusion combustion, however, the gaseous premixed flame is diffusion-kinetic combustion. The solid-surface combustion may be either diffusion combustion (under high pressure, high temperature or low relative velocity) or kinetic combustion (under low pressure, low temperature or high relative velocity).

The word "flame" in its scientific sense can be defined as a zone with a sharp change of temperature and concentration. One of important flame properties is that the flame zone can propagate automatically. The flame propagation velocity is the flame velocity relative to the cold fresh combustible mixture which is equal to the difference between the flame displacement velocity and the mixture flow velocity, as that observed in flame propagation in a long tube with one closed end. For stationary flames as those in Bunsen burners or flat burners, the displacement velocity is zero and the propagation velocity should be equal to the flow velocity. Two possible regimes of flame propagation were observed in experiments: deflagration with flame velocity of 0.2—1 m/s and detonation with flame velocity near 3000 m/s. Other flame properties are carbon formation, flame radiation (due to carbon or soot particles), and ionization with concentration up to 10^{12} ions/cm^3 in laminar premixed flames.

The adiabatic flame temperature is an ideal final temperature of a premixed adiabatic combustion system, when either fuel (for lean mixture with coefficient of excess air greater than unity) or oxygen (for rich mixture with coefficient of excess air less than unity) is totally burned. The coefficient of excess air is defined as the ratio of the actual air/fuel ratio to its stoichiometric ratio. The adiabatic flame temperature will reach its maximum value at the stoichiometric air/fuel ratio.

3. COMBUSTION THEORY

Although fire was discovered and utilized by human beings almost a million years ago, and electricity was discovered and utilized only three hundreds years ago, but the theory of electricity and magnetism had been already established in the 19th century, and combustion theory developed much more slowly. Little was known about combustion till the middle of the 18th century, this is because combustion is a very complex physical and chemical process. Before Lavoisier it was thought that anything containing "combustible element" can burn. In the period from 1756 to 1777 Lavoisier and Lomonosov first pointed out that combustion is an oxidation reaction, on the basis of their experimental observation. In the 19th century owing to the development of thermochemistry and chemical thermodynamics, the combustion equipment had been taken as a thermodynamic system, and its initial state and final state were considered, the heat of combustion, the equilibrium concentration and adiabatic combustion temperature were predicted. In the early years of the 20th century Semenov, Lewis and Von Elbe developed the combustion kinetics which indicates that combustion is a chain reaction which can be accelerated by active centers. Perhaps, the quantitative or mathematical theory of combustion based on the concept of interaction between reaction and heat and mass transfer, including ignition and extinction, flame propagation, turbulent combustion was first developed by Semenov, Zeldovich, Frank-Kamenetskii, Damköhler, Karlovitz, Shelkin, Shetinkov, Summerfield and others in the period from the thirties to forties of this century. Subsequently, theory of combustion of droplets and carbon particles was developed by Varshavskii, Godsave, Penner, Spalding, Paleev, Predvoditelev, Khitrin, Hottel, Pomerantsev and others in the years from the 1950's to 1960's. From the end of the 1950's to the biginning of the 1960's Von Karman and H. S. Tsien introduced a terminology called "Aerothermochemistry" or "Dynamics of reacting fluids" which is a combination of combustion theory with continuum mechanics. This trend was subsequently developed by Penner, Williams, Marble, Toong, Cheng and the author of this book. Starting from the biginning of the 1970's the computer simulation or numerical modeling of combustion was developed on the basis of the achievements in computational fluid dynamics and heat transfer. In this field contribution was made by Spalding, Launder, Patankar, Crowe, Pratt, Swithenbank, Smoot and others. In China the numerical modeling of turbulent gas-particle flows and combustion has been systematically studied for more than 10 years by the author of this book and his colleagues.

It should be pointed out that the development of laser diagnostics of combustion opened a new era in the history of the development of combustion theory. It allows to non-intrusively measure the velocity, temperature and concentration of gas species in flames and to compare these results with numerical predictions. Obviously, using the basic theory, numerical modeling and laser diagnostics together, we can have a better understanding of the combustion phenomena. It can be expected that the combustion theory will develop from a pure descriptive or semiempirical science into an exact quantitative science which can better serve the purpose of engineering design.

4. NUMERICAL MODELING OF TURBULENT GAS-PARTICLE FLOWS AND COMBUSTION

In turbulent gas-particle flows with or without combustion the mass, momentum and energy are transported inside the phases, and there are mass, momentum and energy interactions between phases. Besides, owing to the complex geometric configuration the flows are not symmetrical, but three-dimensional, such as flows in cylindrical combustors with secondary-air inlets or tangential inlets. So, the object to be studied in engineering problems is three-dimensional turbulent two-phase flows and combustion.

The above discussed features of turbulent gas-particle flows and combustion indicate that it is impossible to develop a rigorous scientific design method by using classical analytic solution. The only way to solve this problem is computer simulation or numerical modeling based on modern computational fluid mechanics and large-capacity high-speed digital computers, that is, to establish basic conservation equations under various conditions, to close them, and to solve these nonlinear simultaneous partial differential equations directly by numerical method without simplifications. Indeed, this method has been successfully developed in the last twenty years since the end of 1960's, and it has already been used in engineering design. Nowadays it is possible to predict such complex flows as three-dimensional turbulent recirculating swirling reacting two-phase flows in large scale furnaces and the predicted results are plausible as compared with experiments. Thoroughly incorporating the multi-dimensional modeling of turbulent gas-particle flows and combustion into engineering design will certainly be realized not in the far future.

Before the sixties of this century, the method of optimization or scale-up design was mainly empirical, based on experiments, intuition or semiempirical calculations by using similarity laws and model tests. For example, the presently used thermodynamic calculation for boiler furnaces or the engineering calculations in water intake and discharge design are of the last kind. Although these empirical methods are useful up to now and even in the future because of their simplicity and feasibility in qualitative evaluation, but due to their strong empirical nature and narrow-range applicability, the predictions cannot be applied in widely varied conditions. The numerical modeling of gas-particle flows and combustion is a result of recent development of modern digital computers, numerical analysis, computational fluid dynamics, heat transfer and combustion since the 1970's. It is a powerful tool for engineering design, which allows to predict the real processes in fluid machines, heat exchangers, combustors, furnaces, plasma generators, rivers and atmospheric environment, and to obtain the detailed data for engineering design. Therefore. it can reduce the experiments to minimum, saving significantly the expenses, time and research efforts. Without any exaggeration it can be considered as a revolution in engineering design.

There has been different understanding of the word "modelling" or "simulation". Thirty years ago, even still ealier, modeling was considered as an experimental method in model-apparatus according to certain similarity laws, or even visualization experiments of water or gas flows which has only qualitative similarity.

Undoubtedly, these methods are useful not only in the past and present days, but also in the future. However, there are serious shortcomings of such methods. For example, at least more than ten similarity parameters are needed for the similarity of turbulent reacting two-phase flows in models with their prototypes. This is the conclusion given by the present author 25 years ago from deriving the basic conservation equations of turbulent reacting two-phase flows. Actually it is impossible to keep the equality of all these parameters. Hence we have to retain only those parameters which are assumed to be predominant, and neglect the parameters assumed to be less important, and the model can have at most only qualitative similarity, and in many situations it completely losses the similarity.

The method of mathematical modeling was developed a long time ago. Earlier developed models are zero-dimensional models, one-dimensional models and so-called "modular modeling" (0D + 1D). The zero-dimensional model is a thermodynamic model which does not consider fluid mechanics. Uniform temperature and concentration distributions in space are assumed. It is used in internal combustion engines, and in chemical engineering it is called WSR (well-stirred reactor) model. These models can give only the final state based on the initial state, such as the total heat transfer and combustion performance as functions of given conditions, for example, the engine output as a function of the crank angle, but it cannot predict the detailed field distributions of velocity, temperature and concentration. Therefore, it can give only qualitative information, or semiempirical quantitative data based largely on experiments. The one-dimensional models can predict variation of different properties along the axial direction by assuming uniform distributions in cross sections. These models allow to give more information than that given by the zero-dimensional models, but still the flow, heat transfer and combustion processes are simplfied. These models are sometimes called "plug flow" models. Further developed semiempirical models are "modular models" in which the whole flow field is divided into several PFR (plug-flow reactor) parts in terms of forward flows and WSR (well-stirred reactor) parts in terms of recirculating flows. These models are actually zero-D plus 1-D models. Obviously, neither one-dimensional models nor modular models can give quantitative information of the whole flow field. For some engineering processes near to very simple flows, such as boundary layers or jets, sometimes the approximate integral models by using semiempirical profiles are used. This last kind of model can be considered as "quasi-two-dimensional" or "one and half dimensional".

What we would like to discuss in this book are multi-dimensional, two-dimensional or three-dimensional differential models, which is called numerical modeling, based on the principles of computational fluid mechanics, heat transfer and combustion, that is, to directly solve the non-linear simultaneous partial differential conservation equations of mass, momentum, energy, species and other quantities by a numerical procedure without any simplification. These models can give the detailed information of flow, heat transfer and combustion characteristics, that is the temporal and spatial distributions of different properties of gas and particle phases. Hence, they can serve the purpose of optimization and scale-up design. As a matter of fact, the numerical modeling has already become, and will be a powerful tool in engineering design.

From the experience of exploration and investigation for many years, the procedures of numerical modeling can be stated as follows:

(1) Establishing basic conservation equations

The first step in numerical modeling is to establish the basic conservation equations based on the principles of fluid mechanics, heat transfer, combustion, plasma dynamics, and other sciences, that is, the continuity equation, momentum equation, energy equation, species equation, turbulent kinetic energy equation and other equations. These equations constitute a system of non-linear simultaneous partial differential equations which cannot be solved by the classical analytical method, and can only be solved by numerical methods. There is little argument about the basic equations of laminar single-phase flows. However, for turbulent flows, in particular turbulent two-phase flows, there are different forms of basic conservation equations based on different modeling theories. So, how to establish the basic equations is an important part of the modeling theory in the first place.

(2) Specifying boundary conditions

The second step is to determine the computed domain and to specify the boundary conditions at the inlet, exit, axis (or symmetrical plane), walls and free surfaces (if they exist) of the domain according to the given geometrical configuration and sizes. For turbulent two-phase flows it is necessary to specify the boundary conditions of time-averaged and RMS values of each variable of gas and particle phases respectively. It is important to properly specify these boundary conditions, however, this is not an easy task. For example, we need to know, how to give the values of turbulent kinetic energy and Reynolds stress components at the inlet, walls and free surface, the particle velocity and concentration at the walls, and the conditions at the exit with limited length for strongly swirling or recirculating flows.

(3) Selecting models or closure methods

Usually, the basic equations of turbulent two-phase flows are not a closed system. The correlation terms of stresses, heat fluxes, and reaction source terms for gas and particle phases are unknown. To solve these problems, it is necessary to develop modeling theory based on physical concepts or some hypotheses. These models include turbulence models of gas and particle phases, global models of particle phases, gas-phase reaction models in turbulent flows, radiative heat transfer models, and pollutant formation models. The attention of this book is focused on the first three models.

(4) Establishing finite difference equations

In numerical methods the partial differential equations must be discretized. In most cases the finite difference method is used in the modeling of turbulent two-phase flows. The finite difference equations or FDE's are obtained either by Taylor's series expansion, or by integration in the control volume, by using certain difference scheme, such as central difference scheme, upwind difference scheme and power-law difference scheme. Other discretization method, such as finite element method and finite analysis method can also be used.

(5) Selecting solution procedure

There are different solution procedures for single-phase flows. For example, the methods developed by Patankar and Spalding are GENMIX marching integration method for parabolic flows (boundary layers, jets, pipe and nozzle flows), vorticity-stream function algorithm and SIMPLE (pressure-velocity correction) algorithm for elliptic flows (recirculating flows). In SIMPLE method the TDMA, line-by-line iterations and under-relaxations are usually adopted. Special solution procedures have been proposed for turbulent reacting two-phase flows. Among them the typical algorithms are PSIC, IPSA, GEMCHIP, PCGC-2 and LEAGAP.

(6) Developing solution technique

There is a need to develop the special detailed numerical method or solution technique for different problems, accounting for their special features. For example. besides the basic numerical procedure, we need to make a reasonable and economical grid arrangement, sometimes even a grid system floating with time or space location, to treat the irregular boundaries, to choose the underrelaxation factors and iteration or sweeping methods. For two-phase flows the iterations between gas and particle phases, between turbulence and chemical kinetics must be done by choosing an appropriate method. Besides, the problems such as how to correct the mass continuity of particle phase, how to calculate the particle trajectories, have to be solved.

(7) Developing the computer code

The next step is to develop the computer code based on the above obtained finite difference equations and developed numerical procedures. The code must include the main program and subroutines, and be flexible and convenient for utilization and possible further modifications. A user's manual is always needed. The computer code for turbulent reacting two-phase flows is much more complex than that for single-phase non-reacting flows.

(8) Debugging the computer code

The so-called "debugging the code" is to eliminate the accidental and systematic errors in the computer code, including the errors in the algorithm, in order to properly operate the code and to give the convergent and reasonable results. Frequently, the subroutines of gas flow field are the most important parts in the comprehensive code. Debugging the code is a cumbersome task, it needs careful and hard work. It can be successful only after computational practice.

(9) Comparing predictions with experiments

After the code has been debugged, the computation can be carried out for different flow parameters and geometrical configurations. To assess the feasibility, the merits and shortcomings of the models and numerical procedures, predictions must be compared with measurements, in particular accurate laser measurements. For turbulent two-phase flows comparison has to be made not only for the gas phase, but also for the particle phases.

(10) Refining models and numerical procedures

After assessing the models and numerical procedures they can be refined or further improved to overcome their shortcomings, or new models and numerical procedures will be proposed, until satisfactory results are obtained. Usually it is a long run to achieve this goal, it needs unshakable research efforts for many years.

The above-described is the whole process of numerical modeling. It can be seen that the numerical modeling is a problem of not only numerical methods and numerical techniques, but it includes the theoretical basis of a series of physical and chemical processes, and also detailed measurements. It is a result of mutual verification among theory, computation and experiment. It can be concluded that the numerical modeling developed only in this way can be considered as sufficiently reliable and useful in practice.

PART I

FUNDAMENTALS OF COMBUSTION AND GAS-PARTICLE FLOWS

CHAPTER 1 BASIC EQUATIONS OF MULTICOMPONENT REACTING FLOWS

1.1 BASIC RELATIONS OF MULTICOMPONENT GASES

It is well known that flows with combustion or reaction are always multicomponent. For example, in gas combustion at least there are one kind of fuel, one kind of oxidizer, combustion products, inert gases, various atoms, free redicals, ions and electrons. The thermodynamic and transport properties of multicomponent reacting fluids are functions not only of pressure and temperature, but, also of species concentration. Furthermore, the diffusion equation must arise in basic equations for describing such kind of flows and there will be mass source/sink or heat source/sink terms in diffusion and energy equations. Therefore, before going to discuss the basic equations it is necessary to understand the basic relations of thermodynamic properties and transport processes for multicomponent reacting fluids.

For multicomponent gases the total mass density should be the sum of mass concentration of each species and the total pressure is the sum of partial pressure of each species, that is:

$$\rho = \sum_s \rho_s \tag{1.1}$$

$$p = \sum_s p_s \tag{1.2}$$

The mass fraction and molar fraction of species "s" are

$$Y_s = \rho_s/\rho \qquad X_s = p_s/p \tag{1.3}$$

The relations between mass density and molar density (molar concentration) of gas mixture and species "s" are

$$\rho = nM \tag{1.4}$$

$$\rho_s = n_s M_s \tag{1.5}$$

Except under very high temperature or very high pressure, for most of combusting or reacting flows the gas mixture and its species can be treated as ideal or perfect gases, so we can write

$$p_s = \rho_s RT/M_s = n_s RT$$
$$p = \rho RT/M = nRT \tag{1.6}$$

and hence we have

$$X_s = p_s/p = n_s/n$$

Owing to

$$n = \sum n_s$$

and

$$\rho = \sum \rho_s = \sum n_s M_s = nM$$

the following relations between the mean molecular weight and molecular weight of each species, the mass fraction and molar fraction can be obtained as:

$$\begin{cases} M = \sum X_s M_s & (1.7) \\ X_s = Y_s M / M_s & (1.8) \\ \sum \ (Y_s M / M_s) \ = 1 & (1.9) \\ M = \ (\sum Y_s / M_s)^{-1} & (1.10) \\ M / M_s = (M_s \sum Y_s / M_s)^{-1} & (1.11) \end{cases}$$

If the molecular weights of gas species are nearly the same ($M_1 \approx M_2 \approx M_s \approx \cdots$) or the molecular weight of one species is much greater than those of other species ($Y_s \gg Y_1, Y_s \gg Y_2, \cdots$), then we have

$$M \approx M_s, \qquad X_s \approx Y_s$$

Now, let us consider the motion and diffusion in multicomponent reacting gases. Generally speaking, there are three kinds of macroscopic velocities in multicomponent gases

$\boldsymbol{v}$—Mixture velocity with reference to laboratory coordinate

$\boldsymbol{v}_s$—s-species velocity with reference to laboratory coordinate

$\boldsymbol{V}_s$—s-species velocity with reference to mixture motion, caused by diffusion drift of molecular random motion

Obviously there should be

$$\boldsymbol{V}_s = \boldsymbol{v}_s - \boldsymbol{v} \tag{1.12}$$

Corresponding to the above-mentioned three kinds of linear velocities, there are three kinds of mass fluxes

$\rho \boldsymbol{v} = \boldsymbol{g}$—Total mass flux of the gas mixture

$\rho_s \boldsymbol{v}_s = \boldsymbol{g}_s$—Species mass flux

$\rho_s \boldsymbol{V}_s = \boldsymbol{J}_s$—Species diffusion flux

The species mass flux should be equal to the diffusion flux plus the mass flux of this species carried by the mixture motion, that is

$$\boldsymbol{g}_s = \rho_s \boldsymbol{v}_s = \boldsymbol{J}_s + Y_s \rho \boldsymbol{v} = \rho_s \boldsymbol{V}_s + \rho_s \boldsymbol{v} \tag{1.13}$$

The mixture mass flux is the sum of mass fluxes of all species

$$\boldsymbol{g} = \rho \boldsymbol{v} = \sum \boldsymbol{g}_s = \sum \rho_s \boldsymbol{v}_s = \sum \rho_s \boldsymbol{V}_s + \boldsymbol{v} \sum \rho_s = \sum \rho_s \boldsymbol{V}_s + \rho \boldsymbol{v}$$

and here we have

$$\sum \rho_s \boldsymbol{V}_s = \sum \boldsymbol{J}_s = 0 \tag{1.14}$$

Eqs. (1.13) and (1.14) point out that the species mass flux is not its diffusion flux and the sum of diffusion fluxes is zero, or diffusion has no effect on mixture motion. However, the sum of linear diffusion velocities is not zero

$$\sum \boldsymbol{V}_s \neq 0$$

Next, we like to discuss the basic laws of transport processes. For species diffusion (molecular diffusion) in a binary (two-component) mixture caused by concentration gradient, the phenomenological diffusion law or the Fick's Law can be expressed as

$$J_{1j} = - \rho D_{12} \frac{\partial Y_1}{\partial x_j} \tag{1.15}$$

For a multicomponent mixture the Fick's Law can still have the similar form

$$J_{sj} = - \rho D_s \frac{\partial Y_s}{\partial x_j} \tag{1.16}$$

but the molecular diffusivity is a function of species concentration

$$D_s = D_s (Y_1, Y_2, Y_3, \cdots, Y_s, \cdots)$$

In many cases it is likely to introduce further simplifications, such as equality of diffusivity of each species

$$D_1 = D_2 = \cdots = D_s = D$$

For heat conduction in multicomponent reacting gases the Fourier Law must be modified to account for the enthalpy difference between those entering and leaving the volume element carried by the diffusion fluxes. In this case the total heat flux is

$$q_j = - \lambda \frac{\partial T}{\partial x_j} + \sum \rho_s V_{sj} h_s \tag{1.17}$$

Where h_s is the enthalpy of " s " species, including the chemical enthalpy, that is

$$h_s = h_{0s} + \int_{T_0}^{T} c_{ps} \, dT \tag{1.18}$$

The mixture enthalpy will be

$$h = \sum Y_s h_s = \sum Y_s h_{0s} + \int_{T_0}^{T} \sum Y_s c_{ps} \, dT = h_0 + \int_{T_0}^{T} c_p \, dT \tag{1.19}$$

where h_{0s} is enthalpy of formation of " s " species and c_{ps} is the specific heat of " s " species which is also a function of p and T. It should be noticed that: (1) both the species enthalpy and the mixture enthalpy include two parts, the physical (thermal) enthalpy and the chemical enthalpy; (2) the species enthalpy of formation is a constant, but the mixture enthalpy of formation is a function of species concentration; (3) the species specific heat is independent of species concentration, but the mixture specific heat depends on species concentration; and (4) there may be three kinds of mixture enthalpy to be used in multicomponent reacting flows, thermal enthalpy, thermal enthalpy plus chemical enthalpy, and stagnant enthalpy = thermal enthalpy + chemical enthalpy + kinetic energy.

For momentum transfer the relationship between the stress and the deformation rate can still be described by the generalized Newton-Stokes Law used in single-component non-reacting flows

$$p_{ij} = 2\mu \dot{S}_{ij} - p\delta_{ij} - \frac{2}{3}\mu \left(\frac{\partial v_j}{\partial x_j} \right) \delta_{ij} \tag{1.20}$$

where

$$\dot{S}_{ij} = \frac{1}{2} \left(\frac{\partial v_i}{\partial x_j} + \frac{\partial v_j}{\partial x_i} \right)$$

However here the molecular viscosity is a function not only of temperature and pressure, but also of species concentration.

According to the first-order approximation of the kinetic theory, the molecular diffusivity, viscosity and thermoconductivity for a binary mixture or single-component gases can be given as:

$$[D_{12}]_1 = \frac{3}{8n\sigma_{12}^2}\left[\frac{kT(M_1+M_2)}{2\pi M_1 M_2}\right]^{\frac{1}{2}} \qquad (1.21)$$

$$[\mu]_1 = 5\sqrt{kMT}/(16\pi^{\frac{1}{2}}\sigma^2) \qquad (1.22)$$

$$[\lambda]_1 = \frac{5}{2}[\mu]_1 c_v = \frac{75}{64\sigma^2}\left(\frac{k^3T}{\pi M}\right)^{\frac{1}{2}} \qquad (1.23)$$

where k is the Boltzmann constant and σ is the collisional cross section of molecules.

1.2 BASIC LAWS OF CHEMICAL KINETICS

The stoichiometric relation of a chemical reaction can be expressed by

$$\sum \nu_s A_s \rightarrow \sum \nu_s' A_s'$$

where A_s, A_s' are chemical symbols for reactants and products, and ν_s and ν_s' are corresponding stoichiometric coefficients. The reaction rate of " s " species which is defined as the consumed or produced mass per unit volume per unit time can be expressed by

$$w_s = -\left(\frac{d\rho_s}{dt}\right)_{\text{chem}}$$

Obviously there should be

$$\frac{w_1}{\nu_1} = \frac{w_2}{\nu_2} = \cdots = -\frac{w_1'}{\nu_1'} = -\frac{w_2'}{\nu_2'} = \cdots$$

The law of mass action gives

$$w_s = k_s \prod_{s=1}^{z} C_s^{\nu_s}$$

where C_s may be mass concentration ρ_s or molar concentration $X_s M_s$, k_s is the rate coefficient, z is total number of species, $\nu = \nu_1 + \nu_2 + \cdots = \sum \nu_s$ is the apparent reaction order. For most reactions the actual order is not apparent order, that is

$$w_s = k_s \prod_{s=1}^{z} C_s^{m_s} \qquad (1.24)$$

where $m_s \neq \nu_s$ and $m = \sum m_s \neq \nu = \sum \nu_s$, m is the actual reaction order. If there are several simultaneous element reactions and the r-th reaction rate is

$$\sum_{s=1}^{z} \nu_{sr} A_s \longrightarrow \sum_{s=1}^{z'} \nu_{sr}' A_s'$$

then the total reaction rate will be

$$w_s = \sum_r w_{sr} = \sum_r k_{sr} \prod_{s=1}^{z} C_s^{m_{sr}} \qquad (1.25)$$

For a reversible reaction

$$\sum_s \nu_s A_s \rightleftharpoons \sum \nu_s' A_s'$$

the rate of forward reaction is

$$w_{s+} = k_s \prod_s C_s^{m_s}$$

and the rate of backward reaction is

$$w_{s-} = k_s' \prod_s C_s'^{\,m_s}$$

so the net reaction rate is

$$w_s = w_{s+} - w_{s-} = k_s \prod_s C_s^{m_s} - k_s' \prod_s C_s'^{\,m_s'} \qquad (1.26)$$

The total reaction rate of several simultaneous reversible reactions is

$$w_s = \sum_r w_{sr} = \sum_r (k_{sr}\prod_s C_s^{m_{sr}} - k'_{sr}\prod_s C'_s{}^{m'_{sr}}) \tag{1.27}$$

The rate coefficient k_s is a highly nonlinear function of temperature and it can be expressed by the Arrhenius Law as

$$k_s = k_{0s} \exp(-E/RT) \tag{1.28}$$

where E is the activation energy whose physical significance is an energy barrier for the reaction to take place. R is the universal gas constant, and $k_{0s} = BT^{0.5}$, B is called preexponential factor or frequency factor. In treating the combustion problems it is used to consider the reaction as an equivalent one-step reaction, hence the global reaction rate can be expressed by

$$w_s = k_{0s}\rho^{\sum m_s} \exp(-\frac{E}{RT}) \prod_s Y_s^{m_s} \tag{1.29}$$

where k_{0s}, E, m_s and $\sum m_s$ are pure empirical constants.

1.3 BASIC CONSERVATION EQUATIONS

1.3.1 The Reynolds Transport Theorem

In describing the fluid motion or the change of any property in the flow field, either the Eulerian treatment by considering the change from point to point in space, or the Lagrangian treatment by tracking the change in a moving fluid element can be used. For a fluid element the Lagrangian treatment is related to the Eulerian treatment by

$$\frac{d\varphi}{dt} = \frac{\partial\varphi}{\partial t} + v_j \frac{\partial\varphi}{\partial x_j}$$

A similar relation called Reynolds Transport Theorem can be derived for a control volume. Considering the integral of a generalized fluid property φ over a control volume V moving in the Lagrangian coordinate

$$\Phi(t) = \int_V \varphi(x_i, t)\,\delta V$$

and the change of Φ due to the fluid motion by tracking the trajectory of this control volume

$$\frac{D\Phi}{Dt} = \frac{D}{Dt}\int_V \varphi\delta V = \int_V \frac{D}{Dt}(\varphi\delta V) = \int_V \frac{D\varphi}{Dt}\delta V + \int_V \varphi \frac{D\delta V}{\delta V Dt}\delta V$$

$$= \int_V \left(\frac{D\varphi}{Dt} + \varphi\frac{\partial v_j}{\partial x_j}\right)\delta V = \int_V \left[\frac{\partial\varphi}{\partial t} + \frac{\partial}{\partial x_j}(\varphi v_j)\right]\delta V$$

we have

$$\frac{D}{Dt}\int_V \varphi\delta V = \int_V \left[\frac{\partial\varphi}{\partial t} + \frac{\partial}{\partial x_j}(\varphi v_j)\right]\delta V \tag{1.30}$$

or

$$\frac{D\Phi}{Dt} = \frac{\partial\Phi}{\partial t} + \int_S \varphi v_n dS \tag{1.31}$$

Eq. (1.30) or (1.31) is called Reynolds Transport Theorem, which indicates that the change of any fluid property φ integrated over a control volume along the trajectory of this moving control volume is equal to the static changing rate with time plus the changing rate caused by convection in Eulerian coordinate. The generalized conservation law is

$$\frac{D\Phi}{Dt} + \int_V S_\varphi \delta V = 0$$

where S_φ is the source term of φ, that is the production or destruction of φ per unit volume.

1. 3. 2 Continuity Equation

For mass conservation or cotinuity equation, taking $\varphi = \rho$ and $S_\varphi = 0$, $\frac{D\Phi}{Dt} = 0$, we have

$$\frac{\partial \rho}{\partial t} + \frac{\partial}{\partial x_j}(\rho v_j) = 0 \tag{1. 32}$$

or

$$\frac{D\rho}{Dt} + \rho \frac{\partial v_j}{\partial x_j} = 0 \tag{1. 33}$$

It can be seen that the continuity equation for multicomponent reacting gases has the same form as that for single-component non-reacting flows.

1. 3. 3 Species Conservation Equation (Diffusion Equation)

For species conservation or diffusion equation, taking $\varphi = \rho_s = Y_s \rho$ and considering that the mass change in a moving control volume is due to the chemical reaction and the net diffusion flux through the enclosed surface of the control volume

$$\int_V S_\varphi \delta V = \int_S j_s dS + \int_V w_s \delta V$$

we have

$$\frac{D}{Dt}\int_V \rho Y_s \delta V + \int_V \left(\frac{\partial j_{sj}}{\partial x_j} + w_s\right) \delta V = 0$$

Using the Reynolds Transport Theorem and Fick's Law

$$\frac{D}{Dt}\int_V \rho Y_s \delta V = \int_V \left[\frac{\partial}{\partial t}(\rho Y_s) + \frac{\partial}{\partial x_j}(\rho Y_s v_j)\right] \delta V$$

$$\frac{\partial j_{sj}}{\partial x_j} = - \frac{\partial}{\partial x_j}\left(D\rho \frac{\partial Y_s}{\partial x_j}\right)$$

The species conservation equation can be obtained as

$$\frac{\partial}{\partial t}(\rho Y_s) + \frac{\partial}{\partial x_j}(\rho Y_s v_j) = \frac{\partial}{\partial x_j}\left(D\rho \frac{\partial Y_s}{\partial x_j}\right) - w_s \tag{1. 34}$$

or

$$\rho \frac{\partial Y_s}{\partial t} + \rho v_j \frac{\partial Y_s}{\partial x_j} = \frac{\partial}{\partial x_j}\left(D\rho \frac{\partial Y_s}{\partial x_j}\right) - w_s \tag{1. 35}$$

$$\rho \frac{DY_s}{Dt} = \frac{\partial}{\partial x_j}\left(D\rho \frac{\partial Y_s}{\partial x_j}\right) - w_s \tag{1. 36}$$

The first and second terms on the left-hand side and the first and second terms on the right-hand side of Eq. (1. 34) are called the rate of change in time, convection term, diffusion term and reaction term respectively. The summation of all species conservation equations gives the continuity equation due to $\sum w_s = 0$, so, if there are totally "z" species, only "$z - 1$" species conservation equations will be independent.

1.3.4 Momentum Equation

For momentum equation, taking $\varphi = \rho v_i$, and considering that the change of momentum in a moving control volume is equal to the sum of surface forces and body forces

$$\int_V S_\varphi \delta V = \int_V \left(\frac{\partial p_{ij}}{\partial x_j} + \sum \rho_s F_{si} \right) \delta V$$

the momentum equation can be written as

$$\frac{D}{Dt}\int_V \rho v_i \delta V = \int_V \left(\frac{\partial p_{ij}}{\partial x_j} + \sum \rho_s F_{si} \right) \delta V$$

Using the Reynolds Transport Theorem and the generalized Newton Viscous-force Law

$$\frac{D}{Dt}\int_V \rho v_i \delta V = \int_V \left[\frac{\partial}{\partial t}(\rho v_i) + \frac{\partial}{\partial x_j}(\rho v_j v_i) \right] \delta V$$

$$p_{ij} \approx \mu \left(\frac{\partial v_i}{\partial x_j} + \frac{\partial v_j}{\partial x_i} \right) - p\delta_{ij}$$

the final form of momentum equation takes the form as

$$\frac{\partial}{\partial t}(\rho v_i) + \frac{\partial}{\partial x_j}(\rho v_j v_i) = -\frac{\partial p}{\partial x_i} + \frac{\partial}{\partial x_j}\left[\mu \left(\frac{\partial v_i}{\partial x_j} + \frac{\partial v_j}{\partial x_i} \right) \right] + \sum \rho_s F_{si} \quad (1.37)$$

or

$$\rho \frac{Dv_i}{Dt} = -\frac{\partial p}{\partial x_i} + \frac{\partial}{\partial x_j}\left[\mu \left(\frac{\partial v_i}{\partial x_j} + \frac{\partial v_j}{\partial x_i} \right) \right] + \sum \rho_s F_{si} \quad (1.38)$$

The body forces could be the gravitational force ρg_i, the electric force $\rho_e E_i$ and the magnetic force $(\boldsymbol{J} \times \boldsymbol{B})_i$.

It can be seen that the form of momentum equation for multicomponent reacting gases is almost identical to that for single-component gases, except some slight difference in the body-force term.

1.3.5 Eenrgy Equation

The energy conservation equation is based on the First Law of thermodynamics

$$\sum dQ_i = d\varepsilon + \sum dW_i$$

where $\sum dQ_i$ is the sum of various forms of heat added to the system, such as heat conduction, heat radiation, heat of reaction and Joule's heat; $\sum dW_i$ is the sum of work done by various forces, and $d\varepsilon$ is the energy change of the system. The energy per unit mass of a gas mixture is

$$\varepsilon = e + v^2/2$$

where e is the internal energy per unit mass and

$$e = e_{\text{trans}} + e_{\text{rot}} + e_{\text{vibr}} + e_{\text{spin}} + e_{\text{nucl}} + \cdots$$

$v^2/2$ is the kinetic energy per unit mass. The energy per unit mass of "s" species is

$$\varepsilon_s = e_s + v^2/2$$

It should be noticed that $\varepsilon_s \neq e_s + v_s^2/2$, because e_s includes the kinetic energy of random motion. The energy per unit volume of gas mixture is

$$\rho\varepsilon = \sum_s \rho_s e_s = \sum_s \rho_s (e_s + v^2/2) = \sum_s \rho_s e_s + \rho v^2/2 = \rho (e + v^2/2)$$

Taking $\varphi = \rho\varepsilon$, by using the Reynolds Transport Theorem and the continuity equation, the energy change in a moving control volume is

$$\frac{D}{Dt}\int_V \rho\varepsilon \delta V = \int_V \left[\frac{\partial}{\partial t}(\rho\varepsilon) + \frac{\partial}{\partial x_j}(\rho\varepsilon v_j)\right]\delta V$$

$$= \int_V \left[\frac{D}{Dt}(\rho\varepsilon) + \rho\varepsilon\frac{\partial v_j}{\partial x_j}\right]\delta V = \int_V \rho\frac{D\varepsilon}{Dt}\delta V$$

$$= \int_V \rho\frac{D}{Dt}\left(e + \frac{v^2}{2}\right)\delta V$$

The net heat exchange to the control volume by conduction, radiation and multicomponent diffusion is

$$\int_V \frac{\partial}{\partial x_j}\left(\lambda\frac{\partial T}{\partial x_j} + \sum_s D\rho\frac{\partial Y_s}{\partial x_j}h_s + q_{rj}\right)\delta V$$

The mechanical work done by the body forces is

$$\int_V \sum \rho_s \boldsymbol{F}_s \cdot \boldsymbol{v}_s \delta V = \int_V \left(\boldsymbol{v}\cdot\sum\rho_s\boldsymbol{F}_s + \sum\rho_s\boldsymbol{F}_s\cdot\boldsymbol{V}_s\right)\delta V$$

The mechanical work done by the surface forces is

$$\int_S \boldsymbol{p}_n \cdot \boldsymbol{v} dS = \int_V \frac{\partial}{\partial x_j}(p_{ij}\cdot v_i)\delta V$$

So the energy equation can be written as

$$\rho\frac{D}{Dt}\left(e + \frac{v^2}{2}\right) = \frac{\partial}{\partial x_j}\left(\lambda\frac{\partial T}{\partial x_j}\right) + \frac{\partial}{\partial x_j}\left(\sum_s D\rho\frac{\partial Y_s}{\partial x_j}h_s\right) - \frac{\partial q_{rj}}{\partial x_j}$$

$$+ v_i\sum\rho_s F_{si} + \sum\rho_s F_{si}V_{si} + \frac{\partial}{\partial x_j}(p_{ij}v_i) \qquad (1.39)$$

Now, the work done by the surface force can be divided into two parts

$$\frac{\partial}{\partial x_j}(p_{ij}v_i) = v_i\frac{\partial p_{ij}}{\partial x_j} + p_{ij}\frac{\partial v_i}{\partial x_j}$$

where the first part $v_i\left(\frac{\partial p_{ij}}{\partial x_j}\right)$ is related with the kinetic energy and the work done by the body force in accordance with the mixture velocity through the momentum equation. This relationship can be obtained by multiplying v_i to each term of the momentum equation as

$$v_i\rho\frac{Dv_i}{Dt} = \rho\frac{D}{Dt}\left(\frac{v^2}{2}\right) = v_i\frac{\partial p_{ij}}{\partial x_j} + v_i\sum\rho_s F_{si} \qquad (1.40)$$

On the other hand, the second part $p_{ij}\left(\frac{\partial v_i}{\partial x_j}\right)$, or the work of deformation itself can be divided into two parts

$$p_{ij}\frac{\partial v_i}{\partial x_j} = (-p\delta_{ij} + \tau_{ij})\frac{\partial v_i}{\partial x_j} = -p\frac{\partial v_j}{\partial x_j} + \tau_{ij}\frac{\partial v_i}{\partial x_j}$$

where $-p\left(\frac{\partial v_j}{\partial x_j}\right)$ is compression/expansion work, and $\tau_{ij}\frac{\partial v_i}{\partial x_j}$ is called viscous energy dissipation, or work of shear deformation. We have

$$\Phi \equiv \tau_{ij}\frac{\partial v_i}{\partial x_j} = \left[2\mu S_{ij} - \left(\frac{2}{3}\mu\frac{\partial v_j}{\partial x_j}\right)\delta_{ij}\right]\frac{\partial v_i}{\partial x_j}$$

Owing to

$$\frac{\partial v_i}{\partial x_j} = \frac{1}{2}\left[\left(\frac{\partial v_i}{\partial x_j} + \frac{\partial v_j}{\partial x_i}\right) + \left(\frac{\partial v_i}{\partial x_j} - \frac{\partial v_j}{\partial x_i}\right)\right] = \dot{S}_{ij} + \dot{\xi}_{ij}$$

and

$$\dot{S}_{ij} \cdot \dot{\xi}_{ij} = 0,\ \delta_{ij} \cdot \dot{\xi}_{ij} = 0,\ \tau_{ij} \cdot \dot{\xi}_{ij} = 0,\ \delta_{ij} \frac{\partial v_i}{\partial x_j} = \frac{\partial v_j}{\partial x_j}$$

finally the energy dissipation can be expressed by

$$\Phi = 2\mu \dot{S}_{ij}^2 - \frac{2}{3}\mu\left(\frac{\partial v_j}{\partial x_j}\right)^2 \tag{1.41}$$

or

$$\Phi = 2\mu(\dot{S}_{11}^2 + \dot{S}_{22}^2 + \dot{S}_{33}^2 + \dot{S}_{12}^2 + \dot{S}_{23}^2 + \dot{S}_{31}^2) - \frac{2}{3}\mu(\dot{S}_{11} + \dot{S}_{22} + \dot{S}_{33})^2 \tag{1.42}$$

Substituting the expression

$$p_{ij}\frac{\partial v_i}{\partial x_j} = -p\frac{\partial v_j}{\partial x_j} + \Phi$$

and Eq. (1. 40) into Eq. (1. 39), finally the energy equation in the form of internal energy change can be obtained as

$$\rho\frac{De}{Dt} + p\frac{\partial v_j}{\partial x_j} = \frac{\partial}{\partial x_j}\left(\lambda\frac{\partial T}{\partial x_j}\right) - \frac{\partial q_{rj}}{\partial x_j} + \frac{\partial}{\partial x_j}\left(\sum_s D\rho\frac{\partial Y_s}{\partial x_j}h_s\right) + \Phi + \sum_s \rho_s F_{si} V_{si} \tag{1.43}$$

Notice that compared with the energy equation for non-reacting single-component flows, the energy equation for multicomponent reacting flows has two additional terms: the net enthalpy flux caused by the species diffusion fluxes and the work done by the body forces with the diffusion drift velocity

$$\frac{\partial}{\partial x_j}\left(\sum_s D\rho\frac{\partial Y_s}{\partial x_j}h_s\right), \qquad \sum_s \rho_s F_{si} V_{si}$$

Although the summation of diffusion fluxes is zero, but the net enthalpy flux carried by the diffusion fluxes and the work done by the body forces related with diffusion fluxes are not zero.

It should also be mentioned that Eq. (1. 39) is a full form of energy conservation equation which includes the change of internal energy, as well as the change of kinetic energy, whereas Eq. (1. 43) includes only the change of internal energy which is related to heat transfer and part of mechanical work, and Eq. (1. 40) includes only the change of kinetic energy, which is related to another part of mechanical work and is not related with the heat transfer.

Conventionally, in theoretical studies and numerical models we prefer to use the energy equation in the form of enthalpy change. By using the relation between the enthalpy and the internal ennergy and the continuity equation we have

$$\frac{Dh}{Dt} = \frac{De}{Dt} + \frac{D}{Dt}\left(\frac{p}{\rho}\right) = \frac{De}{Dt} + \frac{1}{\rho}\frac{Dp}{Dt} - \frac{p}{\rho^2}\frac{D\rho}{Dt}$$

and

$$p\frac{\partial v_j}{\partial x_j} = -\frac{p}{\rho}\frac{D\rho}{Dt}$$

Thus, the energy equation in the form of enthalpy change is

$$\rho \frac{Dh}{Dt} - \frac{Dp}{Dt} = \frac{\partial}{\partial x_j}\left(\lambda \frac{\partial T}{\partial x_j}\right) - \frac{\partial q_{rj}}{\partial x_j} + \frac{\partial}{\partial x_j}\left(\sum_s D\rho \frac{\partial Y_s}{\partial x_j} h_s\right) + \Phi + \sum_s \rho_s F_{si} V_{si} \tag{1.44}$$

For low Mach number flows ($Ma \approx 0$), if neglecting body forces and radiation, we have

$$\rho \frac{Dh}{Dt} - \frac{\partial p}{\partial t} = \frac{\partial}{\partial x_j}\left(\lambda \frac{\partial T}{\partial x_j}\right) + \frac{\partial}{\partial x_j}\left(\sum_s D\rho \frac{\partial Y_s}{\partial x_j} h_s\right) \tag{1.45}$$

Considering

$$h = \sum Y_s h_s, \quad h_s = \int c_{ps} dT + h_{0s}$$

and

$$\frac{\partial h}{\partial x_j} = \sum_s h_s \frac{\partial Y_s}{\partial x_j} + \sum_s Y_s c_{ps} \frac{\partial T}{\partial x_j} = \sum_s h_s \frac{\partial Y_s}{\partial x_j} + c_p \frac{\partial T}{\partial x_j}$$

Eq. (1.45) can be rewritten as

$$\rho \frac{Dh}{Dt} - \frac{\partial p}{\partial t} = \frac{\partial}{\partial x_j}\left[\frac{\mu}{\Pr} \frac{\partial h}{\partial x_j} + \mu\left(\frac{1}{\mathrm{Sc}} - \frac{1}{\Pr}\right)\sum_s h_s \frac{\partial Y_s}{\partial x_j}\right] \tag{1.46}$$

In case of $\Pr = \mathrm{Sc} = \mathrm{Le} = 1$, it will be reduced to

$$\frac{\partial}{\partial t}(\rho h) + \frac{\partial}{\partial x_j}(\rho v_j h) = \frac{\partial p}{\partial t} + \frac{\partial}{\partial x_j}\left(\mu \frac{\partial h}{\partial x_j}\right) \tag{1.47}$$

In classical combustion theory it is accustomed to using the energy equation in the form of temperature change. The enthalpy of a gas mixture is

$$h = \sum_s Y_s h_s = \sum_s Y_s h_{0s} + \int_{To}^{T} \sum_s Y_s c_{ps} dT = h_0 + \int_{To}^{T} c_p dT$$

and its substantial derivative is

$$\rho \frac{Dh}{Dt} = \rho \frac{D}{Dt}\left(\sum_s Y_s h_{0s} + \int c_p dT\right) = \rho c_p \frac{dT}{dt} + \sum_s h_{0s} \rho \frac{DY_s}{Dt}$$

$$= \rho c_p \frac{dT}{dt} + \sum h_{0s}\left[\frac{\partial}{\partial x_j}\left(D\rho \frac{\partial Y_s}{\partial x_j}\right) - w_s\right] = \rho c_p \frac{dT}{dt} + \frac{\partial}{\partial x_j}\left(\sum_s D\rho \frac{\partial Y_s}{\partial x_j} h_{0s}\right) - \sum w_s h_{os}$$

Owing to

$$h_s - h_{0s} = \int c_{ps} dT$$

and $\sum w_s h_{0s} = w_s Q_s$ is the heat of reaction, the energy equation will take the form

$$\rho c_p \frac{DT}{Dt} - \frac{Dp}{Dt} = \frac{\partial}{\partial x_j}\left(\lambda \frac{\partial T}{\partial x_j}\right) - \frac{\partial q_{rj}}{\partial x_j}$$
$$+ \frac{\partial}{\partial x_j}\left(\sum_s D\rho \frac{\partial Y_s}{\partial x_j} \int c_{ps} dT\right) + w_s Q_s + \Phi + \sum_s \rho_s F_{si} V_{si} \tag{1.48}$$

In case the Mach number is much smaller than unity, the pressure change and energy dissipation will be negligible compared with the enthalpy change, and if the body force and radiative heat transfer are negligible, assuming

$$c_{p1} = c_{p2} = c_{p3} = \dots = c_{ps} = c_p$$

then the energy equation can be reduced to

$$\rho c_p \frac{DT}{Dt} = \frac{\partial}{\partial x_j}\left(\lambda \frac{\partial T}{\partial x_j}\right) + w_s Q_s \tag{1.49}$$

Eq. (1.49) is called the Zeldovich-Shvab form of energy equation by some investigators. The basic equations for this simplest case can be summarized as follows

$$
\begin{cases}
\dfrac{\partial \rho}{\partial t} + \dfrac{\partial}{\partial x_j}(\rho v_j) = 0 \\
\dfrac{\partial}{\partial t}(\rho v_i) + \dfrac{\partial}{\partial x_j}(\rho v_j v_i) = -\dfrac{\partial p}{\partial x_i} + \dfrac{\partial}{\partial x_j}\left[\mu\left(\dfrac{\partial v_j}{\partial x_i} + \dfrac{\partial v_i}{\partial x_j}\right)\right] + \rho g_i \\
\dfrac{\partial}{\partial t}(\rho Y_s) + \dfrac{\partial}{\partial x_j}(\rho v_j Y_s) = \dfrac{\partial}{\partial x_j}\left(D\rho \dfrac{\partial Y_s}{\partial x_j}\right) - w_s \\
\dfrac{\partial}{\partial t}(\rho c_p T) + \dfrac{\partial}{\partial x_j}(\rho v_j c_p T) = \dfrac{\partial}{\partial x_j}\left(\lambda \dfrac{\partial T}{\partial x_j}\right) + w_s Q_s \\
w_s = B\rho^2 Y_F Y_{ox} \exp(-E/RT) \\
p = \rho RT \sum Y_s / M_s
\end{cases} \tag{1.50}
$$

1.4 SIMILARITY PARAMETERS OF REACTING FLOWS

On the basis of the dimensional analysis using the Momentum Equation (1.38) and Energy Equation (1.48), we can obtain the following similarity parameters for fluid flow

$$\mathrm{Eu} = \frac{\text{Pressure head}}{\text{Velocity head}} = \frac{p_\infty / L}{\rho_\infty u_\infty^2 / L} = \frac{p_\infty}{\rho_\infty u_\infty^2}$$

$$\mathrm{Re} = \frac{\text{Inertial force}}{\text{Viscous force}} = \frac{\rho_\infty u_\infty^2 / L}{\mu_\infty u_\infty / L} = \frac{u_\infty L}{\gamma_\infty}$$

$$\mathrm{Ma} = \frac{\text{Kinetic energy}}{\text{Thermal energy}} = \sqrt{\frac{u_\infty^2/2}{c_p T_\infty}} \sim \frac{u_\infty}{a}$$

heat transfer

$$\mathrm{Pe} = \frac{\text{Heat convection}}{\text{Heat conduction}} = \frac{\rho_\infty u_\infty c_p T_\infty / L}{\lambda_\infty T_\infty / L} = \frac{u_\infty L}{\lambda_\infty / (c_p \rho_\infty)} = \mathrm{Re} \cdot \mathrm{Pr}$$

and combustion

$$D_I = \frac{\text{Heat of reaction}}{\text{Heat convection}} = \frac{\text{Flow time}}{\text{Reaction time}} = \frac{w_{s\infty} Q_s}{\rho_\infty u_\infty c_p T_\infty / L} = \frac{\tau_f}{\tau_c}$$

$$D_{II} = \frac{\text{Heat of reaction}}{\text{Heat conduction}} = \frac{\text{Diffusion time}}{\text{Reaction time}} = \frac{w_{s\infty} Q_s}{\lambda_\infty T_\infty / L^2} = \frac{\tau_d}{\tau_c}$$

It can be seen that these two Damköhler numbers are specific parameters for the similarity of combusting flows. Therefore, the similarity of two low-Mach-number combustion systems can be expressed by the identity of the following independent similarity parameters:

$$
\begin{cases}
\mathrm{Re} = \text{const} \\
\mathrm{Pr} = \text{const} \\
D_I = \text{const}
\end{cases} \tag{1.51}
$$

because D_{II} is the combination of D_I with Pe.

1.5 ZELDOVICH TRANSFORMATION AND GENERALIZED REYNOLDS ANALOGY

Zeldovich proposed a transformation by combining the energy equation with the diffusion equation, or the diffusion equation of a certain species with that of other species, to eliminate the reaction terms in the conservation equations of the combined functions. Later it can be shown that this transformation is useful not only in the classical analytical studies in combustion theory, but also in numerical modeling of diffusion-controlled gas-phase combustion.

Considering the similar form of the diffusion equation and the energy equation

$$\begin{cases} \rho v_j \dfrac{\partial Y_s}{\partial x_j} = \dfrac{\partial}{\partial x_j}\left(D\rho \dfrac{\partial Y_s}{\partial x_j}\right) - w_s \\ \rho v_j c_p \dfrac{\partial T}{\partial x_j} = \dfrac{\partial}{\partial x_j}\left(\lambda \dfrac{\partial T}{\partial x_j}\right) + w_s Q_s \\ \rho v_j \dfrac{\partial h}{\partial x_j} = \dfrac{\partial}{\partial x_j}\left(\dfrac{\lambda}{c_p} \dfrac{\partial h}{\partial x_j}\right) \end{cases}$$

and introducing the combined functions

$$X = Y_1 - Y_2/\beta, \qquad Z = c_p T + Y_s Q_s$$

in case of Pr=Sc=Le=1 and c_p=const, we have the following equations

$$\begin{cases} \rho v_j \dfrac{\partial X}{\partial x_j} = \dfrac{\partial}{\partial x_j}\left(\dfrac{\lambda}{c_p} \dfrac{\partial X}{\partial x_j}\right) \\ \rho v_j \dfrac{\partial Z}{\partial x_j} = \dfrac{\partial}{\partial x_j}\left(\dfrac{\lambda}{c_p} \dfrac{\partial Z}{\partial x_j}\right) \\ \rho v_j \dfrac{\partial h}{\partial x_j} = \dfrac{\partial}{\partial x_j}\left(\dfrac{\lambda}{c_p} \dfrac{\partial h}{\partial x_j}\right) \end{cases}$$

For the boundary conditions

at the surface: $u = 0, v = v_w, X = X_w, Z = Z_w, h = h_w$

and at the infinity: $u = u_\infty, v = 0, \ X = X_\infty, Z = Z_\infty, h = h_\infty$

there should be

$$\frac{X - X_w}{X_\infty - X_w} = \frac{Z - Z_w}{Z_\infty - Z_w} = \frac{h - h_w}{h_\infty - h_w} \tag{1.52}$$

which implies that there is a similarity of h profile with X profile and Z profile. In a particular case, for a laminar boundary layer over a flat plate, denoting

$$\overline{X} = \frac{X - X_w}{X_\infty - X_w}, \qquad \overline{Z} = \frac{Z - Z_w}{Z_\infty - Z_w}$$

$$\overline{u} = u/u_\infty, \qquad \overline{h} = \frac{h - h_w}{h_\infty - h_w}$$

we have

$$\begin{cases} \frac{\partial}{\partial x}(\rho u) + \frac{\partial}{\partial y}(\rho v) = 0 \\ \rho u \frac{\partial \bar{u}}{\partial x} + \rho v \frac{\partial \bar{u}}{\partial y} = \frac{\partial}{\partial y}\left(\mu \frac{\partial \bar{u}}{\partial y}\right) \\ \rho u \frac{\partial \bar{X}}{\partial x} + \rho v \frac{\partial \bar{X}}{\partial y} = \frac{\partial}{\partial y}\left(\mu \frac{\partial \bar{X}}{\partial y}\right) \\ \rho u \frac{\partial \bar{Z}}{\partial x} + \rho v \frac{\partial \bar{Z}}{\partial y} = \frac{\partial}{\partial y}\left(\mu \frac{\partial \bar{Z}}{\partial y}\right) \\ \rho u \frac{\partial \bar{h}}{\partial x} + \rho v \frac{\partial \bar{h}}{\partial y} = \frac{\partial}{\partial y}\left(\mu \frac{\partial \bar{h}}{\partial y}\right) \end{cases}$$

$$(\bar{u})_w = (\bar{X})_w = (\bar{Z})_w = (\bar{h})_w = 0, \quad (\bar{u})_\infty = (\bar{X})_\infty = (\bar{Z})_\infty = (\bar{h})_\infty = 1$$

and

$$\bar{u} = \bar{X} = \bar{Z} = \bar{h}$$

or

$$\frac{u}{u_\infty} = \frac{X - X_w}{X_\infty - X_w} = \frac{Z - Z_w}{Z_\infty - Z_w} = \frac{h - h_w}{h_\infty - h_w} \tag{1.53}$$

These similar distributions of velocity, enthalpy, X-function and Z-function are called "Generalized Reynolds analogy" which is the extension of the Reynolds analogy of flow, heat and mass transfer for single-component flows to multicomponent reacting flows.

1.6 BOUNDARY CONDITIONS AT THE INTERFACE AND STEFAN FLUX

1.6.1 Interfaces

In reacting flow systems the interfaces between the gas and solid/liquid phases may be the channel wall, ablating surface, catalytic surface, solid-fuel or liquid-fuel surface. Frequently, there are physical or chemical changes on the interfaces, such as evaporation or condensation, sublimation, devolatilization and heterogeneous (solid-gas) reactions. No physical and chemical change (phase change and chemical reactions) take place only at the inert surface, such as the channel wall made of non-catalytic ceramic materials. These physical or chemical processes at the interfaces will determine the basic features of the heat and mass transfer to the surface. To properly specify the boundary conditions, it is necessary to have a better understanding of what happens on the interfaces.

1.6.2 Stefan Flux

It is well-known that in case of single-component non-reacting viscous flows over an inert surface, there will be no velocity slip for the fluid element immediately adjacent to the solid wall, or in other words, the fluid will be fully adherent to the wall. However, the situation is found to be different in multicomponent reacting flows. Each gas species will have a diffusion flux at the surface, and these diffusion fluxes must have interaction among themselves. On the other hand, the phase change or the chemical reaction will consume or yield a total mass flux at the interface. Therefore, under the effects of physical or chemical processes and diffusion interaction, there will be a net mass flux normal to the surface, called Stefan flux, which was first discovered by Stefan in studying water evaporation.

Let us take some examples to illustrate the origin and behavior of Stefan flux. The simplest example is water evaporation in air. In this case there are only two species—water vapor and air in the space above the water surface (Fig. 1. 1).

Obviously, the concentration of water vapor at the water surface is higher than that in air. If Y_1, Y_2 denote the mass fractions of water vapor and air respectively and y is the distance normal to the surface, and assume that the evaporation rate is totally determined by molecular diffusion, then the boundary condition at the surface would

$$\begin{cases} u = 0, \quad v = 0 \\ g_{1w} = J_{1w} = -D\rho\left(\dfrac{\partial Y_1}{\partial y}\right)_w \end{cases}$$

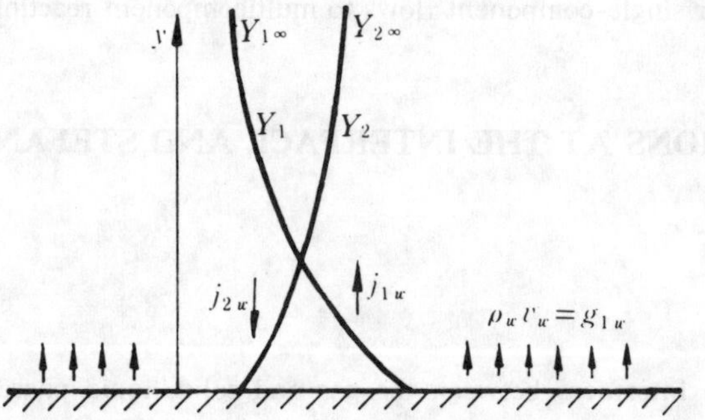

Fig. 1. 1

However, this statement is not the actual case, because there is also an air diffusion flux at the surface

$$J_{2w} = -D\rho\left(\frac{\partial Y_2}{\partial y}\right)_w \tag{1. 54}$$

and owing to $Y_1 + Y_2 = 1$ and

$$\left(\frac{\partial Y_1}{\partial y}\right)_w < 0, \qquad J_{1w} > 0$$

we have

$$\left(\frac{\partial Y_2}{\partial y}\right)_w = -\left(\frac{\partial Y_1}{\partial y}\right)_w > 0$$

or

$$J_{2w} = -J_{1w} < 0$$

This is an air diffusion flux to the surface, equal in magnitude but opposite in direction to the water-vapor diffusion flux. Now, there is only water evaporation on the surface and the surface neither absorbs nor releases air under the steady state, or the total mass flux of air at the surface must be zero. Hence, under the effects of diffusion interaction and surface evaporation (but not the external flow field) a total net mass flux of gas mixture is caused at the surface. This mass flux $\rho_w v_w$ is called Stefan flux. The species mass fluxes at the surface consist of two parts: diffusion fluxes and fluxes carried by convection of the whole gas mixture, that is

$$g_{1w} = J_{1w} + Y_{1w}\rho_w v_w \tag{1.55}$$

$$g_{2w} = J_{2w} + Y_{2w}\rho_w v_w \tag{1.56}$$

so, we have

$$g_{2w} = -D\rho\left(\frac{\partial Y_2}{\partial y}\right)_w + Y_{2w}\rho_w v_w = 0 \tag{1.57}$$

$$g_{1w} = -D\rho\left(\frac{\partial Y_1}{\partial y}\right)_w + Y_{1w}\rho_w v_w = D\rho\left(\frac{\partial Y_2}{\partial y}\right)_w + Y_{1w}\rho_w v_w$$

$$= (Y_{2w} + Y_{1w})\,\rho_w v_w = \rho_w v_w$$

and

$$-D\rho\left(\frac{\partial Y_1}{\partial y}\right)_w = (1 - Y_{1w})\,\rho_w v_w \tag{1.58}$$

It can be concluded from these equations that: (1) the water evaporation rate is the total mass flux of water vapor but not its diffusion flux; (2) the total mass flux of water vapor is equal to its diffusion flux plus the flux carried by the convection of whole gas mixture; and (3) the evaporation rate is equal to the total mass flux of the gas mixture at the surface, that is the Stefan flux $\rho_w v_w$.

The second example is the combustion of solid carbon in pure oxygen environment (Fig. 1.2). For simplicity, assume that there is only one reaction on the surface

$$\mathrm{C} + \mathrm{O}_2 \longrightarrow \mathrm{CO}_2$$

$$12\mathrm{kg} \quad 32\mathrm{kg} \quad 44\mathrm{kg}$$

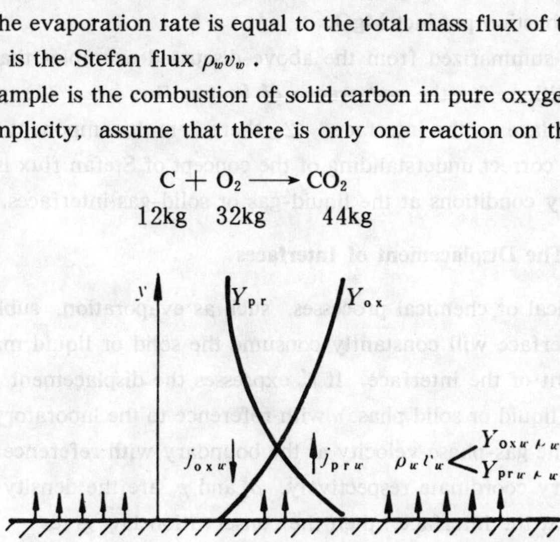

Fig. 1.2

The gas mixture consists of O_2 and CO_2, hence we have

$$J_{ox,w} = -D\rho\left(\frac{\partial Y_{ox}}{\partial y}\right)_w$$

$$J_{pr,w} = -D\rho\left(\frac{\partial Y_{pr}}{\partial y}\right)_w$$

$$\sum Y_s = Y_{ox} + Y_{pr} = 1$$

Obviously, owing to diffusion interaction the diffusion fluxes of these two species are equal in magnitude but opposite in direction, that is

$$\left(\frac{\partial Y_{ox}}{\partial y}\right)_w = -\left(\frac{\partial Y_{pr}}{\partial y}\right)_w, \qquad J_{ox,w} = -J_{pr,w}$$

However, the reaction of carbon oxidation requires

$$g_{ox,w} = -\frac{32}{44} g_{pr,w} \tag{1.59}$$

so, a net mass flux or the Stefan flux arises at the surface, and we have

$$g_{ox,w} = -D\rho\left(\frac{\partial Y_{ox}}{\partial y}\right)_w + Y_{ox,w}\rho_w v_w \tag{1.60}$$

$$g_{pr,w} = -D\rho\left(\frac{\partial Y_{pr}}{\partial y}\right)_w + Y_{pr,w}\rho_w v_w \tag{1.61}$$

$$g_{ox,w} + g_{pr,w} = \rho_w v_w \tag{1.62}$$

$$\rho_w v_w = -\frac{12}{32} g_{ox,w} = g_c \tag{1.63}$$

Notice that in this case the mass flux of each species is not zero and is not equal to the Stefan flux, only the summation of mass fluxes of two species gives the Stefan flux.

In the above-mentioned examples the direction of Stefan flux is positive (from the surface to gas space). Alternatively, the direction of Stefan flux may be negative (from gas space to the surface) as in the case of combustion of magnesium in air with solid-state combustion product MgO.

It can be summarized from the above-discussed examples that the necessary and sufficient conditions for the occurrence of Stefan flux are: (1) physical or chemical processes take place on the interface; (2) there is multicomponent diffusion to or from the surface. A correct understanding of the concept of Stefan flux is important to specify the boundary conditions at the liquid-gas or solid-gas interfaces.

1.6.3 The Displacement of Interfaces

The physical or chemical processes, such as evaporation, sublimation or combustion on the interface will constantly consume the solid or liquid material which causes the displacement of the interface. If v''_w expresses the displacement velocity of the condensed phase (liquid or solid phase) with reference to the laboratory coordinate, and v'_w and v_w denote the gas-phase velocity at the boundary with reference to the interface and to the laboratory coordinate respectively, ρ_1 and ρ_w are the density of condensed phase and gas phase at the interface, then the conservation law gives

$$\rho_1 v''_w = \rho_w v'_w = \rho_w (v_w + v''_w) \tag{1.64}$$

In most cases the solid or liquid density is much greater than the gas density, and the following approximations can be made

$$\rho_1 \gg \rho_w, \quad v''_w \ll v_w$$

Hence we have

$$\rho_1 v''_w \approx \rho_w v_w \tag{1.65}$$

Under this condition we can use the so-called "quasi-stationary" treatment, that is the coordinate fixed with the interface can be approximately considered as the laboratory coordinate. However, in some cases, such as under very high pressure, when ρ_1 and ρ_w are of the same order of magnitude, the quasi-stationary treatment will be inappropriate.

1. 6. 4 Boundary Conditions at the Liquid-gas Interface

On the basis of the concept of Stefan flux, the boundary conditions at the liquid-gas interface can be given as follows (Fig. 1. 3) :

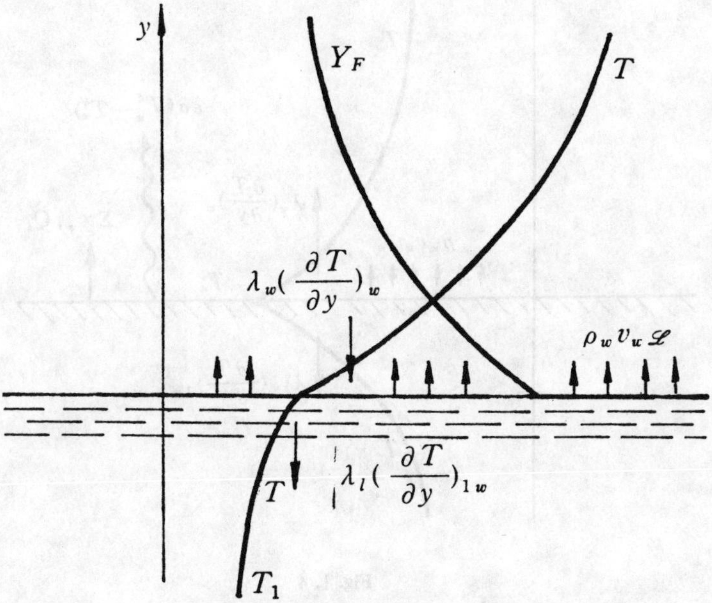

Fig. 1. 3

for momentum equation

$$u = 0, \qquad v = v_w \neq 0 \tag{1.66}$$

for species conservation equations

$$\begin{cases} g_{sw} = -D\rho\left(\dfrac{\partial Y_s}{\partial y}\right)_w + Y_{sw}\rho_w v_w = \alpha\rho_w v_w \\ (s = F, \alpha = 1; s = \text{ox, pr, iner, ...}, \alpha = 0) \\ \sum g_{sw} = \rho_w v_w = g_{Fw} \\ \sum Y_s = Y_F + Y_{\text{ox}} + Y_{\text{pr}} + Y_{\text{iner}} = 1 \end{cases} \tag{1.67}$$

and for energy equation

$$\begin{cases} -\lambda\left(\dfrac{\partial T}{\partial y}\right)_w + \varepsilon\sigma(T_\infty^4 - T_w^4) = -\rho_w v_w \mathscr{L} - \lambda_l\left(\dfrac{\partial T}{\partial y}\right)_{lw} & (1.68) \\ Y_{Fw} = B_w \exp\left(-\dfrac{E_w}{RT_w}\right) & (1.68a) \end{cases}$$

When neglecting radiation and heat conduction into the liquid, we have

$$\lambda\left(\frac{\partial T}{\partial y}\right)_w = \rho_w v_w \mathscr{L} = \rho_w v_w q_e \tag{1.69}$$

1. 6. 5 Boundary Conditions at the Solid-gas Interface

For the solid-gas interface we can give the boundary conditions for momentum equation (Fig. 1. 4)

$$u = 0, \qquad v = v_w \neq 0$$

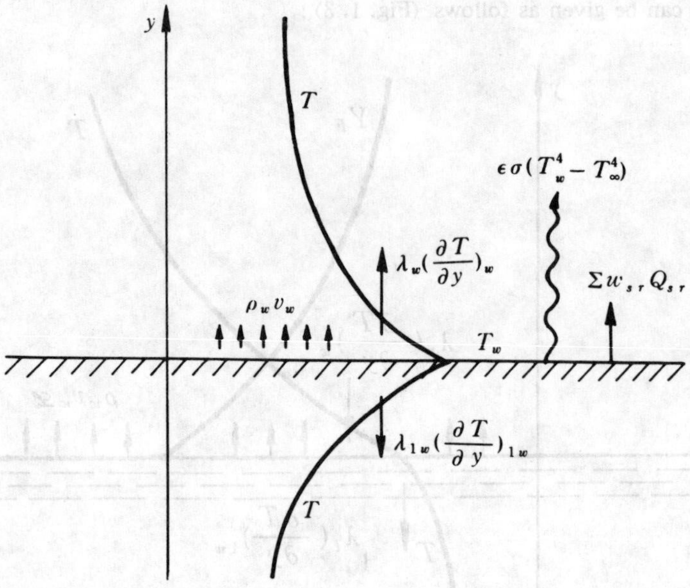

Fig. 1. 4

for species conservation equations

$$g_{sw} = -D\rho\left(\frac{\partial Y_s}{\partial y}\right)_w + Y_{sw}\rho_w v_w = \sum_r w_{sr} \tag{1.70}$$

$$g_{\text{iner}, w} = -D\rho\left(\frac{\partial Y_{\text{iner}}}{\partial y}\right)_w + Y_{\text{iner}, w}\rho_w v_w = 0 \tag{1.71}$$

$$\sum g_{sw} = \rho_w v_w \tag{1.72}$$

$$\alpha g_{Fw} = \beta g_{\text{ox}, w} = \cdots \tag{1.73}$$

and for energy equation

$$\sum_r w_{sr} Q_{sr} = \lambda\left(\frac{\partial T}{\partial y}\right)_w + \lambda_1\left(\frac{\partial T}{\partial y}\right)_{1w} + \varepsilon\sigma(T_w^4 - T_\infty^4) \tag{1.74}$$

1. 6. 6 The Generalized Reynolds Analogy at the Liquid-gas Interface

In conventional heat and mass transfer problems for non-reacting single-component flows the Reynolds analogy indicates the similar form of shear stress, mass flux and heat flux as

$$\begin{cases} \tau = \mu\left(\dfrac{\partial u}{\partial y}\right) \\ g = J = -D\rho\left(\dfrac{\partial Y}{\partial y}\right) \\ q = -\lambda\left(\dfrac{\partial T}{\partial y}\right) \end{cases}$$

For multicomponent reacting flows, owing to the presence of Stefan flux the heat and mass fluxes at the interface are

$$q_w = -\lambda\left(\frac{\partial T}{\partial y}\right)_w$$

$$g_{sw} = -D\rho\left(\frac{\partial Y_s}{\partial y}\right)_w + Y_{sw}\rho_w v_w$$

and the Reynolds analogy will not hold even when Pr=Sc=Le=1. However, if we take

$$F = \frac{Y_F}{1 - Y_{Fw}}, \qquad \theta = \frac{c_p(T_\infty - T)}{q_e}$$

then from Eq. (1.67) and Eq. (1.69) we have

$$\rho_w v_w = -D\rho\left(\frac{\partial F}{\partial y}\right)_w = -\frac{\lambda}{c_p}\left(\frac{\partial \theta}{\partial y}\right)_w \tag{1.75}$$

In case there is no reaction in the gas phase, and if Le=1, the diffusion equation and energy equation and their boundary conditions will be identical to each other, and we obtain

$$F = \theta, \quad F_w = \theta_w \tag{1.76}$$

or

$$\frac{c_p(T_\infty - T_w)}{q_e} = \frac{Y_{Fw}}{1 - Y_{Fw}} \tag{1.77}$$

Therefore, the temperature and fuel vapor concentration at the surface can be determined by Eq. (1.77) together with the saturation condition (1.68a) before solving the diffusion and energy equations. Eq. (1.75) is called the generalized Reynolds analogy at the liquid-gas interface.

CHAPTER 2 IGNITION AND EXTINCTION

2.1 BACKGROUND

Ignition and extinction are the typical critical phenomena in combustion which indicate the transition from slow reaction to high-temperature fast reaction and vice versa respectively. Conventionally it is considered that there are two types of ignition: self ignition—including thermal ignition in closed vessels, well-stirred reactors, diesel engines etc. by heating the entire system and ignition in a coal pile or a stack without external heating; forced ignition—ignition by a hot surface, spark, flamelet or other local heating sources, such as ignition in gas-turbine combustors and Otto engines. However, the mechanism of all ignition phenomena is the interaction between heat release by chemical reaction and heat loss by conduction, convection or radiation, no matter whether the combustible mixture is entirely or locally heated.

A long time ago, and even not far from now, it was thought that the ignition temperature is a constant property of each kind of fuel. Now we know that the ignition temperature is a function of pressure, velocity, geometrical size, fuel/air ratio, and fuel/oxidizer properties. There exists a functional relationship expressing the ignition or extinction condition. The initial condition or boundary condition, at which starting from a certain instant or a certain location the reaction can automatically accelerate and will rapidly reach the high-temperature state, is called the ignition condition.

Let's consider two important features of combustion reactions. The first is the high heating effect of combustion reactions. For example, the carbon, hydrocarbon and hydrogen combustion (oxidation) will give the heating effect of 3.68×10^4 kJ/kg, 4.18×10^4 kJ/kg and 1.38×10^5 kJ/kg respectively. The second is the rapid increase of reaction rate with temperature due to high activation energy. The activation energy of hydrocarbon fuel combustion is in the range of $E = 8.36 \times 10^4 - 1.67 \times 10^5$ kJ/ (kg • mol). Assuming that the reaction rate w_s is proportional to $\exp(-E/RT)$ and $E = 8.36 \times 10^4$ kJ/ (kg • mol), we have

$$\text{at} \quad T_1 = 500\text{K:} \qquad \exp(-E/RT_1) \approx 2 \times 10^{-9}$$

$$\text{at} \quad T_2 = 1000\text{K:} \qquad \exp(-E/RT_2) \approx 4 \times 10^{-5}$$

that is, $T_2 = 2T_1$, $w_{s2} = w_{s1} \times 20000$!

In the region of $0 < RT/E < 0.5$, which corresponds to $T < 5000$K, $\exp(-E/RT)$ always rapidly increases with the increase of RT/E. Conventional combustion processes take place in the temperature range far below 5000K.

Therefore, the critical phenomena—ignition and extinction are caused by the interaction between chemical reactions (kinetics) with high heating effect and high activation energy and heat-mass transfer (fluid flow).

2.2 DIMENSIONAL ANALYSIS

The ignition and extinction theory or theory of critical phenomena is the analysis of the interaction between flow and chemical reactions. Generally speaking, for given inlet and boundary conditions the numerical simulation can predict the velocity, temperature and concentration distributions in space and their change in time for reacting flows, from which the presence or absence of critical phenomena can be determined. However, if we need to determine only qualitatively the condition for occurrence of these critical phenomena and where and when they happen, then the ignition and extinction theory can serve this purpose without carrying out the numerical computations.

Our starting point is the energy equation for low Mach number, constant specific heat, laminar reacting flows with neglected radiation and body forces

$$\rho c_p \frac{dT}{dt} = \frac{\partial}{\partial x_j}\left(\lambda \frac{\partial T}{\partial x_j}\right) + w_s Q_s \tag{2.1}$$

After Frank-Kamenetskii, assuming that the fuel and oxidizer concentration, gas-mixture density and thermoconductivity are kept almost unchanged before ignition, taking

the reaction time $\tau_c = \dfrac{RT_1^2}{E}\dfrac{\rho c_p}{w_{s1}Q_s}$

the diffusion time $\tau_d = \rho c_p L^2/\lambda$

the characteristic temperature RT_1^2/E

the dimensionless time $\bar{t} = t/\tau_c$

the dimensionless temperature $\theta = (T - T_1)E/(RT_1^2)$

the dimensionless coordinate $\xi_j = x_j/L$

and multiplying each term of Eq. (2.1) by $E/(RT_1^2)$, the dimensionless form of energy equation can be obtained as:

$$\frac{d\theta}{d\bar{t}} = \frac{\tau_c}{\tau_d}\frac{\partial}{\partial \xi_j}\left(\frac{\partial \theta}{\partial \xi_j}\right) + w_s/w_{s1} \tag{2.2}$$

where T_1 is a typical temperature which may be T_∞ for self ignition, T_w for ignition by a hot surface and T_f for ignition by a flamelet.

Following Frank-Kamenetskii, we can further assume that the reaction has obvious effect only in a narrow temperature range near T_1, or

$$(T - T_1)/T_1 = \Delta T/T_1 \ll 1$$

hence the exponential term can be approximately expanded as

$$\exp\left(-\frac{E}{RT}\right) = \exp\left[-\frac{E}{RT_1(1 + \Delta T/T_1)}\right]$$
$$\approx \exp\left(-\frac{E}{RT_1}\right)\exp\left[\frac{E}{RT_1^2}(T - T_1)\right] = \exp\left(-\frac{E}{RT_1}\right)e^\theta$$

and we have

$$w_s \approx w_{s1}e^\theta$$

Finally, Eq. (2.2) becomes

$$\frac{d\theta}{d\bar{t}} = \frac{\partial}{\partial \xi_j}\left(\frac{\partial \theta}{\partial \xi_j}\right) + e^\theta \cdot \frac{\tau_d}{\tau_c} \tag{2.3}$$

The general solution of Eq. (2. 3) will have the form

$$\theta = \theta\left(\frac{\tau_d}{\tau_c}, \xi_j, \bar{t}\right) \tag{2. 4}$$

in which the only parameter is τ_d/τ_c . Therefore, in the Lagrangian treatment, for a certain τ_d/τ_c the temperature development in space or in time will have the critical phenomena, and the ignition condition should be

$$\frac{\tau_d}{\tau_c} = \text{const} \tag{2. 5}$$

For a steady flowing system the energy equation is

$$\frac{d\theta}{dt} = v_j \frac{\partial \theta}{\partial x_j} = \frac{1}{\tau_f}\bar{v}_j \frac{\partial \theta}{\partial \xi_j} = \frac{1}{\tau_d}\frac{\partial}{\partial \xi_j}\left(\frac{\partial \theta}{\partial \xi_j}\right) + \frac{e^\theta}{\tau_c}$$

or

$$\bar{v}_j \frac{\partial \theta}{\partial \xi_j} = \frac{1}{\text{Pe}}\frac{\partial}{\partial \xi_j}\left(\frac{\partial \theta}{\partial \xi_j}\right) + e^\theta \frac{\tau_f}{\tau_c} \quad \left(\tau_f = \frac{L}{u_\infty},\ \bar{v}_j = \frac{v_j}{u_\infty}\right) \tag{2. 6}$$

The general solution of Eq. (2. 6) will have the form

$$\theta = \theta\left(\text{Pe}, \frac{\tau_f}{\tau_c}, \bar{v}_j, \xi_j\right) \tag{2. 7}$$

Obviously the ignition condition in this case is

$$\frac{\tau_f}{\tau_c} = f(\text{Pe}) \tag{2. 8}$$

For a steady and stagnant system owing to $\frac{d\theta}{dt} = 0$, the energy equation is reduced to

$$\frac{\partial}{\partial \xi_j}\left(\frac{\partial \theta}{\partial \xi_j}\right) + e^\theta \frac{\tau_d}{\tau_c} = 0 \tag{2. 9}$$

The general solution of Eq. (2. 9) is

$$\theta = \theta\left(\frac{\tau_d}{\tau_c}, \xi_j\right) \tag{2. 10}$$

and once again the ignition condition should be

$$\frac{\tau_d}{\tau_c} = \text{const}$$

The final form of functions (2. 5) and (2. 8) will be determined by the analytical solutions of Eq. (2. 3) with different simplifications, since Eq. (2. 3) in its full form can be solved only by numerical methods.

2. 3 UNSTEADY AND LAGRANGIAN TREATMENT

2. 3. 1 Thermal Ignition in a Closed Vessel—Unsteady Treatment

Consider the simplest case—a closed spherical vessel of diameter d, filled with combustible gas mixture (Fig. 2. 1) , inside which there is a uniform temperature field (sufficiently large thermoconductivity) but changeable with time, or $T = T(t)$. The concentration of gas species in this vessel is also uniform and kept almost unchanged before ignition, and there is only heat conduction between the wall and the surrounding environment.

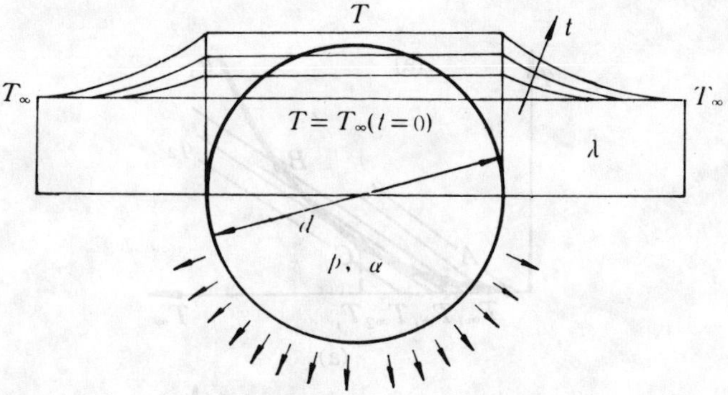

Fig. 2. 1

For the sake of simplicity, assume that the initial state is $T = T_\infty$ at $t = 0$ both inside the sphere and in the surrounding environment by a certain method of heating. After $t > 0$, the temperature inside the vessel rises due to chemical reaction, while the temperature in the surrounding environment is still T_∞ all the time. The simplified energy equation for this closed system is

$$\rho V c_v \frac{dT}{dt} = V w_s Q_s - 2\pi d \lambda (T - T_\infty)$$

or

$$\rho c_v \frac{dT}{dt} = q_1 - q_2 = w_s Q_s - 12\lambda (T - T_\infty)/d^2 \qquad (2.11)$$

Following Semenov, the curves $q_1(T)$, $q_2(T)$ and $T(t)$ can be qualitatively shown in Figs. 2. 2 and 2. 3. It can be seen from Figs. 2. 2 (a) and 2. 3 that if the initial temperature of gas mixture is $T_{\infty 2}$, the heat release q_1 is always larger than the heat loss q_2, or dT/dt is always greater than zero, then a rapid temperature rise will certainly occur sooner or later, or combustion will certainly take place. If the initial temperature is $T_{\infty 1}$, there will be two equilibrium points A and B. Point A is a steady slow reaction state and no combustion will take place in this case. Point B is physically unrealistic or false equilibrium state. If the initial temperature is just $T_{\infty i}$, then $q_1(T)$ and $q_2(T)$ will have a tangential point at C, which is a critical equilibrium point expressing the transition from steady slow reaction to unsteady state with rapid temperature rise. Therefore, such an initial condition coincides with the ignition condition, and it can be expressed by

$$\begin{cases} q_1 = q_2 \\ \dfrac{\partial q_1}{\partial T} = \dfrac{\partial q_2}{\partial T} \end{cases} \qquad (2.12)$$

Not only increasing the initial temperature T_∞, but also increasing the pressure p or changing the air /fuel ratio (Fig. 2. 2b), reducing the thermoconductivity λ and increasing the diameter d (Fig. 2. 2c) can lead to ignition. It should be mentioned that the ignition temperature is the initial temperature $T_{\infty i}$ corresponding to the ignition condition, but not the temperature T_i at the tangential point C. Furthermore, combustion or steady fast reaction will take place not at T_i, but at a temperature much higher than T_i, which will be discussed later.

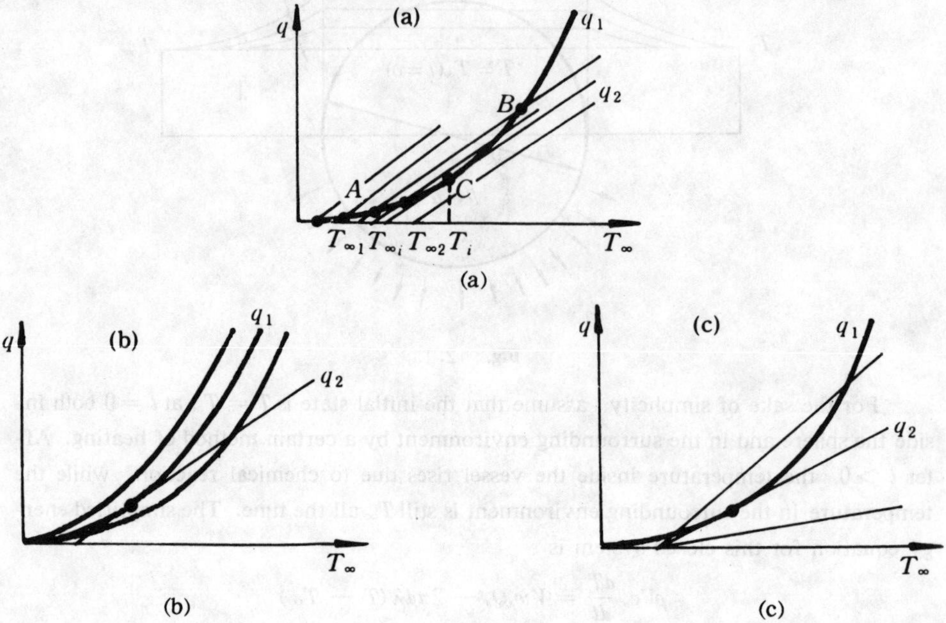

Fig. 2. 2

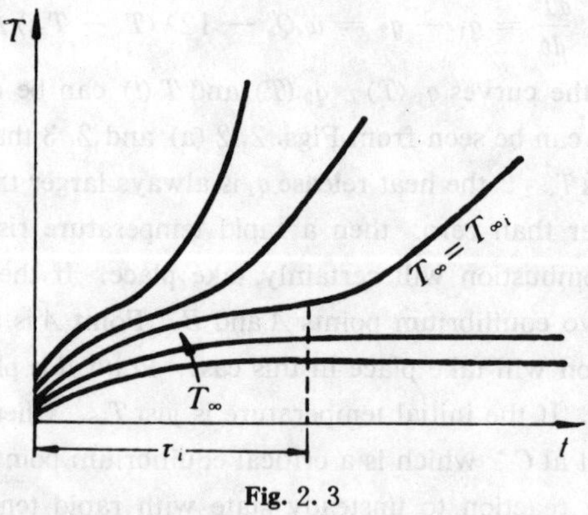

Fig. 2. 3

To derive the functional relationship of ignition, by using Eqs. (2. 11) and (2. 12), we have

$$w_s Q_s = \frac{12\lambda}{d^2}(T_i - T_{\infty i}) \tag{2. 13}$$

$$\frac{E}{RT_i^2} w_s Q_s = 12\,\frac{\lambda}{d^2} \tag{2. 14}$$

Equation (2. 13) divided by Eq. (2. 14) gives

$$T_i - T_{\infty i} = RT_i^2/E$$

and solving for T_i gives the first-order approximation

$$T_i \approx T_{\infty i} \tag{2. 15}$$

and the second-order approximation

$$T_i - T_{\infty i} \approx RT_{\infty i}^2 / E \tag{2.16}$$

Substituting Eq. (2.15) into Eq. (2.14) or (2.16) into (2.13), the final form of ignition relationship can be obtained as (dropping the subscript "i")

$$\frac{E}{RT_\infty^2} w_{s\infty} Q_s = 12 \frac{\lambda}{d^2} \tag{2.17}$$

By denoting

$$\tau_c = \frac{RT_\infty^2}{E} \frac{\rho c_v}{w_{s\infty} Q_s}, \qquad \tau_d = \frac{\rho c_v d^2}{\lambda}$$

we have

$$\frac{\tau_d}{\tau_c} = 12 \tag{2.18}$$

Equation (2.18) is an alternative form of the functional relationship for thermal ignition in a closed vessel, which is in agreement with Eq. (2.5), if taking 12 as the constant on the right-hand side of Eq. (2.5). Obviously, $\tau_d/\tau_c < 12$ and $\tau_d/\tau_c > 12$ correspond to steady slow reaction and fast reaction (combustion) respectively.

Assuming a reaction-rate expression as

$$w_{s\infty} = B\sqrt{T_\infty}\, p_\infty^n \varphi(\alpha) \exp\left(-\frac{E}{RT_\infty}\right)$$

and substituting this expression into Eq. (2.17), the ignition relationship becomes

$$d^2 p^n T_\infty^{-n-1.5} \varphi(\alpha) \exp\left(-\frac{E}{RT_\infty}\right) = A \tag{2.19}$$

where A, n, E and $\varphi(\alpha)$ are determined by experiments. The physical significance of the ignition relationship may be discussed on the basis of Eq. (2.19). First, the ignition condition is a functional relationship in the form of

$$f(d, p, T_\infty, \alpha) = 0 \tag{2.20}$$

and clearly the concept of $T_i = \text{const}$ as an ignition condition is not true. Actually the ignition condition is a functional relationship including reaction kinetics and heat conduction factors. For constant λ, d and α we have

$$\lg \frac{p}{T_\infty^{1+1.5/n}} = \frac{B}{T_\infty} + C \tag{2.21}$$

which means that the ignition temperature will monotonously decrease with the increase of pressure. This can be shown in Fig. 2.4. For constant p, d and λ there are two limits of flammability—lean limit and rich limit—and a minimum ignition temperature, which is confirmed by experiments (Fig. 2.5). For constant p, λ and α the ignition temperature will decrease with increasing diameter of the vessel, which is also in agreement with experiments (Fig. 2.6).

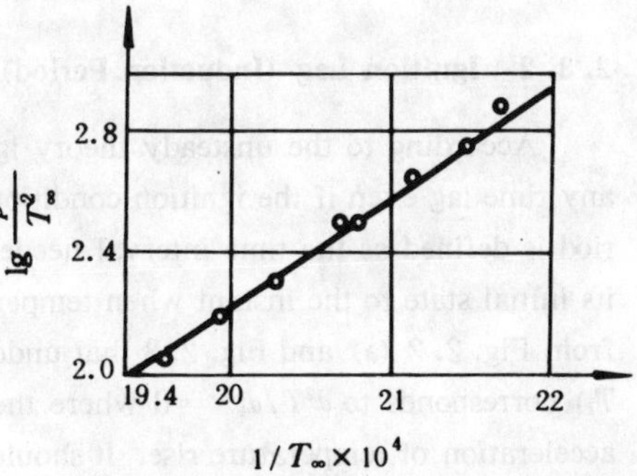

Fig. 2.4

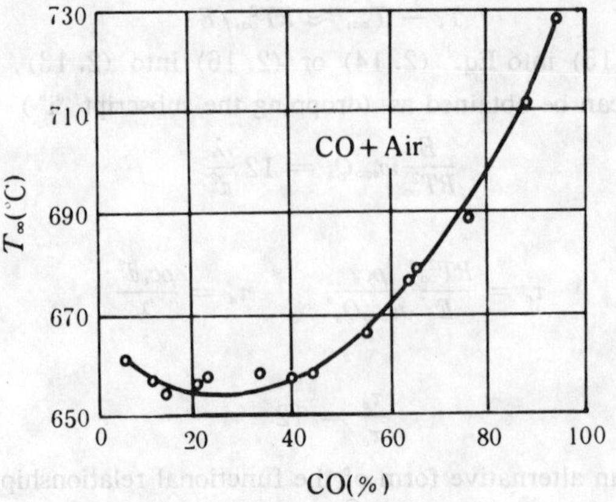

Fig. 2. 5

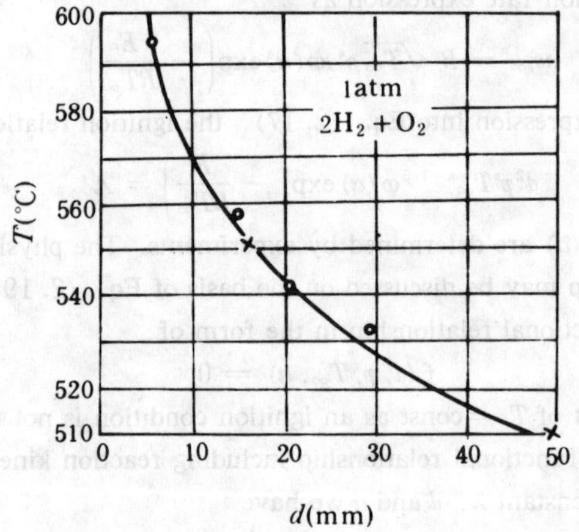

Fig. 2. 6

2. 3. 2 Ignition Lag (Induction Period)

According to the unsteady theory ignition does not occur instantaneously without any time lag even if the ignition condition is satisfied. The ignition lag or induction period is defined as the time interval needed for the combustible mixture to change from its initial state to the instant when temperature starts to increase rapidly. It can be seen from Fig. 2. 2 (a) and Fig. 2. 3 that under the ignition condition the point ($t = \tau_i, T = T_i$) corresponds to $d^2T/dt^2 = 0$ where the deceleration of temperature rise turns to the acceleration of temperature rise. It should be noticed that if $T_\infty < T_{\infty i}$ or the initial condition does not satisfy the ignition condition, there will be an infinite induction period $\tau_i = \infty$. Under the ignition condition $T_\infty = T_{\infty i}$, τ_i will take a maximum finite value. If $T_\infty > T_{\infty i}$ or the initial condition is superior to the ignition condition, τ_i will not be zero, and will decrease with the increase of temperature. In the last situation the curve of temperature rise has also a turning point corresponding to $(dT/dt)_{\min}$, but not $(dT/dt)=0$.

For deriving the expression of ignition lag, introducing

$$\tau_c = \frac{RT_\infty^2}{E}\,\frac{\rho c_v}{w_{s\infty}Q_s},\quad \tau_d = \rho c_v d^2/\lambda,\quad \theta = (T - T_\infty)E/(RT_\infty^2),\quad \bar{t} = t/\tau_c$$

the simplified energy equation (2.11) can be transformed into the following dimensionless form

$$\frac{d\theta}{d\bar{t}} = e^\theta - 12\theta\tau_c/\tau_d \tag{2.22}$$

The idea of finding an adiabatic ignition lag is first to solve $\theta = \theta(\bar{t})$ for adiabatic or zero heat-loss case and then to obtain $t = \tau_i$ by substituting $\theta = \theta_i$. For the adiabatic case Eq. (2.22) becomes

$$\frac{d\theta}{dt} = e^\theta/\tau_c \tag{2.23}$$

The solution of Eq. (2.23) with the initial condition of $\theta = 0$ at $t = 0$ gives

$$\frac{t}{\tau_c} = 1 - e^{-\theta} \tag{2.24}$$

Substituting $\theta_i = 1$ into Eq. (2.24), the adiabatic ignition lag can be obtained as

$$\tau_i = \left(1 - \frac{1}{e}\right)\tau_c$$

or

$$\tau_i = \left(1 - \frac{1}{e}\right)\frac{RT_\infty^2}{E}\,\frac{\rho c_v}{w_{s\infty}Q_s} \tag{2.25}$$

By using above proposed reaction-rate expression, the ignition lag will take a more explicit form

$$\tau_i = Kp^{1-n}[\varphi(\alpha)]^{-1}T_\infty^{0.5+n}\exp\left(\frac{E}{RT_\infty}\right) \tag{2.26}$$

where K, n, E and $\varphi(\alpha)$ are determined by experiments. Obviously, for constant p and α the ignition lag as a function of temperature is

$$\ln \tau_i \approx \frac{E}{RT_\infty} + \text{const} \tag{2.27}$$

For constant temperature and α the ignition lag as a function of pressure is

$$\tau_i \sim p^{1-n} \tag{2.28}$$

Equations (2.27) and (2.28) indicate that the ignition lag rapidly decreases with the increase of temperature and also decreases with the increase of pressure. The linear functions of $\ln \tau_i$ vs. as $1/T_\infty$ and $\ln \tau_i$ vs. as $\ln p$ are really observed in experiments (Fig. 2.7 and 2.8), from which the activation energy E and reaction order n can be determined.

For constant T_∞ and p there exist lean and rich limits for the ignition lag which can be concluded from Eq. (2.26) and from the following empirical formula

$$\tau_i X_{C_3H_8} X_{ox}^{0.25} = 0.03 + 5.4 \times 10^{-9}\exp\left(\frac{14500}{T_\infty}\right) \tag{2.29}$$

where X is the molar fraction.

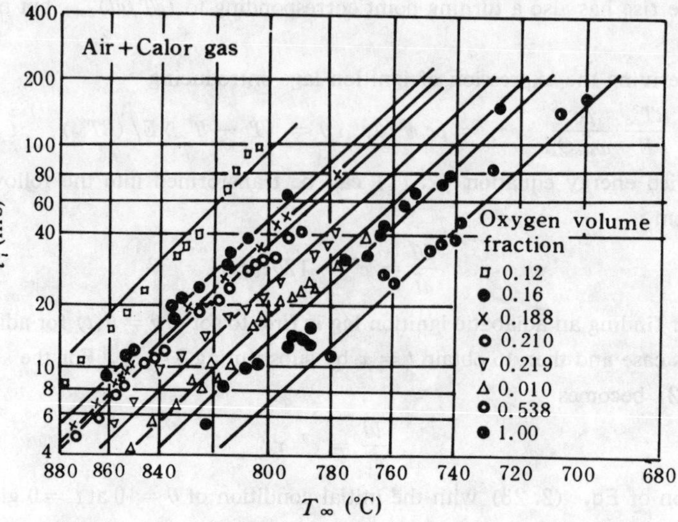

Fig. 2.7

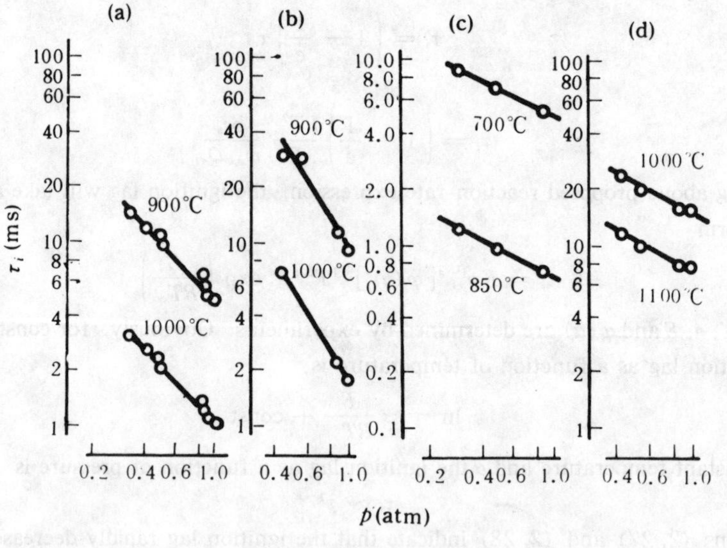

Fig. 2.8

2.3.3 Ignition and Extinction in a Simple Flow System—a Well-Stirred Reactor

Consider a simple flow system—a well-stirred reactor with an inlet and an exit (Fig. 2.9). The inlet temperature and concentration of a certain species are T_∞ and Y_∞ respectively, the total mass flux is M. The temperature T and concentration Y inside the reactor and of the products are uniform and have the same value. The system is

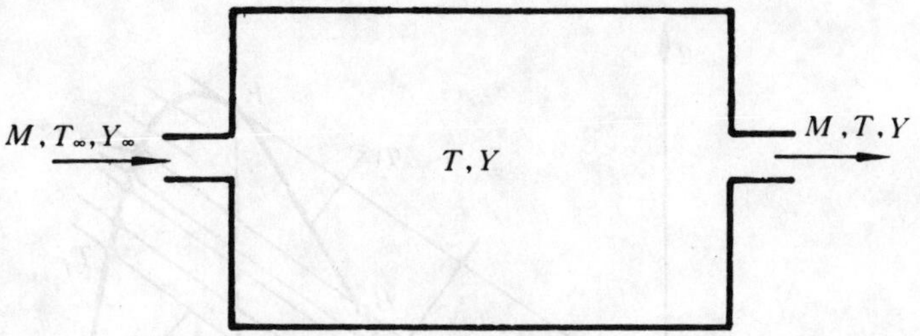

Fig. 2. 9

adiabatic and with a first-order reaction. For such a system the temperature and concentration change due to reaction and heat and mass loss can be written as:

$$\rho V c_v \frac{dT}{dt} = VQB\rho_\infty Y \exp\left(-\frac{E}{RT}\right) - Mc_p(T - T_\infty) = Q_1 - Q_2 \qquad (2.30)$$

$$\rho V \frac{dY}{dt} = VB\rho_\infty Y \exp\left(-\frac{E}{RT}\right) - M(Y_\infty - Y) = G_1 - G_2 \qquad (2.31)$$

For any steady state of slow or fast reaction, we have

$$G_1 = G_2, \qquad Q_1 = Q_2$$

or

$$VQB\rho_\infty Y \exp\left(-\frac{E}{RT}\right) = Mc_p(T - T_\infty) \qquad (2.32)$$

$$VB\rho_\infty Y \exp\left(-\frac{E}{RT}\right) = M(Y_\infty - Y) \qquad (2.33)$$

Hence the relation between the temperature and concentration during adiabatic combustion is

$$Q(Y_\infty - Y) = c_p(T - T_\infty) \qquad (2.34)$$

or

$$\frac{Y}{Y_\infty} = \frac{T_m - T}{T_m - T_\infty} \qquad (2.35)$$

where $T_m = T_\infty + Y_\infty Q/c_p$ is the adiabatic combustion temperature. The heat release and heat loss per unit volume should be

$$q_1 = \frac{Q_1}{V} = QB\rho_\infty Y_\infty\left(\frac{T_m - T}{T_m - T_\infty}\right)\exp\left(-\frac{E}{RT}\right) \qquad (2.36)$$

$$q_2 = \frac{Q_2}{V} = \frac{M}{V}c_p(T - T_\infty) \qquad (2.37)$$

It can be seen from curves of q_1 and q_2 vs. as T (Fig. 2. 10) that unlike the case discussed in Secs. 2. 3. 1 and 2. 3. 2, owing to the concentration change in this simple flow system there are three intersection points with two steady states: slow steady reaction "A" and fast steady reaction "A'" (steady combustion). Variation of the inlet temperature can lead to the occurrence of two critical states: ignition "I" and extinction "E". For $T_\infty > T_{\infty I}$ there is only steady combustion. For $T_\infty < T_{\infty E}$ there is only steady slow reaction. For $T_{\infty E} < T_\infty < T_{\infty I}$ there will be no ignition but also no extinction, if combustion already exists.

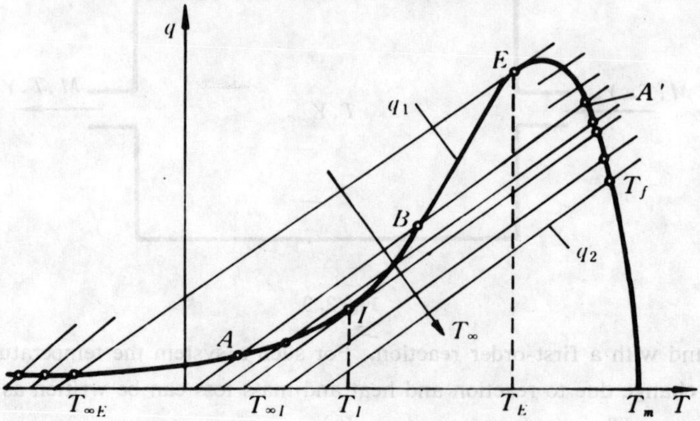

Fig. 2. 10

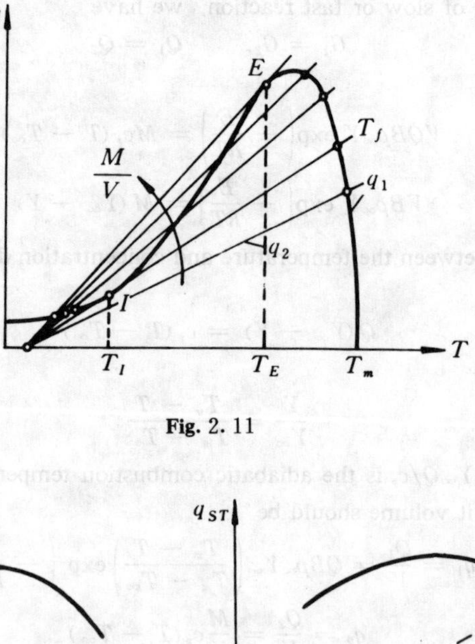

Fig. 2. 11

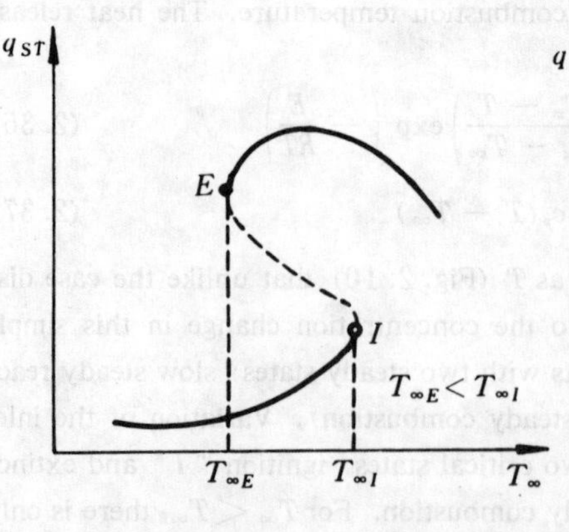

Fig. 2. 12

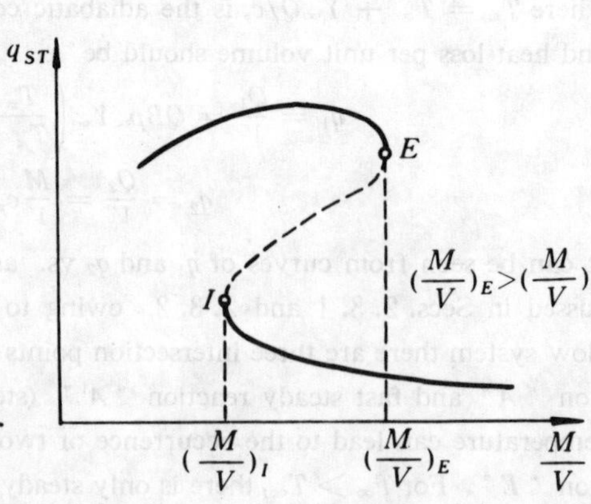

Fig. 2. 13

Variation of M/V can also lead to ignition or extinction (Fig. 2. 11). It can be seen that ignition and extinction are similar phenomena, but they do not happen at the same critical point. The relation between the ignition point " I " and the extinction point " E " can also be seen from curves of the heat release of steady states or equilibrium states q_{st} (the intersection points of curves q_1 with q_2) vs. as parameters T_∞ and M/V in Figs. 2.12 and 2.13. The ignition-extinction criterion can be expressed as:

$$Q_1 = Q_2, \qquad G_1 = G_2, \qquad \frac{\partial Q_1}{\partial T} = \frac{\partial Q_2}{\partial T} \tag{2.38}$$

For further analysis, introducing

reactiveness $\varepsilon_1 = 1 - Y/Y_\infty$, $\varepsilon_2 = c_p(T - T_\infty)/Y_\infty Q$

characteristic times $\tau_f = V\rho_\infty/M \sim L/V$, $\tau_c = 1/B$

dimensionless temperature $\theta = RT/E$

dimensionless heat release $\zeta = Ec_p/(RY_\infty Q)$

the steady-state expressions become

$$\frac{\tau_f}{\tau_c}(1-\varepsilon_1)\exp\left(-\frac{1}{\theta}\right) = \varepsilon_1$$

$$\frac{\tau_f}{\tau_c}(1-\varepsilon_1)\exp\left(-\frac{1}{\theta}\right) = \varepsilon_2 = \zeta(\theta - \theta_\infty)$$

or

$$\begin{cases} \varepsilon_1 = \varepsilon_2 \\ \varepsilon_1 = \left[1 + \dfrac{\tau_c}{\tau_f}\exp\left(\dfrac{1}{\theta}\right)\right]^{-1} \\ \varepsilon_2 = \zeta(\theta - \theta_\infty) \end{cases} \tag{2.39}$$

The ignition-extinction criterion becomes

$$\varepsilon_1 = \varepsilon_2, \qquad \frac{\partial \varepsilon_1}{\partial \theta} = \frac{\partial \varepsilon_2}{\partial \theta} \tag{2.40}$$

$$\begin{cases} \left[1 + \dfrac{\tau_c}{\tau_f}\exp\left(\dfrac{1}{\theta}\right)\right]^{-1} = \zeta(\theta - \theta_\infty) \\ \dfrac{\dfrac{1}{\theta^2}\dfrac{\tau_c}{\tau_f}\exp\left(\dfrac{1}{\theta}\right)}{\left[1 + \dfrac{\tau_c}{\tau_f}\exp\left(\dfrac{1}{\theta}\right)\right]^2} = \zeta \end{cases} \tag{2.41}$$

Obviously, the general form of ignition or extinction relation will be

$$f(\tau_f/\tau_c, \theta_\infty, \zeta) = 0$$

On the basis of Eqs. (2.39) and (2.41), the ignition and extinction caused by variation of D$= \tau_f/\tau_c$ can be shown in Figs. 2.14 and 2.15. which shows $\varepsilon \approx 0$ for ignition and $\varepsilon \approx 1$ for extinction. By using these approximations, the final form of ignition and extinction relations can be obtained as:

$$\begin{cases} \dfrac{\tau_f/\tau_c}{\tau_f/\tau_c + \exp\left(\dfrac{1}{\theta_\infty}\right)} = \zeta\theta_\infty^2 \\ \dfrac{\dfrac{1}{\theta_m^2}\dfrac{\tau_c}{\tau_f}\exp\left(\dfrac{1}{\theta_m}\right)}{\left[1 + \dfrac{\tau_c}{\tau_f}\exp\left(\dfrac{1}{\theta_m}\right)\right]^2} = \zeta \end{cases} \tag{2.42}$$

or

$$\begin{cases} 1+\dfrac{M}{\rho_\infty VB}\exp\left(\dfrac{E}{RT_\infty}\right)=\dfrac{Y_\infty QE}{c_p RT_\infty^2} \\ \dfrac{M}{\rho_\infty VB}\exp\left(\dfrac{E}{RT_m}\right)\left[1+\dfrac{M}{\rho_\infty VB}\exp\left(\dfrac{E}{RT_m}\right)\right]^{-2}=\dfrac{Rc_pT_m^2}{EY_\infty Q} \end{cases} \tag{2.43}$$

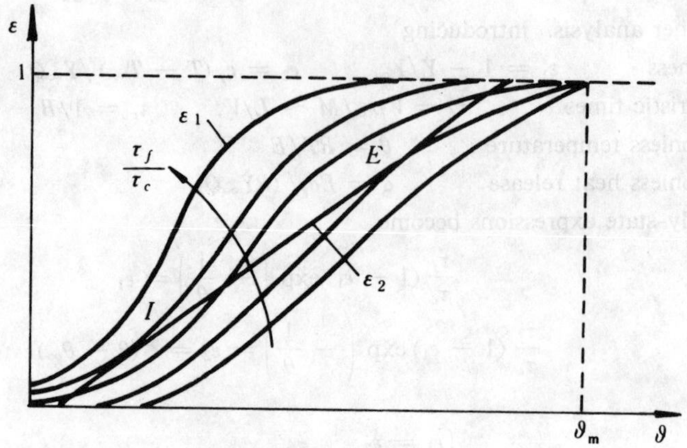

Fig. 2. 14

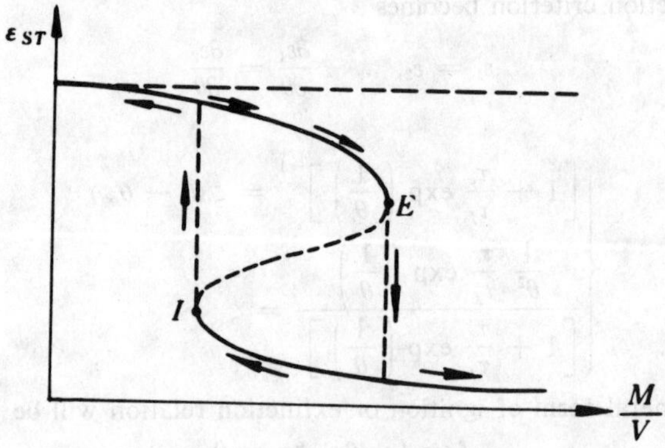

Fig. 2. 15

It can be concluded from this analysis that: (1) ignition and extinction are closely related phenomena, affected by the velocity, diameter, inlet temperature, inlet concentration and fuel/air ratio in a similar way; (2) however, ignition and extinction are not the same critical point, ignition takes place under more favorable conditions; and (3) the initial temperature has stronger effect on ignition than on extinction, and the initial concentration has stronger effect on extinction than on ignition.

2. 3. 4 Lagrangian Treatment of a General Flow System

The Lagrangian treatment can be used for a general flow system. For a fluid element the energy equation is

$$\rho c_p \frac{dT}{dt} = \frac{\partial}{\partial x_j}\left(\lambda_e \frac{\partial T}{\partial x_j}\right) + w_s Q_s$$

where the first and second terms on the right hand side can be considered as the heat loss and heat release of the fluid element. So, the ignition or extinction of any fluid element shoud be

$$\begin{cases} \dfrac{\partial}{\partial x_j}\left(\lambda_e \dfrac{\partial T}{\partial x_j}\right) = - w_s Q_s \\ \dfrac{\partial}{\partial T}\left[\dfrac{\partial}{\partial x_j}\left(\lambda_e \dfrac{\partial T}{\partial x_j}\right)\right] = - Q_s \dfrac{\partial w_s}{\partial T} \end{cases} \tag{2.44}$$

and its dimensionless form is

$$\begin{cases} \dfrac{\partial}{\partial \xi_j}\left(\dfrac{\partial \theta}{\partial \xi_j}\right) = - \mathrm{D}e^{\theta} \\ \dfrac{\partial}{\partial \theta}\left[\dfrac{\partial}{\partial \xi_j}\left(\dfrac{\partial \theta}{\partial \xi_j}\right)\right] = - \mathrm{D}e^{\theta} \end{cases} \tag{2.45}$$

where

$$\mathrm{D} = \tau_d / \tau_c$$

By using an approximate relation

$$\frac{\partial}{\partial \xi_j}\left(\frac{\partial \theta}{\partial \xi_j}\right) \approx \psi(\theta)$$

on the basis of the temperature distribution of nonreacting flows, the critical condition can be expressed as

$$\begin{cases} \psi(\theta) = - \mathrm{D}e^{\theta} \\ \psi'(\theta) = - \mathrm{D}e^{\theta} \end{cases} \tag{2.46}$$

where $\psi(\theta)$ is a function of Re_∞, Pr_∞. By solving

$$\psi(\theta_i) = \psi'(\theta_i)$$

we can find θ_i, and the critical condition becomes

$$\mathrm{D} = - \frac{\psi(\theta_i)}{\exp(\theta_i)} \tag{2.47}$$

In most cases T_i is slightly different from T_1 or $\theta_i \approx 0$, so

$$\mathrm{D} = - \psi(0) \tag{2.48}$$

can be taken as the critical condition.

As an example, consider the ignition of a cold gas mixture with temperature T_∞ in a laminar boundary layer over a hot flat plate with temperature T_w (Fig. 2. 16). Ignition must occur at first in the region immediately adjacent to the hot plate, so the ignition condition is

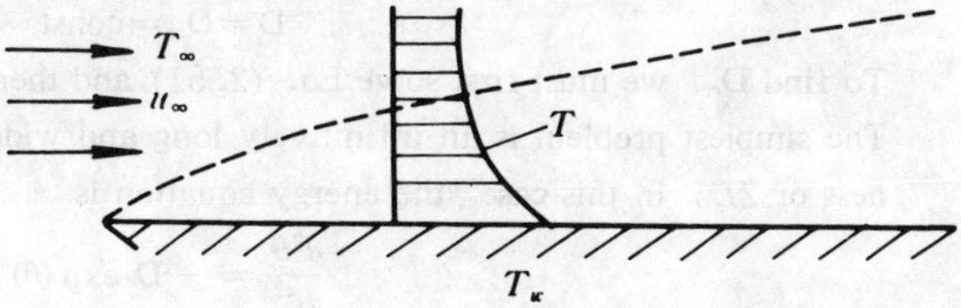

Fig. 2. 16

$$\left[\frac{\partial}{\partial y}\left(\lambda\frac{\partial T}{\partial y}\right)\right]_w = - w_{sw}Q_s \tag{2.49}$$

In case of Pr=1, we have

$$\left[\frac{\partial}{\partial y}\left(\lambda\frac{\partial T}{\partial y}\right)\right]_w \approx \lambda_w\left(\frac{\partial^2 T}{\partial y^2}\right)_w = -\lambda_w(T_w - T_\infty)\frac{f'''(0)}{2}\frac{u_\infty}{\nu_\infty x}\left(\frac{T_\infty}{T_w}\right)$$

where $f'''(0)$ is the third-order derivative of the Blasius Function at the wall. Hence the ignition condition is

$$\left(\frac{T_\infty}{T_w}\right)\lambda_w(T_w - T_\infty)\frac{f'''(0)}{2}\frac{u_\infty}{\nu_\infty x} = w_{sw}Q_s$$

or

$$\frac{\tau_f}{\tau_{cw}} = \frac{f'''(0)}{2}\left(\frac{T_\infty}{T_w}\right)\frac{E(T_w - T_\infty)}{RT_w^2} \tag{2.50}$$

where

$$\tau_f = \frac{x}{u_\infty}, \qquad \tau_{cw} = \frac{\rho_w c_p R T_w^2}{w_{sw}Q_s E}$$

2.4 STEADY AND EULERIAN TREATMENT

2.4.1 Thermal Ignition in a Closed Vessel—Steady Treatment

The basic idea of steady treatment is to consider only the steady equililbrium states of slow reaction in Fig. 2.2. (a). There will be different steady equilibrium states corresponding to different temperature distribution inside the vessel under different boundary conditions. If under certain boundary condition the steady temperature distribution becomes impossible, this boundary condition is just the ignition condition. It has been shown in Sec. 2.2 that the dimensionless form of energy equation for a stagnant reacting gas mixture is

$$\frac{\partial}{\partial \xi_j}\left(\frac{\partial\theta}{\partial \xi_j}\right) = -\,\mathrm{D}\exp(\theta) \tag{2.51}$$

where $\mathrm{D} = \tau_d/\tau_c$. The general form of its solution is

$$\theta = \theta(\mathrm{D}, \xi_j) \tag{2.52}$$

and the ignition condition is

$$\mathrm{D} = \mathrm{D}_{cr} = \mathrm{const} \tag{2.53}$$

To find D_{cr}, we must first solve Eq. (2.51) and then seek for the limiting value of D. The simplest problem is an infinitively long and wide slab reaction zone with a thickness of $2L$. In this case, the energy equation is

$$\frac{d^2\theta}{d\xi^2} = -\,\mathrm{D}\exp(\theta) \tag{2.54}$$

with the boundary condition of

$$\begin{cases} \text{at} \quad \xi = 0: & \dfrac{d\theta}{d\xi} = 0 \\ \text{at} \quad \xi = L: & \theta = 0 \end{cases} \tag{2.55}$$

The analytical solution of Eq. (2.54) with boundary condition (2.55) is

$$\theta = \ln \left[a \Big/ \left(\mathrm{ch}^2 \sqrt{\frac{aD}{2}}\xi \right) \right], \quad a = \mathrm{ch}^2 \sqrt{\frac{aD}{2}} \tag{2.56}$$

Denoting $\sigma = \sqrt{\frac{aD}{2}}$, the function $\theta = \theta(D, \xi)$ can be transformed into $\theta = \theta(\sigma, \xi)$, namely

$$\theta = \ln \frac{\mathrm{ch}^2 \sigma}{\mathrm{ch}^2 (\sigma \xi)}$$

and the relation between σ and D is

$$\frac{\mathrm{ch}\sigma}{\sigma} = \left(\frac{D}{2} \right)^{-\frac{1}{2}}$$

Figure 2.17 gives two possible values of σ for a given D corresponding to two possible steady temperature distribution. The higher value of σ has no physical significance, since the concentration change has not been taken into account in this analysis. It can be seen from Fig. 2.17 that there will be no solution of σ or a steady temperature will not be possible, if D is greater than 0.88, Therefore, the ignition condition is

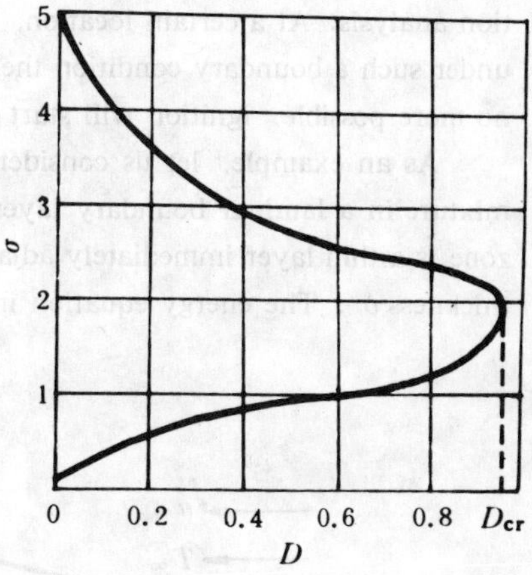

Fig. 2.17

$$D_{cr} = 0.88 \tag{2.57}$$

The maximum temperature rise at the center line before ignition is

$$\theta_i = \frac{E}{RT_\infty^2}(T_i - T_\infty) = \ln \frac{\mathrm{ch}^2 \sigma_{cr}}{\mathrm{ch}^2 (0)} = 1.25 \tag{2.58}$$

By using the similar method from analytical or numerical solution we can find

for cylindrical reactor $\qquad D_{cr} = 2.00, \quad \theta_i = 1.44 \qquad$ (2.59)

for spherical reactor $\qquad D_{cr} = 3.32, \quad \theta_i = 1.67 \qquad$ (2.60)

It is interesting to note that by setting $L = r = d/2$, the result obtained in Sec. 2.3.1 based on the unsteady treatment

$$\frac{E}{RT_\infty^2} w_{s\infty} Q_s = 12 \frac{\lambda}{d^2} \tag{2.17}$$

is equivalent to $D_{cr} = 3$ $\qquad$ with $\qquad \theta_i \approx 1$

It can also be shown that the unsteady treatment leads to

for cylindrical reactor $\qquad D_{cr} = 2 \qquad$ (2.61)

for slab reactor $\qquad D_{cr} = 1 \qquad$ (2.62)

Obviously, the unsteady and steady treatments give the results in good agreement.

2.4.2 Eulerian Treatment of a General Flow System

For a general flow system with chemical reaction, the whole flow domain may be divided into a reaction zone and a nonreacting zone with their corresponding simplified energy equations

$$\frac{\partial}{\partial x_j}\left(\lambda_e \frac{\partial T}{\partial x_j} \right) = - w_s Q_s \tag{2.63}$$

and

$$\rho v_j c_p \frac{\partial T}{\partial x_j} = \frac{\partial}{\partial x_j}\left(\lambda_e \frac{\partial T}{\partial x_j}\right) \tag{2.64}$$

For the reaction zone, the boundary temperature gradient for the limiting case of steady slow reaction can be found on the basis of the steady treatment. For the nonreacting zone, the boundary temperature gradient can be taken from the ordinary heat convection analysis. At a certain location, if the former exceeds the latter, which means that under such a boundary condition the steady temperature distribution of slow reaction is no more possible, ignition will start at that location.

As an example, let us consider once again the problem of ignition of a cold gas mixture in a laminar boundary layer over a hot flat plate (Fig. 2.18). The reaction zone is a thin layer immediately adjacent to the wall with boundary temperature T_w and thickness δ_r. The energy equation in the reaction zone is:

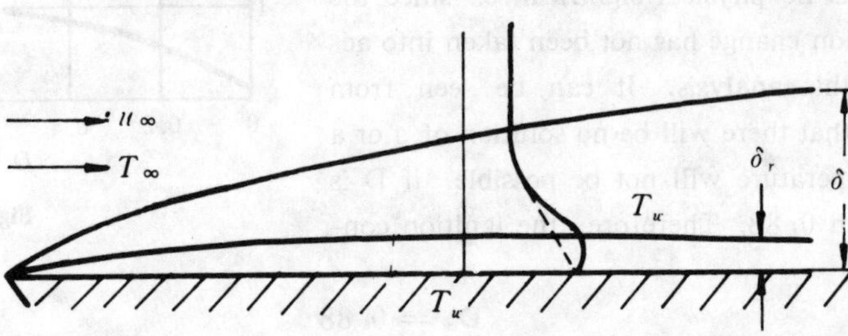

Fig. 2.18

$$\frac{\partial^2 \theta}{\partial \xi^2} = -\mathrm{D}e^{\theta} \tag{2.65}$$

where

$$\theta = \frac{E}{RT_w^2}(T - T_w), \qquad \xi = \frac{y}{\delta_r}, \qquad \mathrm{D} = \frac{E}{RT_w^2}\frac{w_{sw}Q_s}{\lambda_w}\delta_r^2$$

This equation can be transformed into

$$\frac{\partial \theta}{\partial \xi} = \sqrt{2\mathrm{D}\,(a - e^{\theta})} \tag{2.66}$$

where "a" is a constant. The limiting case of steady slow reaction corresponds to a maximum temperature rise with

$$\theta \approx 1 \qquad \text{at} \qquad \frac{\partial \theta}{\partial \xi} = 0$$

Hence we have

$$\frac{\partial \theta}{\partial \xi} = \sqrt{2\mathrm{D}\,(e - e^{\theta})} \tag{2.67}$$

Taking $\theta = 0$, we obtain the temperature gradient at the boundary of the reaction zone

$$\left(\frac{\partial \theta}{\partial \xi}\right)_2 = \sqrt{2\mathrm{D}\,(e - 1)}$$

or

$$\left(\frac{\partial T}{\partial y}\right)_2 = \sqrt{2\,(e-1)\,\frac{RT_w^2}{E}\,\frac{w_{sw}Q_s}{\lambda_w}} \tag{2.68}$$

The temperature gradient at the wall for nonreacting case is

$$\left(\frac{\partial T}{\partial y}\right)_1 = \frac{f''(0)}{2}(T_w - T_\infty)\sqrt{\frac{u_\infty}{\nu_\infty x}}\left(\frac{T_\infty}{T_w}\right) \tag{2.69}$$

The critical condition is

$$\left(\frac{\partial T}{\partial y}\right)_1 = \left(\frac{\partial T}{\partial y}\right)_2$$

or

$$\frac{f''(0)}{2}(T_w - T_\infty)\sqrt{\frac{u_\infty}{\nu_\infty x_i}} = \sqrt{2(e-1)}\frac{RT_w^2}{E}\frac{w_{sw}Q_s}{\lambda_w} \tag{2.70}$$

It can alse be expressed by

$$\frac{\tau_f}{\tau_{cw}} = \frac{[f''(0)]^2}{8(e-1)}\left(\frac{T_\infty}{T_w}\right)^{0.5}\left(\frac{E}{RT_w}\right)^2\left(\frac{T_w - T_\infty}{T_w}\right)^2 \tag{2.71}$$

where

$$\tau_f = \frac{x_i}{u_\infty}, \quad \tau_{cw} = \frac{\rho_w c_p R T_w^2}{w_{sw} Q_s E}$$

Considering $T_\infty / T_w \ll 1$, Eq. (2.50) and Eq. (2.71) can be simplified into

$$\frac{\tau_f}{\tau_{cw}} = \frac{f'''(0)}{2}\left(\frac{T_\infty}{T_w}\right)\frac{E}{RT_w} \tag{2.72}$$

$$\frac{\tau_f}{\tau_{cw}} = \frac{[f''(0)]^2}{8(e-1)}\left(\frac{T_\infty}{T_w}\right)^{0.5}\left(\frac{E}{RT_w}\right)^2 \tag{2.73}$$

It can be seen that these equations are qualitatively similar but quantitatively different in their right-hand side.

2.5 ZERO-GRADIENT ANALYSIS

2.5.1 Basic Phenomena

Conventionally, ignition of a cold combustible mixture by a hot surface, such as a hot wire, a hot plate, a hot cylinder, a hot ball, or even a spark, a flamelet, is called "forced ignition". Such kind of ignition often takes place in gasoline engines and gas turbine combustors. In Secs. 2.3.4 and 2.4.2 we have discussed the ignition of a cold mixture by a hot plate by using the Lagrangian and Eulerian treatment of a general flow system. In this paragraph, we will analyze the phenomena by using the so-called "zero-gradient" analysis proposed by Zeldovich, Khitrin and Goldenberg.

Consider the case of pure cenvective heat transfer from a hot inert plate to inert or single-component gases. The temperature gradient at the wall

$$\left(\frac{\partial T}{\partial y}\right)_w = \frac{\mathrm{Nu}_x}{x}(T_\infty - T_w) = 0.332\sqrt{\frac{u_\infty}{\nu_\infty x}}\mathrm{Pr}^{0.33}(T_\infty - T_w)$$

decreases monotonously with the increase of x -distance, as shown by the dashline temperature profiles in Fig. 2.19 (a). In case of heat transfer with slow reaction in the gas mixture with the wall temperature slightly higher than the temperature of oncoming flow, the temperature profiles in the boundary layer will have a slight deformation, as shown by the solid lines in Fig. 2. 19 (a). Theoretically speaking, if the plate is sufficiently long, eventually at a certain distance, the temperature gradient at the wall may become zero due to the combined effect of heat-release accumulation and thickening boundary layer. However, no ignition will take place in this case. If the wall temperature is sufficiently high (e. g. 1273 K), under certain conditions ignition may take place at a finite distance x_i smaller than the plate length L and the temperature profile will deform so remarkably, that there will be a local temperature peak, as shown by the solid lines in Fig. 2. 19 (b). In this case, the wall temperature gradient will increase from negative values through zero to positive values along the x direction, as shown in Fig. 2. 19 (c).

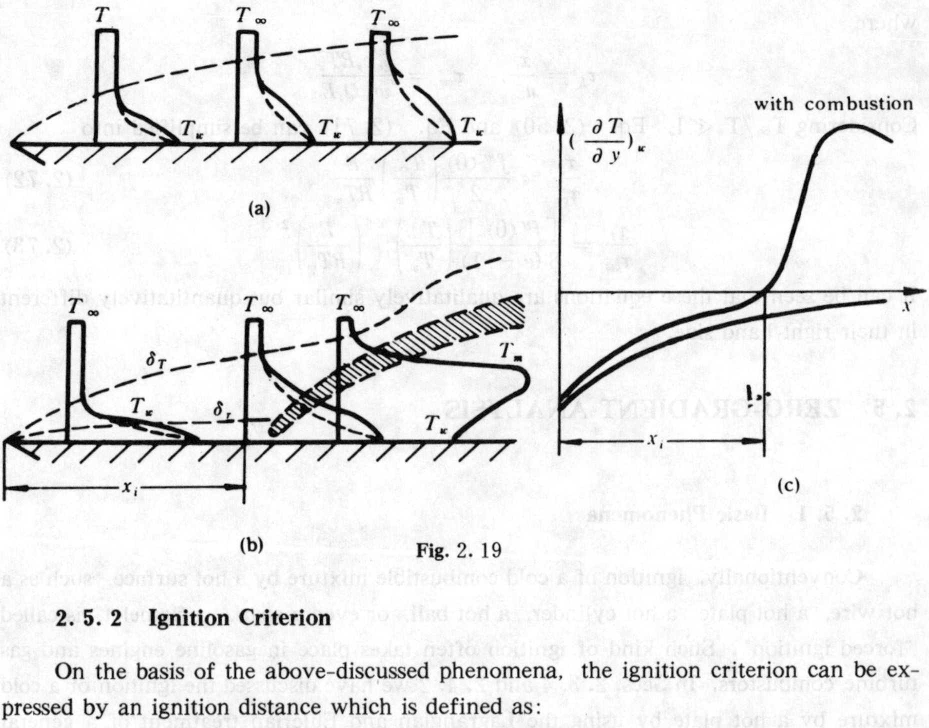

Fig. 2. 19

2. 5. 2 Ignition Criterion

On the basis of the above-discussed phenomena, the ignition criterion can be expressed by an ignition distance which is defined as:

$$\text{at} \quad x = x_i \quad : \quad \left(\frac{\partial T}{\partial y}\right)_w = 0 \qquad (2.74)$$

This concept is similar to what we have used in flow separation, where the velocity-profile deformation due to negative pressure gradient leads to

$$\left(\frac{\partial u}{\partial y}\right)_w = 0$$

at the separation point. The "Zero temperature gradient" analysis was proposed by Zeldovich, and was used by Khitrin and Goldenberg, and T. Y. Toong in their simplified solution and exact solution respectively. Although this criterion is only a relative condition and actually flame may arise not just from this distance, but the physical meaning is that starting from $x = x_i$ the heat transfer will change its direction: from gas to wall.

Obviously, if the ignition distance is greater than the plate length, or $x_i > L$, there will be ignition failure.

2.5.3 Simplfied Two-zone Theory

The whole boundary layer can be divided into two zones (Fig. 2.19 (d)): a thin reaction zone immediately adjacent to the wall whose thickness δ_r is much smaller than the thickness of the thermal boundary layer δ_T; a heat transfer zone where reaction is negligible. In the high-temperature and low-velocity reaction zone the convective term in the energy equation is negligible, so the energy equation

$$\rho u c_p \frac{\partial T}{\partial x} + \rho v c_p \frac{\partial T}{\partial y} = \frac{\partial}{\partial y}\left(\lambda \frac{\partial T}{\partial y}\right) + w_s Q_s$$

can be simplified into

$$\lambda_w \frac{\partial^2 T}{\partial y^2} = - w_s Q_s \tag{2.75}$$

In the heat-transfer zone where the reaction term can be neglected the energy equation can be simplified into

$$\rho u c_p \frac{\partial T}{\partial x} + \rho v c_p \frac{\partial T}{\partial y} = \frac{\partial}{\partial y}\left(\lambda \frac{\partial T}{\partial y}\right) \tag{2.76}$$

To solve Eq. (2.75), we can change it into

$$\frac{\partial^2 T}{\partial y^2} = \frac{\partial T}{\partial y}\frac{\partial}{\partial T}\left(\frac{\partial T}{\partial y}\right) = \frac{\partial}{\partial T}\left[\frac{1}{2}\left(\frac{\partial T}{\partial y}\right)^2\right] = -\frac{w_s Q_s}{\lambda_w}$$

Integrating this equation gives

$$\left(\frac{\partial T}{\partial y}\right)_w^2 - \left(\frac{\partial T}{\partial y}\right)_{\text{I}}^2 = \frac{2Q_s}{\lambda_w}\int_{T_w}^{T_r} w_s dT \approx \frac{2Q_s}{\lambda_w}\int_{T_w}^{T_\infty} w_s dT$$

Using $\left(\frac{\partial T}{\partial y}\right)_w = 0$ at $x = x_i$, we obtain

$$\left(\frac{\partial T}{\partial y}\right)_{\text{I}} = \sqrt{\frac{2Q_s}{\lambda_w}\int_{T_\infty}^{T_w} w_s dT} \tag{2.77}$$

The solution of Eq. (2.76) gives the temperature gradient at the wall for nonreacting cases

$$\left(\frac{\partial T}{\partial y}\right)_{\text{I}} = \left(\frac{\partial T}{\partial y}\right)_{*w} = \frac{\text{Nu}_x^*}{x}(T_w - T_\infty)$$

where Nu_x^* is the Nusselt number for convective heat transfer of nonreacting flows. The coupling of two zones requires

$$\left(\frac{\partial T}{\partial y}\right)_{\text{I}} = \left(\frac{\partial T}{\partial y}\right)_{\text{I}} \tag{2.78}$$

Hence we obtain the Khitrin-Goldenberg's Expression for ignition by a hot plate

$$\frac{(\text{Nu}^*)^2}{x_i^2} = \frac{1}{(T_w - T_\infty)^2}\frac{2Q_s}{\lambda_w}\int_{T_\infty}^{T_w} w_s dT \tag{2.79}$$

After approximate integration we have

$$\int_{T_\infty}^{T_w} w_s dT \approx \frac{RT_w^2}{E} w_{sw}$$

For the heat transfer in a laminar boundary layer with $\text{Pr}=1$, we have

$$\mathrm{Nu}_x^* = \frac{f''(0)}{2}\sqrt{\frac{u_\infty x}{\nu_\infty}\left(\frac{T_\infty}{T_w}\right)} \qquad (2.80)$$

Therefore, the ignition expression becomes

$$\frac{x_i}{u_\infty} = \frac{[f''(0)]^2}{8}\left(\frac{T_\infty}{T_w}\right)^{0.5}\left(\frac{E}{RT_w}\right)^2\left(\frac{T_w - T_\infty}{T_w}\right)^2 \frac{RT_w^2}{E}\frac{c_p\rho_w}{w_{sw}Q_s} \qquad (2.81)$$

or

$$\frac{\tau_f}{\tau_{cw}} = \frac{[f''(0)]^2}{8}\left(\frac{E}{RT_w}\right)^2\left(\frac{T_w - T_\infty}{T_w}\right)^2\left(\frac{T_\infty}{T_w}\right)^{0.5} \qquad (2.82)$$

where

$$\tau_f = \tau_i = x_i/u_\infty, \qquad \tau_{cw} = \frac{RT_w^2}{E}\frac{c_p\rho_w}{w_{sw}Q_s}$$

x_i is called the ignition distance, and τ_i is called the ignition lag, but the latter is different from that defined in the case of ignition in a closed vessel discussed in Sec. 2. 3. 2.

Comparison of Eq. (2. 82) with Eq. (2. 71) shows that these equations are quite similar with only the difference in a factor of $(e-1)$. For constant pressure and air/fuel ratio, the ignition relation can be approximately expressed by

$$\ln\left(\frac{x_i}{u_\infty}\right) = \frac{E}{RT_w} + \text{const} \qquad (2.83)$$

The linear functions of $\ln x_i$ vs as $\frac{1}{T_w}$ and x_i vs as u_∞ have been verified by the experiments on ignition by a hot plate (Fig. 2. 20 and 2. 21). Both increasing and decreasing

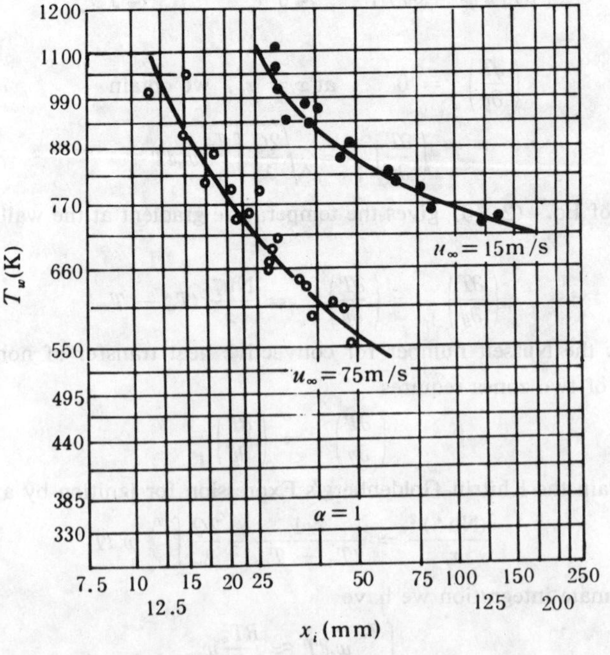

Fig. 2. 20

air/fuel ratio can lead to increasing the ignition distance x_i. This is verified by the experiments on ignition by a hot plate (Fig. 2. 22). All this indicates that the simplified theory is qualitatively in agreement with experiments.

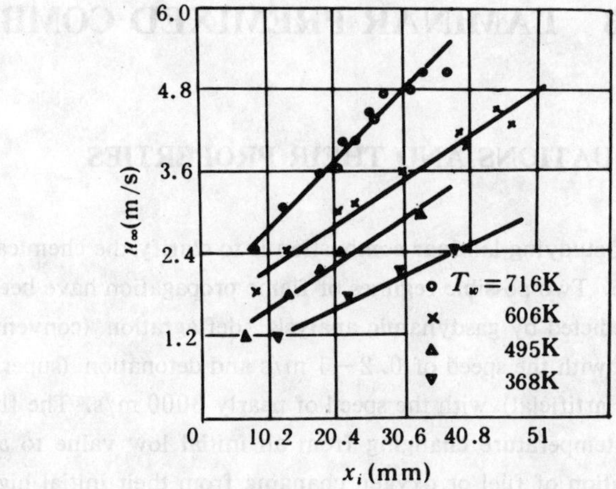

Fig. 2. 21

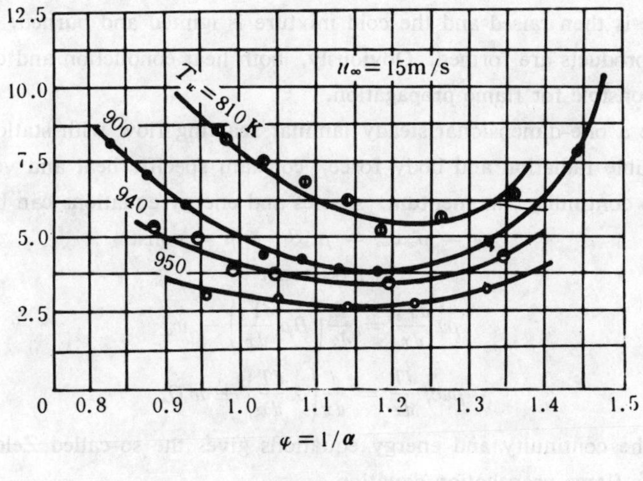

Fig. 2. 22

CHAPTER 3 LAMINAR PREMIXED COMBUSTION

3.1 BASIC EQUATIONS AND THEIR PROPERTIES

The purpose of studying laminar combustion is to clarify the chemical kinetics of a combustible mixture. Two possible regimes of flame propagation have been observed by experiments and predicted by gasdynamic analysis: deflagration (conventional laminar or turbulent flame) with the speed of 0.2—1 m/s and detonation (supersonic combustion, spontaneous or artificial) with the speed of nearly 3000 m/s. The flame is a finite reaction zone with temperature changing from an initial low value to a final highest value and concentration of fuel or oxygen changing from their initial highest values to zero. The flame propagation is a process, in which the heat is conducted from the hot combustion products to their neighbouring cold fresh mixture. The temperature of the cold mixture is then raised and the cold mixture is ignited and burned. Finally, new combustion products are formed. Obviously, both heat conduction and chemical reaction are responsible for flame propagation.

Assume a one-dimensional steady laminar reacting flow with stationary reaction zone, negligible radiation and body force, constant specific heat and very low Mach number, the continuity, momentum, species and energy equations can be written as:

$$\rho u = \rho_\infty u_\infty = \rho_\infty S_l = m = \text{const} \tag{3.1}$$

$$p \approx \text{const} \tag{3.2}$$

$$\rho u \frac{dY_s}{dx} = \frac{d}{dx}\left(D\rho \frac{dY_s}{dx}\right) - w_s \tag{3.3}$$

$$\rho u c_p \frac{dT}{dx} = \frac{d}{dx}\left(\lambda \frac{dT}{dx}\right) + w_s Q_s \tag{3.4}$$

Combining the continuity and energy equations gives the so-called Zeldovich-Frank-Kamenetsky's flame propagation equation

$$\rho_\infty S_l c_p \frac{dT}{dx} = \frac{d}{dx}\left(\lambda \frac{dT}{dx}\right) + w_s Q_s \tag{3.5}$$

The adiabatic (without heat losses to the wall, cold boundary and hot boundary) boundary conditions are:

$$x \to -\infty: \quad T \to T_\infty,\ Y_s \to Y_{s\infty},\ \frac{dT}{dx} \to 0,\ \frac{dY_s}{dx} \to 0$$

$$x \to +\infty: \quad \frac{dT}{dx} \to 0,\ \frac{dY_s}{dx} \to 0 \tag{3.6}$$

We are interested not in the profiles $T(x)$ or $Y_s(x)$ themself, but in finding out $m = \rho_\infty S_l$ or S_l —the finite velocity at which the steady T-profile or Y-profile moves.

The solution of a second-order differential equation with three boundary conditions exists only for a certain S_l. This is called an eigen-value problem. Another problem is the so-called "Cold-boundary difficulty". Setting

$$p = \frac{dT}{dx}, \qquad \lambda = \text{const}$$

we have

$$\frac{\lambda}{c_p \rho_\infty} p \frac{dp}{dT} - S_l p + \frac{w_s Q_s}{c_p \rho_\infty} = 0 \tag{3.7}$$

It can be seen from the cold-boundary condition $p\,(-\infty) = o$ that if $w_s\,(T_\infty) \neq 0$, S_l will approach infinity, or there will be no solution. The existence of a definite solution requires

$$w_s\,(T_\infty) = 0, \qquad \text{at} \quad x \to -\infty$$

whose physical meaning is no ignition at the cold boundary. However, the practical cases are

$$w_s\,(T) \sim \exp\,(-E/RT)\,,\; w_s\,(T_\infty) \neq 0$$

This is called the "Cold-boundary difficulty". Fortunately, in case of flame propagation T_∞ is much smaller than the ignition temperature $T_{\infty i}$ and $w_s\,(T_\infty)$ is really negligibly small, so the physical process can be approximately described by the above-stated mathematical formulation.

Next, let us consider the similarity of temperature and concentration profiles and the conservation of enthalpy across the flame zone. Integrating the Energy Equation (3. 4) and Diffusion Equation (3. 3) from $x = -\infty$ to $x = +\infty$ by using the boundary condition gives

$$\rho_\infty S_l c_p\,(T_m - T_\infty) = \int_{-\infty}^{+\infty} w_s Q_s dx = \rho_\infty S_l Y_{s\infty} Q_s$$

$$c_p\,(T_m - T_\infty) = Y_{s\infty} Q_s$$

or

$$c_p T_m = c_p T_\infty + Y_{s\infty} Q_s \tag{3.8}$$

It is obvious that the enthalpy at the cold boundary is always equal to that at the hot boundary, no matter the Lewis number $\text{Le} = D\rho/\lambda/c_p$ is unity or not, and the temperature at the hot boundary is the adiabatic flame temperature T_m. Taking

$$\text{Le} = \frac{D\rho}{\lambda/c_p} = 1$$

$$\theta = c_p\,(T_m - T)\,/\,(Y_{s\infty} Q_s) = (T_m - T)\,/\,(T_m - T_\infty)$$

$$F = Y_s / Y_{s\infty}$$

we have

$$S_l \frac{d\theta}{dx} = \frac{d}{dx}\left(\frac{\lambda}{\rho_\infty c_p}\frac{d\theta}{dx}\right) - w_s / Y_{s\infty}$$

$$S_l \frac{dF}{dx} = \frac{d}{dx}\left(\frac{\lambda}{\rho_\infty c_p}\frac{dF}{dx}\right) - w_s / Y_{s\infty}$$

with

$$\theta\,(o) = F\,(o) = 1,\; \theta\,(\infty) = F\,(\infty) = 0$$

and hence

$$\theta = F \tag{3.9}$$

$$Y_s/Y_{s\infty} = (T_m - T)/(T_m - T_\infty) = c_p(T_m - T)/(Y_{s\infty}Q_s)$$

or

$$c_pT + Y_sQ_s = c_pT_\infty + Y_{s\infty}Q_s = c_pT_m \tag{3.10}$$

The conclusions are: (A) when Le=1, T-profile and Y-profile are similar across the flame; (B) when Le=1 and only when Le=1, the enthalpy will be conserved across the flame everywhere.

3.2 TWO-ZONE APPROXIMATE SOLUTION

The classical two-zone model was proposed by Zeldovich and Frank-Kamenetsky. The two-zone concept can be illustrated in Fig. 3.1. Considering that the heat-source term is a highly nonlinear function of temperature, we can divide the flame into two zones: the preheating zone and the reaction zone. In the preheating zone the reaction term in the energy equation can be neglected due to negligible heat release under lower temperature, and in the reaction zone the convection term can be neglected, compared with other two terms.

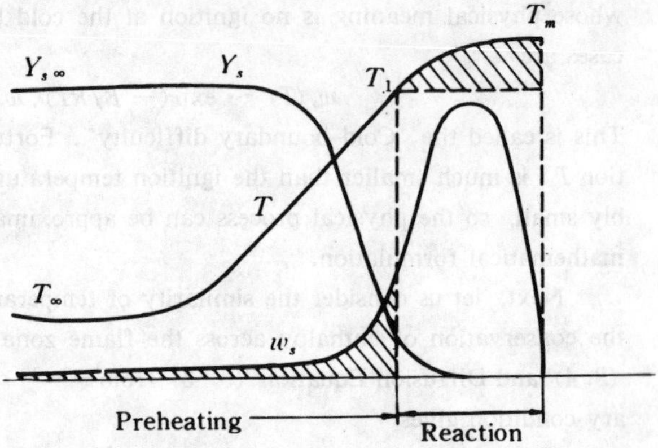

Fig. 3.1

Therefore, the simplified energy equations for these two zones can be written as:

$$\rho_\infty S_l c_p \frac{dT}{dx} = \frac{d}{dx}\left(\lambda \frac{dT}{dx}\right) \tag{3.11}$$

$$\frac{d}{dx}\left(\lambda \frac{dT}{dx}\right) + w_sQ_s = 0 \tag{3.12}$$

Integrating Eq. (3.11) from $x = -\infty$ to the location of $T = T_1$ (where T_1 is the temperature of preheating zone at the hot boundary) by using the cold-boundary condition gives

$$\rho_\infty S_l c_p (T_1 - T_\infty) = \lambda \left(\frac{dT}{dx}\right)_1 \tag{3.13}$$

Integrating Eq. (3.12) from the location of $T = T_2$ (where T_2 is the temperature at the cold boundary of the reaction zone) to $x = \infty$ by using the hot-boundary condition gives

$$\left(\frac{dT}{dx}\right)_2 = \sqrt{\frac{2Q_s}{\lambda}\int_{T_2}^{T_m} w_s dT} \tag{3.14}$$

Assume that these two zones are directly coupled, that is:

$$T_1 = T_2, \quad \left(\frac{dT}{dx}\right)_1 = \left(\frac{dT}{dx}\right)_2 \tag{3.15}$$

the physical meaning of this assumption is the equality of heat release by reaction with the heat loss by conduction. Then, we have

$$S_l = \sqrt{\frac{2\lambda Q_s \int_{T_1}^{T_m} w_s dT}{\rho_\infty^2 c_p^2 (T_1 - T_\infty)^2}} \tag{3.16}$$

Further approximations are made by taking

$$T_1 - T_\infty \approx T_m - T_\infty$$

and

$$\int_{T_1}^{T_m} w_s dT \approx \int_{T_\infty}^{T_m} w_s dT$$

These approximations seem to be contradictionary, however they are based on the consideration that the fraction of the reaction-rate integral over temperature range of T_∞ to T_1 will make negligible contribution to the whole integral, and the temperature of starting reaction T_1 is near the temperature T_m. Therefore the final expression of the laminar flame-propagation velocity is

$$S_l = \sqrt{\frac{2\lambda Q_s \int_{T_\infty}^{T_m} w_s dT}{\rho_\infty^2 c_p^2 (T_m - T_\infty)^2}} \tag{3.17}$$

The physical meaning of Eq. (3.17) can be explained as follows. The term

$$\frac{1}{T_m - T_\infty} \int_{T_\infty}^{T_m} w_s dT$$

is a mean reaction rate over the temperature range of $(T_m - T_\infty)$, and

$$\frac{\rho_\infty c_p (T_m - T_\infty)}{\int_{T_\infty}^{T_m} w_s dT / (T_m - T_\infty)} = \bar{\tau}_c$$

is a mean reaction time. Denoting

$$\bar{\tau}_f = \frac{a}{S_l^2} = \frac{\delta}{S_l}, \quad \delta = \frac{a}{S_l}, \quad a = \frac{\lambda}{c_p \rho_\infty}$$

we have

$$S_l \sim \sqrt{\frac{a}{\bar{\tau}_c}} \tag{3.18}$$

and

$$\bar{\tau}_f / \bar{\tau}_c = 2 = \text{const} \tag{3.19}$$

where δ is the flame thickness and $\bar{\tau}_f$ is a mean flame propagation time. Clearly, the flame propagation velocity depends on heat release by reaction and heat conduction, but it is only a function of physico-chemical properties of the gas mixture and is independent of flow parameters, such as gas velocity and geometric size. Eq. (3.17) gives

$$S_l \sim \sqrt{w_s / \rho_\infty^2} \sim \sqrt{p^n / p^2}$$

Hence we have

$$S_l \sim p^{\frac{n}{2} - 1} \tag{3.20}$$

and

$$\delta \sim p^{-\frac{n}{2}} \tag{3.21}$$

For hydrocarbons the reaction order is $n=1.5-2$, therefore the laminar flame propagation velocity slightly increases or keeps unchanged with the decrease of pressure, and the flame thickness increases with the decrease of pressure.

Some modifications have been made to the Z. F. K. 's two-zone model. The numerical solution of Eq. (3. 5) is shown in Fig. 3. 2. The results indicate

$$\int_{T_\infty}^{T_m} w_s dT \approx (1.6 \sim 2.5)\int_{T_1}^{T_m} w_s dT$$

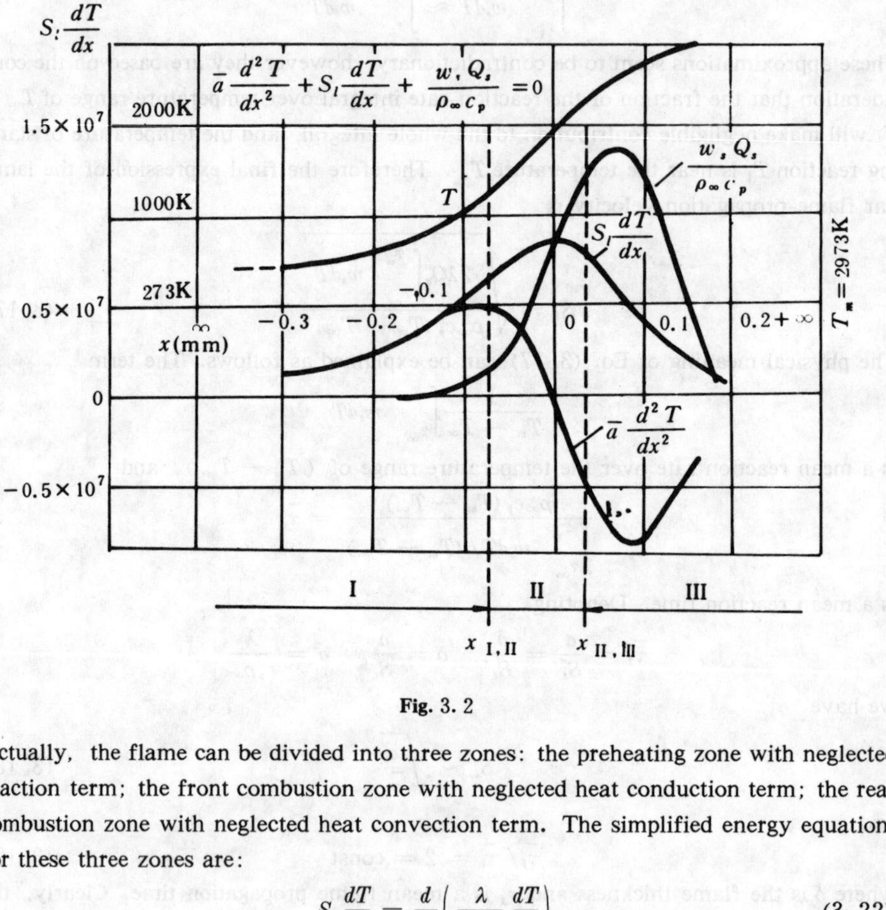

Fig. 3. 2

Actually, the flame can be divided into three zones: the preheating zone with neglected reaction term; the front combustion zone with neglected heat conduction term; the rear combustion zone with neglected heat convection term. The simplified energy equations for these three zones are:

$$S_l\frac{dT}{dx}=\frac{d}{dx}\left(\frac{\lambda}{c_p\rho_\infty}\frac{dT}{dx}\right) \tag{3. 22}$$

$$S_l\frac{dT}{dx}=\frac{w_s Q_s}{\rho_\infty c_p} \tag{3. 23}$$

$$\frac{d}{dx}\left(\frac{\lambda}{\rho_\infty c_p}\frac{dT}{dx}\right)=-\frac{w_s Q_s}{\rho_\infty c_p} \tag{3. 24}$$

Actually, the Z. F. K. model takes

$$\left(\frac{dT}{dx}\right)_{\mathrm{I,II}}=\left(\frac{dT}{dx}\right)_{\mathrm{II,I}}$$

as the coupling condition. Alternatively, the modified two-zone model takes

$$\left(\frac{dT}{dx}\right)_{\text{I, II}} = \left(\frac{dT}{dx}\right)_{\text{II, I}} \tag{3. 25}$$

as the coupling condition. Owing to

$$\left(\frac{dT}{dx}\right)_{\text{I, II}} = \frac{\rho_\infty c_p S_l (T_1 - T_\infty)}{\lambda_1}$$

$$\left(\frac{dT}{dx}\right)_{\text{II, I}} = \frac{w_{s1} Q_s}{\rho_\infty c_p S_l}$$

we have

$$S_l = \sqrt{\frac{\lambda_1 w_{s1} Q_s}{\rho_\infty^2 c_p^2 (T_1 - T_\infty)}} \tag{3. 26}$$

However, T_1 is an unknown quantity. The second condition may be determined from an ignition concept

$$\frac{\partial}{\partial T}\left[\left(\frac{dT}{dx}\right)_{\text{I, II}}\right] = \frac{\partial}{\partial T}\left[\left(\frac{dT}{dx}\right)_{\text{II, I}}\right]$$

or

$$\left(\frac{d^2T}{dx^2}\right)_{\text{I, II}} = \left(\frac{d^2T}{dx^2}\right)_{\text{II, I}} \tag{3. 27}$$

that is

$$\frac{S_l c_p \rho_\infty}{\lambda_1} = \frac{Q_s}{\rho_\infty c_p S_l} w_s' (T_1)$$

Finally the modified two-zone model gives

$$\begin{cases} S_l = \sqrt{\dfrac{\lambda_1 w_{s1} Q_s}{c_p^2 \rho_\infty^2 (T_1 - T_\infty)}} \\ \dfrac{1}{T_1 - T_\infty} = \dfrac{w_s' (T_1)}{w_s (T_1)} \end{cases} \tag{3. 28}$$

where $w_s (T_1)$, $w_s' (T)$ are the reaction rate and its derivative at the temperature T_1 respectively.

3. 3 EXACT SOLUTION

Toong and Klein used different iteration methods to find the exact solution of the flame propagation equation. Integrating Eq. (3. 5) from $x = -\infty$ to $x = \infty$ gives

$$\rho_\infty c_p S_l (T_m - T_\infty) = Q_s \int_{-\infty}^{\infty} w_s dx = Q_s \int_{T_\infty}^{T_m} w_s \left(\frac{dx}{dh}\right) dh$$

or

$$S_l = \frac{Q_s}{\rho_\infty c_p (T_m - T_\infty)} \int_{T_\infty}^{T_m} \frac{w_s}{\left(\dfrac{dh}{dx}\right)} dh \tag{3. 29}$$

where $dh = c_p dT$. Equation (3. 29) points out that the laminar flame propagation velocity depends on the temperature-averaged reaction rate, which has already been indicated by the approximate solution. If we can express the enthalpy gradient *(dh/dx)* in terms of a function of h, then the integral on the right-hand-side of Eq. (3. 29) can be calculated and S_l can be found. In doing so, the energy equation Eq. (3. 5) may be once again integrated from $x = x$ to $x = \infty$, the result gives

$$\rho_\infty c_p S_l (T_m - T) = -\lambda \frac{dT}{dx} + Q_s \int_T^{T_m} \frac{w_s}{\left(\frac{dh}{dx}\right)} dh$$

or

$$\frac{dh}{dx} = \frac{c_p Q_s}{\lambda} \int_T^{T_m} \frac{w_s}{\left(\frac{dh}{dx}\right)} dh - \frac{S_l \rho_\infty c_p}{\lambda} (h_m - h) \tag{3. 30}$$

Solving Eq. (3. 29) and Eq. (3. 30) with two unknowns S_l and (dh/dx) by using an iteration method can give the exact solution of S_l.

The iteration procedure is:

(1) As a starting step, using the idea of approximate solution, take

$$\left(\frac{dT}{dx}\right)^{(1)} = \sqrt{\frac{2Q_s}{\lambda} \int_T^{T_m} w_s dT} = f_1 (T)$$

or

$$\left(\frac{dh}{dx}\right)^{(1)} = \sqrt{\frac{2Q_s c_p^2}{\lambda} \int_T^{T_m} w_s dT} = f_1 (h) \tag{3. 31}$$

(2) Substitute $(dh/dx)^{(1)}$ into the right-hand-side of Eq. (3. 29) to find out $S_l^{(1)}$.

(3) Substitute $S_l^{(1)}$ and $(dh/dx)^{(1)}$ into the right-hand-side of Eq. (3. 30) to find out $(dh/dx)^{(2)} = f_2 (h)$.

(4) Iterate till convergence, that is

$$\left(\frac{dh}{dx}\right)^{(n-1)} \longrightarrow \left(\frac{dh}{dx}\right)^{n}, \; S_l^{(n-1)} \longrightarrow S_l^{(n)}$$

It can be shown that the first approximation $S_l^{(1)}$ obtained by using Eq. (3. 29) and Eq. (3. 31) is just the approximate solution obtained by the two-zone model. The exact expression is

$$\rho_\infty S_l c_p (T_m - T_\infty) = Q_s \int_{-\infty}^{\infty} w_s dx$$

Taking the approximate expression

$$\left(\frac{dT}{dx}\right)^{(1)} \approx \sqrt{\frac{2Q_s}{\lambda} \int_T^{T_m} w_s dT}$$

in the entire range of $x = -\infty$ to $x = \infty$, we have

$$Q_s \int_{-\infty}^{\infty} w_s dx \approx \sqrt{2\lambda Q_s \int_{T_\infty}^{T_m} w_s dT}$$

or

$$\rho_\infty S_l c_p (T_m - T_\infty) = \sqrt{2\lambda Q_s \int_{T_\infty}^{T_m} w_s dT}$$

hence we have

$$S_l^{(1)} = \sqrt{\frac{2\lambda Q_s}{\rho_\infty^2 c_p^2 (T_m - T_\infty)^2} \int_{T_\infty}^{T_m} w_s dT}$$

It is obvious that the flame velocity S_l obtained from the two-zone model, or from the first approximation is smaller than that given by the exact solution, because the approximate temperature gradient is greater than the exact one in most regions of the flame.

3.4 EXPERIMENTAL RESULTS IN BUNSEN BURNERS

Historically, different methods have been developed to measure the laminar flame propagation velocity, such as long tube, constant-volume bomb, constant-pressure bomb (soap bubble), flat-flame burner and Bunsen burner. Among them the Bunsen burner is most widely used to produce a steady laminar premixed flame. To determine the flame propagation velocity, either the flame-cone angle, or the flame shape is measured by direct photography, shadow-graphy, interferometry, or by thermocouples.

The experimental results indicate that the flame velocity depends on the initial gas temperature, pressure, fuel/air ratio, fuel type (in particular the chemical kinetics properties), but is independent of the gas velocity and tube diameter. The effect of fuel/air ratio and fuel type can be shown in Figs. 3.3, 3.4. S_l will reach its maximum value at a fuel/air ratio slightly richer than the stoichiometric ratio, and the flame velocity of unsaturated hydrocarbons is greater than that of saturated hydrocarbons (Fig. 3.4). The effect of pressure is

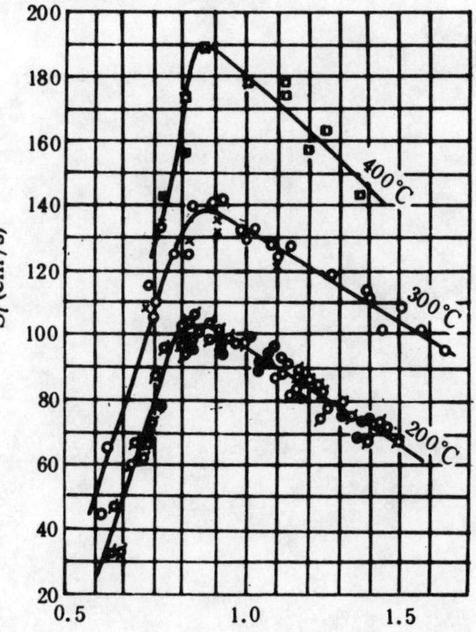

Fig. 3.3

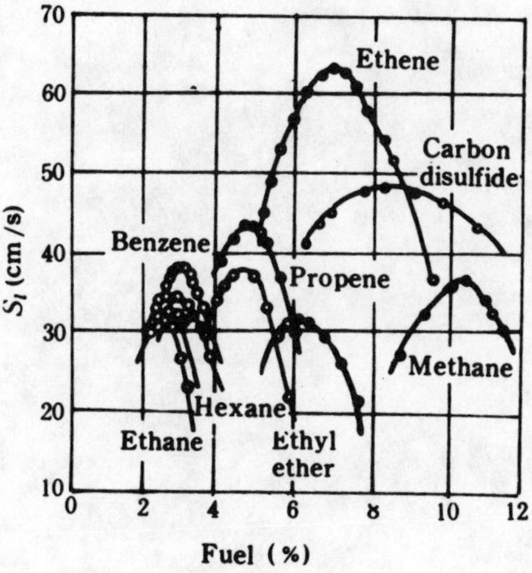

Fig. 3.4

$$S_l \sim p^{(-0.15)-(-0.45)} \tag{3.32}$$

which implies that in most cases S_l slightly increases with the decrease of pressure, or the reaction order is between 1 and 2. The effect of initial temperature is

$$S_l \sim T_\infty^{1.5-2.0} \tag{3.33}$$

One of the ways to combine the theory with experiments is to predict the chemical kinetics properties by using the theory and experimental data. For example, the activation energy E and reaction order n can be determined by

$$\ln S_l^2 = -\frac{E}{RT_m} + C_1 \tag{3.34}$$

and

$$\ln S_l = \left(\frac{n}{2} - 1\right) \ln p + C_2 \tag{3.35}$$

Substituting n and E into the formula of S_l, the preexponential factor B can be found. As an example, by using the flame theory and experimental data in Bunsen burners, the kinetics of $CO\text{-}O_2$ reaction has been found as:

$$w_{CO} = (1.14 - 2.5) \times 10^6 Y_{CO} Y_{ox}^{0.25} \rho^{1.25} T^{-2.25} \exp\left(-\frac{23000}{T}\right)$$

This result is in agreement with that obtained by direct kinetic measurements, which illustrates that the laminar flame theory is plausible.

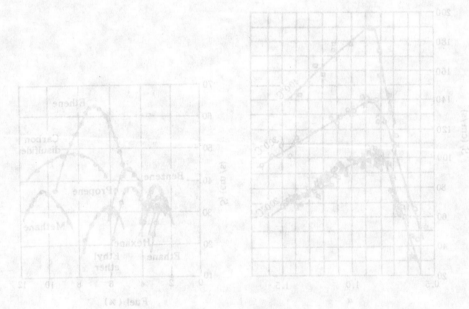

CHAPTER 4 DROPLET AND PARTICLE COMBUSTION

4. 1 DROPLET EVAPORATION WITH AND WITHOUT COMBUSTION

The combustion of liquid fuel may take place in free surface, spray or fuel vapor after prevaporization. One of the striking features of combustion of liquid fuel is that in most cases there is no reaction at the liquid surface and combustion takes place only in gas phase with vaporized fuel, because the temperature of the liquid surface is lower than the boiling point which in turn is much lower than the ignition temperature. Frequently, the ignition temperature of hydrocarbon-air mixtures is greater than 900K. The boiling point of different petroleum products is:

Gasoline	360—380K
Kerosene	420—500K
Transformer oil	560—610K
Lubrication oil	620—670K

Besides, the latent heat for evaporation is much smaller than the activation energy for reaction. Therefore, combustion of liquid fuel always takes place after its evaporation. Furthermore, the combustion of liquid free surface or a liquid droplet is always diffusion-controlled combustion, since the fuel vapor and oxygen will diffuse from opposite directions to the flame zone prior to combustion. It has been found in experimental observation that the burning rate of liquid fuel is inversely proportional to its density. All these indicate that evaporation plays an important role in liquid fuel combustion.

A widely used method of liquid fuel combustion is spray combustion. Depending on the inlet temperature, the droplet size, and the fuel type, spray combustion may take the form of fully vaporized combustion , droplet diffusion combustion or vapor flame plus droplet combustion. In any case, to study the droplet evaporation with and without combustion is of vital importance for understanding the mechanism of liquid fuel combustion and numerical modeling of two-phase flows and combustion.

4. 1. 1 Droplet Evaporation in Low-Temperature Environment

The evaporation is a phase change in which the liquid molecules at the free surface absorb the energy from the gas environment and vaporize. The static behavior of evaporation, such as latent heat, and boiling point are thermodynamic properties. The dynamic behavior of evaporation, such as the evaporation rate, depends on heat and mass transfer. For evaporation in low-temperature environment, the effect of evaporation on flow, heat and mass transfer can be neglected, and the droplet evaporation can be considered as a result of conventional heat and mass transfer of a nonevaporating sphere. In this case the total heat flux is :

$$Q = 4\pi r_p^2 h (T_g - T_w) = \pi d_p \,\mathrm{Nu}\, \lambda (T_g - T_w)$$

and

$$Q = G q_e$$

where G is the evaporation rate, and q_e is the heat of evaporation. Hence, we have

$$G = \pi d_p \,\mathrm{Nu}\, \frac{\lambda}{c_p} B \tag{4.1}$$

where

$$B = \frac{c_p (T_g - T_w)}{q_e}$$

The law of mass transfer gives

$$G = \pi d_p^2 h_d \rho (Y_w - Y_g) = \pi d_p \,\mathrm{Nu}\, D\rho (Y_w - Y_g)$$

or

$$G = \pi d_p \,\mathrm{Nu}\, D\rho Y_w \tag{4.2}$$

The physical meaning of Eq. (4. 1) and Eq. (4. 2) is that the evaporation rate is proportional to the droplet diameter d_p, the thermophysical properties λ/c_p or $D\rho$, and parameter B. From Eq. (4. 1) and Eq. (4. 2) and the saturation relation we can determine the temperature and the vapor concentration at the droplet surface as

$$\begin{cases} B = c_p (T_g - T_w)/q_e = \mathrm{Le}\, Y_w \\ Y_w = B_w \exp(-E_w/RT_w) \end{cases} \tag{4.3}$$

Experiments point out that even for high intensity evaporation in high-temperature environment the linear relation of G vs. d_p, λ/c_p and $D\rho$ is true, however the linear relation of G vs B is questionable owing to the reverse effect of evaporation on heat and mass transfer.

4. 1. 2 Basic Equations in the Gas Environment Around an Evaporating Droplet

Let us consider the case of droplet evaporation in a stagnant gas environment without free or forced convection (Fig. 4. 1). Assume that there is a quasi-steady sphericosymmetrical flow around the droplet caused only by the Stefan flux in evaporation. Oxygen diffuses to the surface and fuel vapor diffuses to the environment. The heat radiation is neglected. The heat released from the high-temperature flame zone (if there is combustion) is transferred on one side to the surface for evaporation and on the other side to the environment. The continuity, momentum, gas species and energy equations can be written as:

$$\begin{cases} 4\pi r^2 \rho v = 4\pi r_p^2 \rho_w v_w = \mathrm{const} = G & (4.4) \\ p \approx \mathrm{const} & (4.5) \\ \rho v \dfrac{dY_s}{dr} = \dfrac{1}{r^2}\dfrac{d}{dr}\left(r^2 D\rho \dfrac{dY_s}{dr}\right) - w_s & (4.6) \\ \rho v c_p \dfrac{dT}{dr} = \dfrac{1}{r^2}\dfrac{d}{dr}\left(r^2 \lambda \dfrac{dT}{dr}\right) + w_s Q_s & (4.7) \end{cases}$$

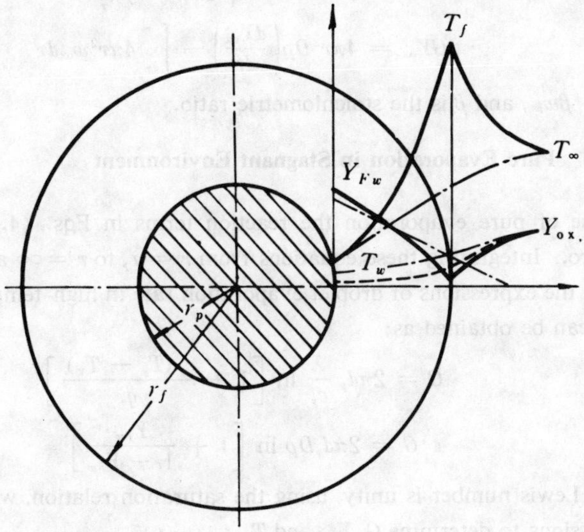

Fig. 4. 1

The boundary conditions are:

at $r = r_p$:

$$\begin{cases} \lambda\left(\dfrac{dT}{dr}\right)_w = \rho_w v_w q_e = G q_e/(4\pi r_p^{\,2}) & (4.8) \\ -D\rho\left(\dfrac{dY_s}{dr}\right)_w + Y_{sw}\rho_w v_w = \alpha\rho_w v_w \quad \begin{matrix}(s = F, \alpha = 1 \\ s \neq F, \alpha = 0)\end{matrix} & (4.9) \\ Y_{Fw} = B_w \exp(-E_w/RT_w) & (4.10) \end{cases}$$

at $r = \infty$:

$$T = T_g, \quad Y_F = 0, \quad Y_{pr} = 0, \quad Y_{ox} = Y_{ox,\infty}, \quad Y_{iner} = Y_{iner,\infty} \tag{4.11}$$

An additional condition is:

$$\Sigma Y_S = Y_F + Y_{ox} + Y_{pr} + Y_{iner} = 1 \tag{4.12}$$

It can be seen that these equations are different from those used in conventional heat conduction or diffusion problems in that there are additional convection and reaction terms and the Stefan flux arises on the surface. Besides, the temperature and fuel vapor concentration at the surface need to be determined.

By multiplying each term with $4\pi r^2$ and using the continuity equation, the diffusion equation and energy equation can be transformed into

$$\frac{d}{dr}(GY_s) = \frac{d}{dr}\left(4\pi r^2 D\rho\frac{dY_s}{dr}\right) - 4\pi r^2 w_s \tag{4.13}$$

$$\frac{d}{dr}(Gc_pT) = \frac{d}{dr}\left(4\pi r^2\lambda\frac{dT}{dr}\right) + 4\pi r^2 w_s Q_s \tag{4.14}$$

After integrating Eqs. (4. 13) and (4. 14) from r_p to r and using boundary conditions at $r = r_p$, we have

$$G[c_p(T - T_w) + q_e] = 4\pi r^2\lambda\frac{dT}{dr} + \int_{r_p}^{r} 4\pi r^2 w_s Q_s dr \tag{4.15}$$

$$G[Y_F - 1] = 4\pi r^2 D\rho\left(\frac{dY_F}{dr}\right) - \int_{r_p}^{r} 4\pi r^2 w_F dr \qquad (4.16)$$

$$GY_{ox} = 4\pi r^2 D\rho\left(\frac{dY_{ox}}{dr}\right) - \int_{r_p}^{r} 4\pi r^2 w_{ox} dr \qquad (4.17)$$

where $w_{ox} = \beta w_F$, and β is the stoichiometric ratio.

4.1.3 Pure Evaporation in Stagnant Environment

In case of pure evaporation the reaction terms in Eqs. (4.15) and (4.16) are equal to zero. Integrating these equations from $r = r_p$ to $r = \infty$ and assuming $\lambda/c_p = D\rho = \text{const}$, the expressions of droplet evaporation rate in high-temperature stagnant environment can be obtained as:

$$G = 2\pi d_p \frac{\lambda}{c_p} \ln\left[1 + \frac{c_p(T_g - T_w)}{q_e}\right] \qquad (4.18)$$

$$G = 2\pi d_p D\rho \ln\left[1 + \frac{Y_{Fw}}{1 - Y_{Fw}}\right] \qquad (4.19)$$

When Lewis number is unity, using the saturation relation, we have the following three expressions to determine G, Y_{Fw} and T_w:

$$\begin{cases} G = 2\pi d_p (\lambda/c_p) \ln(1 + B) & (4.20) \\ B = c_p(T_g - T_w)/q_e = \dfrac{Y_{Fw}}{(1 - Y_{Fw})} & (4.21) \\ Y_{Fw} = B_w \exp\left(-\dfrac{E_w}{RT_w}\right) & (4.22) \end{cases}$$

Equations (4.20) — (4.22) indicate that in case of high-temperature evaporation the evaporation rate is still proportional to droplet diameter and thermoconductivity, but G is proportional to $\ln(1+B)$ instead of B. The ratio of $\ln(1+B)$ over B is smaller than unity, which means that Stefan flux reduces heat and mass transfer in evaporation. Later we will discuss this problem in more detail.

In many practical problems, when the cold droplets are injected into the high-temperature air stream, the heat transfered from gas to droplets will go not only to evaporation but also to preheating the droplets. Therefore we should use the following equations to determine the evaporation rate:

$$\begin{cases} \rho_l \dfrac{\pi d_p^3}{6} c_l \dfrac{dT_w}{dt} = Gq_w - Gq_e & (4.23) \\ G = 2\pi d_p D\rho \ln\left[1 + \dfrac{Y_{Fw}}{1 - Y_{Fw}}\right] & (4.19) \\ G = 2\pi d_p \dfrac{\lambda}{c_p} \ln\left[1 + \dfrac{c_p(T_g - T_w)}{q_w}\right] & (4.24) \\ Y_{Fw} = B_w \exp\left(-\dfrac{E_w}{RT_w}\right) & (4.22) \end{cases}$$

Obviously, in this case, the equations must be solved numerically.

4.1.4 Evaporation with Combustion in Stagnant Environment

In case of droplet evaporation with combustion we can use Zeldovich transformation, taking

$$Z = c_p T + Y_{ox} Q_{ox}$$

and combine the diffusion equation with the energy equation for Le=1 to give

$$G[c_p(T - T_w) + q_e + Y_{ox} Q_{ox}] = 4\pi r^2 \frac{\lambda}{c_p}(c_p T + Y_{ox} Q_{ox})$$

or

$$G[(Z - Z_w) + q_e + Y_{ox} Q_{ox}] = 4\pi r^2 \frac{\lambda}{c_p}\frac{dZ}{dr} \tag{4.25}$$

After integrating from $r = r_p$ to $r = \infty$, we have

$$G = 2\pi d_p \frac{\lambda}{c_p} \ln\left[1 + \frac{Z_\infty - Z_w}{q_e + Y_{ox,w} Q_{ox}}\right]$$

or

$$G = 2\pi d_p \frac{\lambda}{c_p} \ln\left[1 + \frac{c_p(T_g - T_w) + (Y_{ox,\infty} - Y_{ox,w}) Q_{ox}}{q_e + Y_{ox,w} Q_{ox}}\right]$$

For diffusion combustion with $\tau_c/\tau_d \ll 1$, the oxygen concentration at the surface $Y_{ox,w}$ is zero, so we have

$$G = 2\pi d_p \frac{\lambda}{c_p} \ln\left[1 + \frac{c_p(T_g - T_w) + Y_{ox,\infty} Q_{ox}}{q_e}\right] \tag{4.24a}$$

or

$$G = 2\pi d_p \frac{\lambda}{c_p} \ln[1 + B_f] \tag{4.24b}$$

where

$$B_f = \frac{c_p(T_m - T_w)}{q_e}, \qquad T_m = T_g + Y_{ox,\infty} Q_{ox}/c_p$$

It can be noticed that the evaporation rate of a combusting droplet is independent of chemical kinetics. For a combusting droplet we need not find T_w and Y_{Fw}, because T_m is a very high temperature and we can take the boiling point T_b as the surface temperature T_w. The above derived expressions can also be obtained by using an infinitively thin flame-sheet model, which was proposed by Varshavsky, Spalding, Godsave, Paleev and others.

4.1.5 Evaporation of a Combusting or Noncombusting Droplet under Forced Convection

In practical combustors and furnaces there is always relative motion between the gas phase and droplets, that is the droplets are under forced convection. Actually, we need to solve an axisymmetric two-dimensional problem. What we would like to discuss here is a simplified analysis which can explicitely show the effect of different factors on the evaporation rate. This symplified analysis, which is called the "stagnant-film theory" or "hypothetic-film theory", is first to assume a spherical gas layer of finite thickness around the droplet with sphericosymmetric heat conduction equivalent to the real heat convection and with no evaporation, no combustion. This gas layer is called "stagnant film". Then, the next step is to consider sphericosymmetric evaporation and combustion in this film.

The principle of equivalence of a hypothetic heat conduction to the real heat convection for a sphere can be written as:

$$Q = \pi d_p{}^2 h^* (T_g - T_w) = \pi \frac{d_1 d_p \lambda}{r_1 - r_p} (T_g - T_w)$$

so, the radius of the "stagnant film" is

$$r_1 = r_p \frac{\text{Nu}^*}{\text{Nu}^* - 2} \tag{4.26}$$

where Nu^* is the Nusselt number of heat convection for a solid sphere without evaporation and without combustion, which can be expressed by the famous Ranz-Marshall formula:

$$\text{Nu}^* = 2 + 0.6\, \text{Re}^{0.5}\, \text{Pr}^{0.33} \tag{4.27}$$

Now it is easy to consider the effect of forced convection, because it can be done merely by taking $r = r_1$ as the outer boundary of the gas layer instead of originally used $r = \infty$ for a stagnant environment (Fig. 4.2). Therefore, the evaporation rate of a combusting and noncombusting droplet under forced convection can be written as:

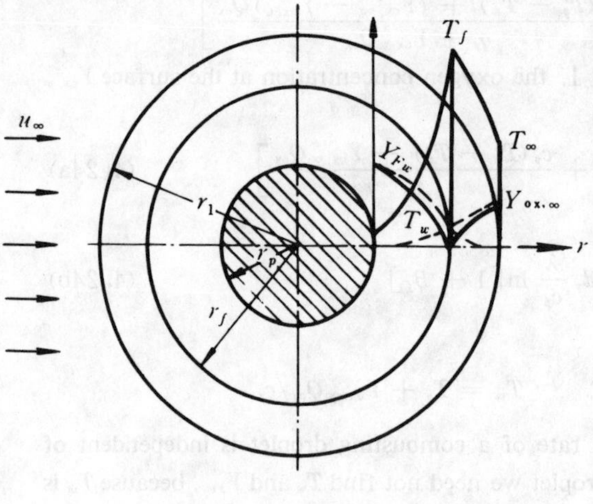

Fig. 4.2

$$G = 4\pi \frac{\lambda}{c_p} \frac{1}{\dfrac{1}{r_p} - \dfrac{1}{r_1}} \times \ln \left[1 + \frac{c_p (T_m - T_w)}{q_e} \right]$$

$$G = 4\pi \frac{\lambda}{c_p} \frac{1}{\dfrac{1}{r_p} - \dfrac{1}{r_1}} \times \ln \left[1 + \frac{c_p (T_g - T_w)}{q_e} \right]$$

or

$$G = \pi d_p \text{Nu}^* (\lambda / c_p) \ln (1 + B_f) \tag{4.28}$$

$$G = \pi d_p \text{Nu}^* (\lambda / c_p) \ln (1 + B) \tag{4.29}$$

At this point, we may compare the real Nusselt number for three cases: pure heat transfer without evaporation or with low-intensity evaporation; high-intensity evaporation; evaporation with combustion. For this comparison a unified Nusselt number is defined as:

$$\text{Nu} = \frac{hd_p}{\lambda} = \frac{qd_p}{\lambda (T_g - T_w)} = \frac{Gq_e}{\pi d_p \lambda (T_g - T_w)}$$

Thus we have the Nusselt number for high-intensity evaporation

$$\text{Nu} = \text{Nu}_e = \text{Nu}^* \ln (1 + B) / B \tag{4.30}$$

and the Nusselt number for evaporation with combustion

$$\text{Nu} = \text{Nu}_f = \text{Nu}^* \ln (1 + B_f) / B \tag{4.31}$$

The conclusion is that the Stefan flux reduces heat and mass tranfer and combustion enhances heat and mass transfer. In general, for droplet evaporation the real Nusselt number is a function not only of Reynolds number and Prandtl number but also of the parameter B or B_f.

On the basis of the above obtained results, we can discuss the so-called "d^2-law" or Sreznevsky Law, according to which the droplet diameter square will decrease linearly with time. From the defination of evaporation rate we can write

$$G = 4\pi r_p{}^2 \rho_l \frac{dr_p}{dt} = \frac{\pi r_p}{2} \rho_l \frac{d\,(d_p{}^2)}{dt}, \qquad K = \frac{d\,(d_p{}^2)}{dt} = \frac{G}{\rho_l \pi r_p/2}$$

so, we have

$$\begin{cases} K = 4\mathrm{Nu}^* \dfrac{\lambda}{\rho_l c_p} \ln(1 + B) \\ K_f = 4\mathrm{Nu}^* \dfrac{\lambda}{\rho_l c_p} \ln(1 + B_f) \end{cases} \tag{4. 32}$$

Obviously, only for the same kind of fuel and oxidizer, if T_g and $Y_{\mathrm{ox},\infty}$ keep unchanged and Nu^* keeps constant, there will be

$$K = \text{const}, \qquad K_f = \text{const}$$

and

$$\begin{cases} d_{p0}{}^2 - d^2 = Kt \\ d_{p0}{}^2 - d^2 = K_f t \end{cases} \tag{4. 33}$$

Equation (4. 33) is called d^2-Law or Sreznevsky Law, and K, K_f are called evaporation constants. The droplet life time is

$$\tau_s = d_{p0}{}^2/K, \qquad \tau_{sf} = d_{p0}{}^2/K_f$$

Finally, we can draw the following conclusions for droplet evapotion: (1) the evaporation rate is always proportional to the droplet diameter, the thermoconductivity and $\ln(1+B)$ or $\ln(1+B_f)$; (2) the evaporation rate of both combusting and noncombusting droplets is independent of chemical kinetics; and (3) the d^2-Law is not universal, it holds only under certain conditions.

4. 2 EXPERIMENTAL RESULTS OF DROPLET EVAPORATION

The droplet evaporation with and without combustion was experimentally studied in spherical burners, porous-sphere burners, suspended droplet apparatus, falling droplet apparatus and other apparatus. The effect of pressure, gas temperature, gas velocity, oxygen concentration and other factors on the evaporation rate of a suspended droplet was studied by measuring the decreasing droplet diameter, by using a cinematograph. When measuring the droplet evaporation with combustion, the droplet was ignited either automatically in high temperature gases, or by an electric spark, if it was suspended in a low-temperature environment.

It was observed that in high-temperature environment the droplet combustion may have three regimes (Fig. 4. 3) : fully enveloped flame, partially enveloped flame and wake flame (parachute flame), as the velocity increases or the droplet diameter decreases during evaporation.

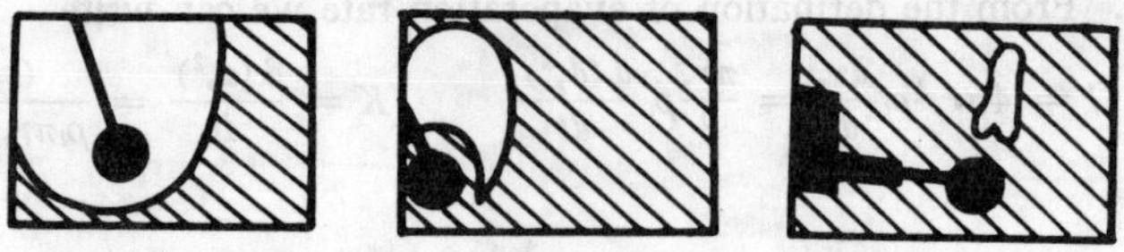

Fig. 4.3

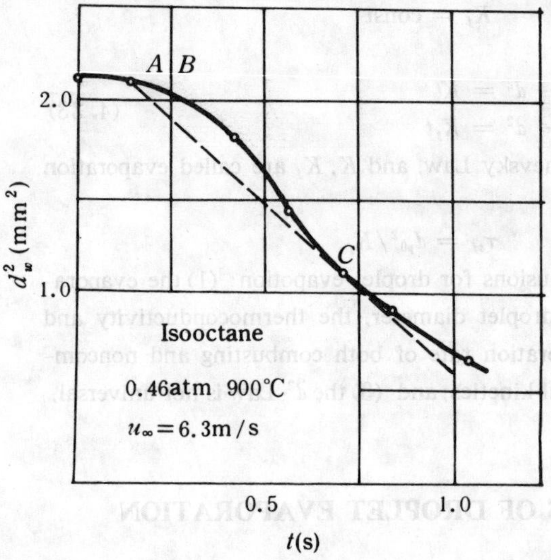

Fig. 4.4

The measured $d_p^2 = d_p^2\,(t)$ curves indicate (Fig. 4.4) that the linear law of decreasing droplet diameter square is only approximately true, because actually there are different stages of droplet behavior, including swelling, preheating, pure evaporation before ignition, evaporation with combustion, pure evaporation after extinction, which leads to different changing law of droplet diameter. The evaporation constants of hydrocarbon droplet combusting in stagnant air are:

Fuel	K_f (mm^2/s)
Kerosene	0.96
Isooctane	0.95
Diesel oil	0.79
$C_{16}H_{34}$	0.70
Heavy oil	0.50

The pressure has almost no effect on the evaporation rate for pressure lower than atmospheric value (Fig. 4.5). For high pressures the effect of pressure is

$$K \sim p^{0.25}$$

Actually, this is a result of the effect of natural convection.

The effect of gas velocity is shown in Fig. 4.6, which shows the increasing velocity leading to different regimes of combustion.

The effect of temperature on K is much stronger than what the theory predicted (Fig. 4.7). It was pointed out by the author that this is caused by the shortcomings of the "stagnant-film" theory.

Although it has been shown theoretically and experimentally that the chemical kinetics has no effect on evaporation rate of combusting and noncombusting droplets, but the chemical kinetics has significant effect on droplet ignition and extinction, which can be analyzed by using the simplified ignition-extinction theory discussed in Chapter 2. Assume that the ignition first occurs at the outer boundary of the stagnant film where the temperature is highest. By taking the zero-gradient analysis and coupling of two zones, that is:

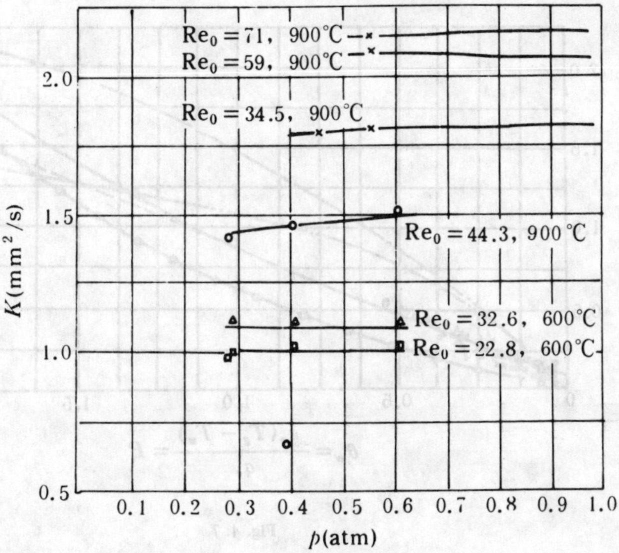

Fig. 4. 5

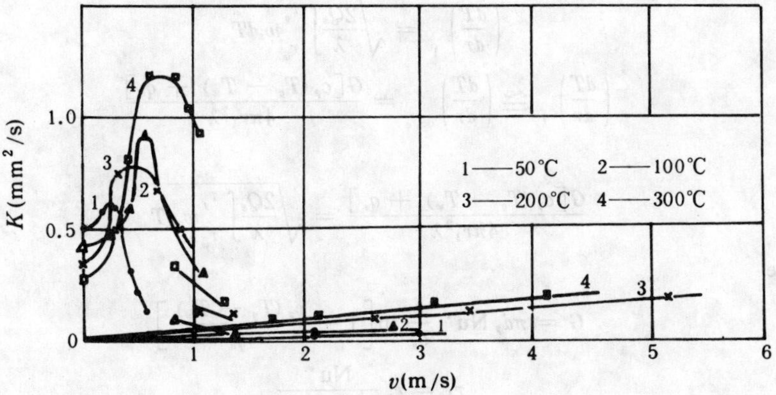

Fig. 4. 6

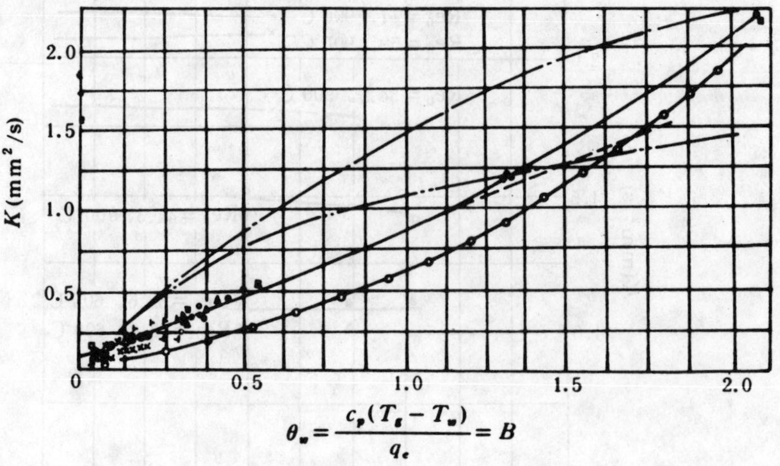

Fig. 4.7

$$\left(\frac{dT}{dr}\right)_\infty = 0, \qquad \left(\frac{dT}{dr}\right)_\mathrm{I} = \left(\frac{dT}{dr}\right)_\mathbf{I}$$

we have

$$\left(\frac{dT}{dr}\right)_\mathrm{I} = \sqrt{\frac{2Q_s}{\lambda}\int_{T_w}^{T_g} w_s dT}$$

$$\left(\frac{dT}{dr}\right)_\mathbf{I} \cong \left(\frac{dT}{dr}\right)_{\infty, e} = \frac{G[c_p(T_g - T_w) + q_e]}{4\pi r_1{}^2\lambda}$$

and

$$\frac{G[c_p(T_g - T_w) + q_e]}{4\pi r_1{}^2\lambda} = \sqrt{\frac{2Q_s}{\lambda}\int_{T_w}^{T_g} w_s dT} \tag{4.34}$$

where

$$G = \pi d_p \mathrm{Nu}^* \frac{\lambda}{c_p} \ln\left[1 + \frac{c_p(T_g - T_w)}{q_e}\right]$$

$$r_1 = r_p \frac{\mathrm{Nu}^*}{\mathrm{Nu}^* - 2}$$

The droplet ignition relation can be expressed as:

$$[(\mathrm{Nu}^* - 2)^2/\mathrm{Nu}^*]^2/d_p{}^2 = AT_g{}^n \exp\left(-\frac{E}{RT_g}\right) \tag{4.35}$$

Similarly, a droplet extinction relation can be obtained as:

$$\frac{(Nu^* - 2)^2}{Nu^*} = AT_f{}^n S_{l\infty} d_p c_p \rho/\lambda \tag{4.36}$$

These relations show the effect of different factors on droplet ignition or extinction. For example, for $\mathrm{Nu}^* \gg 2$ and constant T_g, we have

$$u_\infty/d_p = \mathrm{const} \tag{4.37}$$

and for constant u_∞ we have

$$d_p \sim \exp(E/RT_g) \tag{4.38}$$

Equations (4. 37) and (4. 38) have been verified by experiments of droplet extinction or ignition in laboratory apparatus. It can also be seen from these equations that the gas temperature has stronger effect on ignition than on extinction, and meanwhile the oxygen concentration has stronger effect on extinction than on ignition.

4. 3 SOLID-PARTICLE COMBUSTION

Solid fuels involve metal (Al, Mg, K, Na, W, Mo, Ti, Zr etc.), nonmetal (B, Si, C etc.), solid propellants and fossil fuel—coal. The solid fuel combustion takes different forms, such as solid fuel charge, pulverized fuel (entrained bed), slurry fuel (coal-water mixture, coal-oil mixture, coal-water-oil mixture), fluidized bed, fixed bed and moving bed.

The fossil fuel—coal, including anthracite, bituminous coal, lignite, is solid hydrocarbons. It consists of volatile matter, fixed carbon, moisture and mineral matter (ash). The conventional combustion process of a coal particle can be schematically shown in Fig. 4. 8. It includes preheating, devolatilization, volatile combustion, volatile and char combustion, and char combustion only. Under certain circumstances a first char ignition is also possible, for example, if the coal particle is exposed to very strong radiation. The char combustion often plays major role, but the volatile has strong effect on coal ignition and also on char combustion by changing the char porosity. Unlike the liquid-fuel combustion, the char combustion is frequently heterogeneous combustion where the reaction takes place on the solid surface, because the melting point or sublimation point of many solid fuels is much higher than the ignition temperature. Furthermore, the char combustion is often diffusion-kinetics controlled combustion, instead of pure diffusion combustion of liquid fuel.

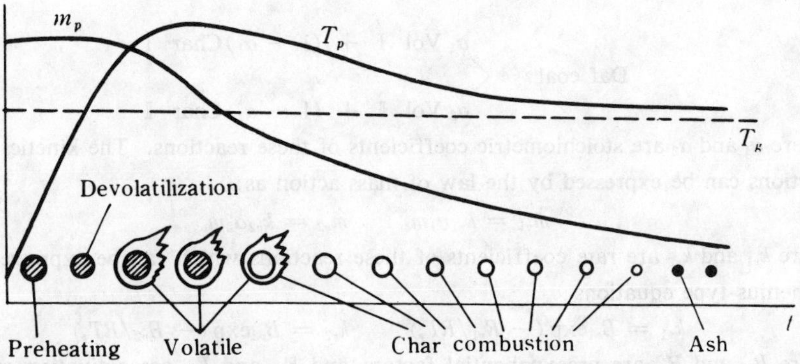

Fig. 4. 8

4.4 COAL DEVOLATILIZATION

It has been observed in experiments that when a cold coal particle is preheated to the temperature higher than 600K, there will be an obvious coal pyrolyzation or devolatilization. The pyrolyzation products or volatiles include CO_2, CO, CH_4, H_2, C_2H_4, H_2O and other higher hydrocarbons C_mH_n. The nitrides in volatiles are major sources of NO_x formation during coal combustion. Owing to very complex physical and chemical structure of coal, the pyrolyzation/devolatilization is a very complex process, including the initial break-down of the chemical key, the formation of unsteady intermediate products and steady final products. Different models of pyrolyzation have been proposed. As an example, a two-reaction model is discussed here.

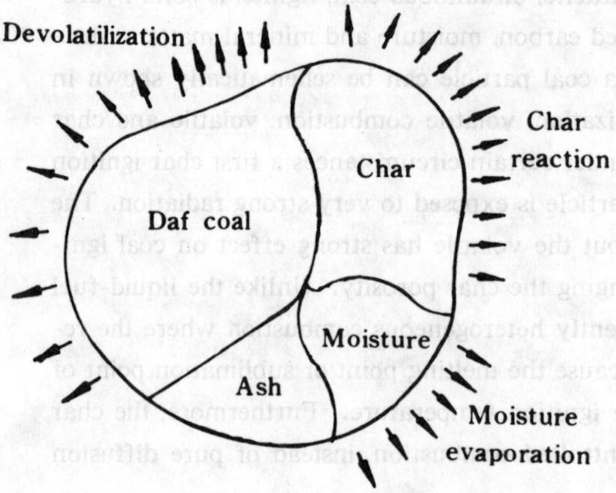

Fig. 4. 9

Considering a coal particle (Fig. 4. 9) consisting of daf coal (dry and ash-free raw coal), char, ash and moisture, we have the particle mass

$$m = m_c + m_h + m_A + m_w \tag{4. 39}$$

The total mass changing rate will be equal to the devolatilization rate plus char reaction rate plus moisture evaporation rate, that is:

$$\dot{m} = \dot{m}_c + \dot{m}_h + \dot{m}_w = \dot{m}_v + \dot{m}_{hr} + \dot{m}_w \tag{4. 40}$$

In the two-reaction model assumption is made with two parallel pyrolyzation reactions as:

$$\text{Daf coal} \begin{cases} \alpha_1 \text{ Vol I} + (1 - \alpha_1)\text{Char I} \\ \alpha_2 \text{ Vol II} + (1 - \alpha_2)\text{Char II} \end{cases} \tag{4. 41}$$

Where α_1 and α_2 are stoichiometric coefficients of these reactions. The kinetics of these reactions can be expressed by the law of mass action as:

$$\dot{m}_{v1} = k_{v1}\alpha_1 m_c \qquad \dot{m}_{v2} = k_{v2}\alpha_2 m_c$$

where k_{v1} and k_{v2} are rate coefficients of these reactions which can be expressed by the Arrhenius-type equations

$$k_{v1} = B_{v1}\exp(-E_{v1}/RT_p) \qquad k_{v2} = B_{v2}\exp(-E_{v2}/RT_p)$$

where B_{v1} and B_{v2} are preexponential factors, and E_{v1} and E_{v2} are activation energy of these reactions. Therefore, the total pyrolyzation/devolatilization rate is

$$\dot{m}_v = m_c[\alpha_1 B_{v1}\exp(-E_{v1}/RT_p) + \alpha_2 B_{v2}(-E_{v2}/RT_p)] \tag{4. 42}$$

The kinetic constants B_{v1}, B_{v2}, E_{v1} and E_{v2} depend on the pressure and coal type. It is considered that the first reaction plays predominant role at lower temperature and the second reaction plays predominant role at higher temperature. According to pyrolyzation experiments α_1 may be taken as the volatile content of proximate analysis and α_2 may be taken as 0.8—1. The typical values of kinetic constants for U.S. bituminous coal are

$$B_{v1} = 3.7 \times 10^5 \ 1/\text{s}, \qquad B_{v2} = 1.46 \times 10^{13} \ 1/\text{s}$$

$$E_{v1} = 7.4 \times 10^4 \ \text{kJ}/(\text{kg} \cdot \text{mol}), \qquad E_{v2} = 2.52 \times 10^5 \ \text{kJ}/(\text{kg} \cdot \text{mol})$$

The mass of daf coal m_c in Eq. (4.42) can be determined from the stoichiometric relation as

$$\dot{m}_c = -(\dot{m}_{v1}/\alpha_1 + \dot{m}_{v2}/\alpha_2) = -m_c[B_{v1}\exp(-E_{v1}/RT_p) + B_{v2}\exp(-E_{v2}/RT_p)] \tag{4.43}$$

Combining Eq. (4.42) with Eq. (4.43) gives

$$\dot{m}_v = m_{co}\exp\left\{-\int_0^t [B_{v1}\exp(-E_{v1}/RT_p) + B_{v2}\exp(-E_{v2}/RT_p)]dt\right\} \times [\alpha_1 B_{v1}\exp(-E_{v1}/RT_p) + \alpha_2 B_{v2}\exp(-E_{v2}/RT_p)] \tag{4.44}$$

It can be seen from Eq. (4.44) that for a given initial particle mass m_{co} and kinetic constants B_{v1}, B_{v2}, E_{v1}, E_{v2}, the coal devolatilization rate is only a function of particle temperature and time. For a cold coal particle entering the high temperature environment with temperature T_g it can be shown by the heat and mass transfer analysis (similar to that used in droplet evaporation) that

$$\dot{m}_v = \pi d_p D\rho \, \text{Nu}^* \ln[1 + Y_{vw}/(1 - Y_{vw})] \tag{4.45}$$

$$\frac{\pi d_p^{\,3}}{6}\bar{\rho}_p c \frac{dT_p}{dt} = \pi d_p \, \text{Nu}^* \lambda \frac{q_v}{c_p} \ln\left[1 + \frac{c_p(T_g - T_p)}{q_v}\right] - \dot{m}_v q_v \tag{4.46}$$

where q_v is the heating effect of pyrolyzation and Y_{vw} is the volatile mass fraction at the particle surface. Thus Eqs. (4.44), (4.45) and (4.46) determine three unknowns T_p, Y_{vw} and $\dot{m}_v$. Obviously, at first the devolatilization rate increases with time due to the increase of temperature, then reaches a maximum value and subsequently goes down with time due to the reduction of daf coal mass.

There are two trends in recent development of devolatilization models. One is to find a generalized one-reaction devolatilization model, the other is to build a multi-reaction model which is based on the pyrolyzation of individual functional groups. Both of them are trying to make the kinetic constants independent of coal type.

4.5 HETEROGENEOUS CARBON REACTIONS

The solid carbon reactions are surface reactions, either on the external surface or on the surface of internal pores. The carbon surface first absorbs the oxidizer and deabsorbs the combustion products after heterogeneous reactions. The basic reactions are carbon oxidation to form CO and CO_2. If the gas environment contains water vapor, carbon will react with water vapor to form H_2 and CH_4. Besides, frequently there is reduction of CO_2 by carbon to form CO. Under certain conditions, CO, H_2 and CH_4 formed by heterogeneous reactions may react with oxygen in the gas layer around carbon particles. Hence, we have the following reactions on the carbon surface or in the gas layer near the surface:

Surface reactions:

(1) $C+O_2 \longrightarrow CO_2+393965$ kJ/mol
(2) $C+0.5O_2 \longrightarrow CO+218614$ kJ/mol
} Primary reactions

(3) $C+CO_2 \longrightarrow 2CO-175351$ kJ/mol
(3a) $C+H_2O \longrightarrow CO+H_2-130165$ kJ/mol
(3b) $C+2H_2O \longrightarrow CO_2+2H_2+45186$ kJ/mol
(3c) $C+2H_2 \longrightarrow CH_4-74700$ kJ/mol
} Secondary reactions

Volume reactions—secondary reactions:

(4) $2CO+O_2 \longrightarrow 2CO+56932$ kJ/mol
(4a) $2H_2+O_2 \longrightarrow 2H_2O+231154$ kJ/mol
(4b) $CH_4+2O_2 \longrightarrow CO_2+2H_2O+889504$ kJ/mol
(4c) $CO+H_2O \longrightarrow CO_2+H_2+40337$ kJ/mol

Experiments have shown that the order of surface reactions is 0.5—1, however, for the convenience of theoretical analysis, the kinetics of first three reactions is considered as an Arrhenius type with first order

$$w_{ox,1} = B_1 \rho_w Y_{ox,w} \exp\left(-\frac{E_1}{RT_p}\right) = k_1 \rho_w Y_{ox,w} \quad (4.47)$$

$$w_{ox,2} = B_2 \rho_w Y_{ox,w} \exp\left(-\frac{E_2}{RT_p}\right) = k_2 \rho_w Y_{ox,w} \quad (4.48)$$

$$w_{co_2,3} = B_3 \rho_w Y_{CO_2,w} \exp\left(-\frac{E_3}{RT_p}\right) = k_3 \rho_w Y_{CO_2,w} \quad (4.49)$$

where the dimension of these reaction rates is kg/ (m^2 • s) and the dimension of preexponential factors B_1, B_2 and B_3 is m/s. The kinetic constants given by different authors are quite different due to different experimental conditions and carbon materials. The range of activation energy given by Aref'ev and Pomerantsev is shown in Tab. 4.1.

Table 4.1 Activation Energy of Carbon Oxidation

Fuel	E_1 (10^3kJ/mol)	E_2	E_3
Electrode Carbon	92 113	100—125	159—184
Anthracite	100—109	105—121	209
Char (or coke)	88—100	96—109	138

It has been discovered by the Russian scientists that there is a relationship between the rate coefficients k_1, k_2, k_3 and the activation energy E_1, E_2, E_3:

$$B_r \exp\left(-\frac{E_r}{RT_p}\right) = k_r = k^* \exp\left[-\frac{E_r}{RT_p}\left(1-\frac{T_p}{T^*}\right)\right] \quad (r = 1, 2, 3) \quad (4.50)$$

When T_p appoaches to T^*, k_r approaches to k^*, where

$$k^* = 10\text{m/s}, \qquad T^* = 2500\text{K}$$

or

$$k^* = 2.5\text{m/s}, \quad T^* = 2200\text{K}$$

However, more appropriate data are considered as:

$$k^* = 100\text{m/s}, \quad T^* = 2600\text{K}$$

$$E_2/E_1 = 1.1, \quad E_3/E_1 = 2.2, \quad B_r = k^* \exp\left(\frac{E_r}{RT^*}\right)$$

These kinetic constants show that in most cases $E_3 > E_2 > E_1$ and $B_3 > B_2 > B_1$, $k_1 > k_2 > k_3$ at low T_p and $k_3 > k_2 > k_1$ at high T_p, which implies that at lower temperature the carbon oxidation reactions to form CO_2 and CO play a major role, while at higher temperature the reduction of CO_2 by carbon to CO plays a predominant role. The kinetics of CO volume reaction is

$$w_{ox4} = 6.6 \times 10^4 \exp\left(\frac{-15000}{RT}\right) \tag{4.51}$$

4.6 CHAR COMBUSTION

Figure 4.10 gives a qualitative picture of temperature and concentration distribution around a char particle with or without CO combustion. There is high CO concentration near the surface and CO diffuses from the surface to the environment, while oxygen diffuses from the environment to the surface. The highest temperature and highest CO_2 concentration are located either at the surface or in the stagnant film. In the latter case, CO_2 will diffuse both to the surface and the environment.

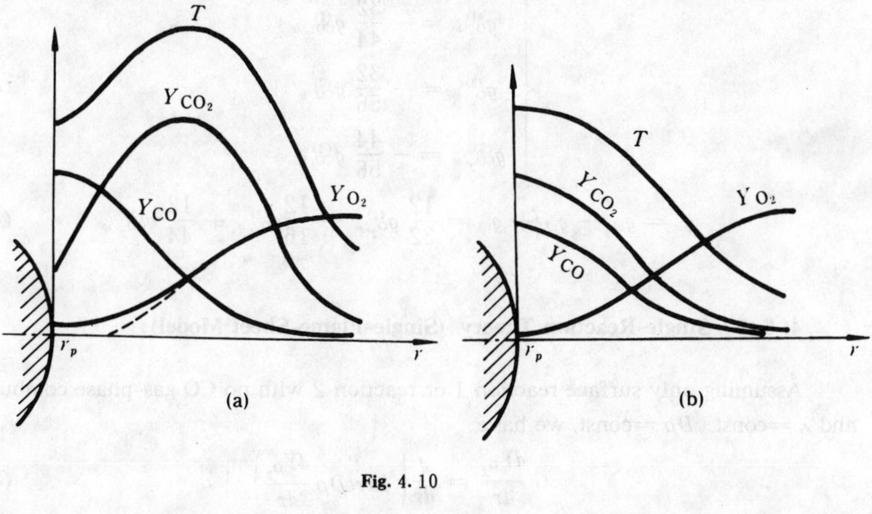

Fig. 4.10

4.6.1 Basic Equations in the Gas Layer Around a Combusting Char Particle

Assuming a sphericosymmetric laminar reacting flow around a combusting char particle caused by the Stefan flux, the continuity, momentum, diffusion and energy equations can be written as:

$$4\pi r^2 \rho v = 4\pi r_p{}^2 \rho_w v_w = \text{const} = G \tag{4.52}$$

$$p \approx \text{const} \tag{4.53}$$

$$\rho v \frac{dY_s}{dr} = \frac{1}{r^2}\frac{d}{dr}\left(r^2 D\rho \frac{dY_s}{dr}\right) - w_s \tag{4.54}$$

$$\rho v c_p \frac{dT}{dr} = \frac{1}{r^2}\frac{d}{dr}\left(r^2 \lambda \frac{dT}{dr}\right) + w_s Q_s \tag{4.55}$$

with boundary conditions at $r = r_p$

$$\begin{cases} g_{sw} = - D\rho\left(\dfrac{dY_s}{dr}\right)_w + Y_{sw}\rho_w v_w = \sum\limits_r B_{sr}\rho_w Y_{sw}\exp\left(-\dfrac{E_r}{RT_p}\right) & (4.56) \\ \Sigma g_{sw} = g_w = \rho_w v_w = g_c & (4.57) \\ \dfrac{\pi d_p{}^3}{6}\rho_c c\dfrac{dT_p}{dt} = \pi d_p{}^2 \sigma\varepsilon (T_g{}^4 - T_p{}^4) - \pi d_p{}^2 \lambda\left(\dfrac{dT}{dr}\right)_w + \sum\limits_r G_{cr}Q_{cr} & (4.58) \end{cases}$$

and at $r = r_1$

$$T = T_g, \qquad Y_s = Y_{s\infty} \tag{4.59}$$

additional conditions

$$\Sigma Y_s = 1 \tag{4.60}$$

$$g_{\text{iner}} = - D\rho\left(\frac{dY_{\text{iner}}}{dr}\right) + Y_{\text{iner}}\rho v = 0 \tag{4.61}$$

and stoichiometric relations

$$\begin{cases} g_{CO,w} = g^{(2)}_{CO,w} + g^{(3)}_{CO,w} \\ g_{O_2,w} = g^{(1)}_{O_2,w} + g^{(2)}_{O_2,w} \\ g_{CO_2,w} = g^{(1)}_{CO_2,w} + g^{(3)}_{CO_2,w} \end{cases} \tag{4.62}$$

$$\begin{cases} g^{(1)}_{O_2,w} = -\dfrac{32}{44}\, g^{(1)}_{CO_2,w} \\ g^{(2)}_{O_2,w} = -\dfrac{32}{56}\, g^{(2)}_{CO,w} \\ g^{(3)}_{CO_2,w} = -\dfrac{44}{56}\, g^{(3)}_{CO,w} \end{cases} \tag{4.63}$$

$$g_c = g_{c1} + g_{c2} + g_{c3} = \frac{12}{32}\, g^{(1)}_{O_2,w} + \frac{12}{16} g^{(2)}_{O_2,w} + \frac{12}{44}\, g^{(3)}_{CO_2,w} \tag{4.64}$$

4.6.2 Single-Reaction Theory (Single-Flame-Sheet Model)

Assuming only surface reaction 1 or reaction 2 with no CO gas-phase combustion, and λ = const, $D\rho$ = const, we have

$$G\frac{dY_{O_2}}{dr} = \frac{d}{dr}\left(4\pi r^2 D\rho \frac{dY_{O_2}}{dr}\right) \tag{4.65}$$

$$G c_p \frac{dT}{dr} = \frac{d}{dr}\left(4\pi r^2 \lambda \frac{dT}{dr}\right) \tag{4.66}$$

Integrating Eq. (4.65) and Eq. (4.66) from $r = r_p$ to $r = r$ gives

$$-4\pi r^2 D\rho\left(\frac{dY_{O_2}}{dr}\right) + GY_{O_2} = G_{O_2}$$

$$4\pi r^2 \lambda \frac{dT}{dr} = G[c_p (T - T_p) + q_w]$$

where

$$q_w = \frac{4\pi r_P^2}{G} \lambda \left(\frac{dT}{dr} \right)_w$$

In the case of only reaction 1, we have

$$G_{O_2} = - \frac{32}{44} G_{CO_2}, \qquad G = G_{O_2} + G_{CO_2} = - \frac{3}{8} G_{O_2}$$

and in the case of only reaction 2, we have

$$G = - \frac{3}{4} G_{O_2}$$

so, the general relation can be expressed as:

$$G = - G_{O_2} / \beta$$

where $\beta = 8/3$ or $4/3$, and the diffusion and energy equations become

$$4\pi r^2 D\rho \frac{dY_{O_2}}{dr} = G(Y_{O_2} + \beta)$$

$$4\pi r^2 \lambda \frac{dT}{dr} = G[c_p (T - T_p) + q_w]$$

Integrating these equations from $r = r_p$ to $r = r_1$ gives

$$G_c = G = \pi d_p D\rho \, \mathrm{Nu}^* \ln[1 + (Y_{O_2,\infty} - Y_{O_2,w}) / (\beta + Y_{O_2,w})] \quad (4.67)$$

$$G_c = G = \pi d_p (\lambda / c_p) \mathrm{Nu}^* \ln[1 + c_p (T_g - T_p) / q_w] \quad (4.68)$$

or

$$G_{O_2} = - \beta G_c = \pi d_p D\rho \, \mathrm{Nu}^* \ln[1 + (Y_{O_2,\infty} - Y_{O_2,w}) / (\beta + Y_{O_2,w})]$$

$$G_{O_2} = - \beta G_c = - \pi d_p (\lambda / c_p) \mathrm{Nu}^* \ln[1 + c_p (T_g - T_p) / q_w]$$

Besides, the reaction kinetics gives

$$G_{O_2} = - \pi d_p^2 B_1 Y_{O_2,w} \rho \exp(- E_1 / RT_p) \quad (4.69)$$

or

$$G_{O_2} = - \pi d_p{}^2 B_2 Y_{O_2,w} \rho \exp(- E_2 / RT_p) \quad (4.70)$$

The particle energy equation is:

$$\frac{\pi d_p^3}{6} \rho_c c \frac{dT_P}{dt} = \pi d_p^2 \sigma \varepsilon (T_g{}^4 - T_p{}^4) - G q_w + G Q_c \quad (4.71)$$

Equations (4.67). (4.68), (4.69) or (4.70), and (4.71) can determine G_c, $Y_{O_2,w}$, q_w and T_p. In case of steady state, rediation equilibrium and Le=1, we have $q_w = Q_c$ and

$$c_p (T_p - T_g) / Q_c = (Y_{O_2,\infty} - Y_{O_2,w}) / (\beta + Y_{O_2,w}) \quad (4.72)$$

so, Eqs. (4.67), (4.69) or (4.70), and (4.72) can give the solution.

The above obtained results indicate that in general cases the char combustion is a kinetics—diffusion controlled combustion, or in other words, it depends on both kinetics and heat and mass transfer. Obviously, the char combustion rate for reaction 2 is larger than that for reaction 1.

For most cases, owing to

$$Y_{O_2,\infty} / \beta \ll 1, \qquad (Y_{O_2,\infty} - Y_{O_2,w}) / (\beta + Y_{O_2,w}) \ll 1$$

we have

$$g_{O_2,w} = \frac{G_{O_2}}{\pi d_p^2} = \frac{\mathrm{Nu}^* D\rho}{d_p}(Y_{O_2,\infty} - Y_{O_2,w}) = - h_d^* \rho (Y_{O_2,\infty} - Y_{O_2,w})$$

$$g_{O_2,w} = - k_{O_2} \rho Y_{O_2,w}, \qquad k_{O_2} = B_1 \exp\left(-\frac{E_1}{RT_p}\right)$$

and

$$\begin{cases} g_{O_2,w} = - K Y_{O_2,\infty} & (4.73) \\ K = \dfrac{1}{\dfrac{1}{h_d^*} + \dfrac{1}{k_{O_2}}} = \dfrac{1}{\dfrac{d_p}{\mathrm{Nu}^* D} + \dfrac{1}{k_{O_2}}} & (4.74) \end{cases}$$

Equations (4.73) and (4.74) dipict clearly the diffusion—kinetics nature of char combustion, which was first pointed out by Frank—Kamenetsky , Blinov and Hottel. For high gas temperature, large particle size and low Nusselt number (or low relative velocity), we have

$$1/h_d^* \gg 1/k_{O_2}, \qquad \mathbf{D_I} = \tau_d/\tau_c = k_{O_2}/h_d^* \gg 1$$

which means that diffusion is much slower than reaction, hence

$$K \approx h_d^*, \qquad Y_{O_2,w} \approx 0$$

$$g_{O_2,w} \approx - h_d^* \rho Y_{O_2,\infty} = - \mathrm{Nu}^* D\rho Y_{O_2,\infty}/d_p \tag{4.75}$$

$$c_p (T_p - T_g) \approx Y_{O_2,\infty} Q_C/\beta$$

This is the limiting case of diffusion—controlled combustion with the lowest oxygen concentration and highest temperature at the particle surface. In this case, the char combustion rate will be proportional to $\sqrt{u_\infty/d_p}$, but is almost independent of T_g .

For low gas temperature, small partcle size and high Nusselt number (or high relative velocity), we have

$$1/k_{O_2} \gg 1/h_d^*, \qquad \mathbf{D_I} \ll 1$$

which means that reaction is much slower than diffusion, hence

$$K \approx k_{O_2}, \qquad Y_{O_2,w} \approx Y_{O_2,\infty}$$

$$g_{O_2,w} = - k_{O_2} Y_{O_2,\infty} \tag{4.76}$$

$$T_p \approx T_g$$

This is the limiting case of kinetics-controlled combustion with the highest oxygen concentration and lowest temperature at the particle surface. In this case, the char combustion rate will increase with the increase of temperature, but is almost independent of particle size and relative velocity. The above-discussed conclusions can be schematically illustrated in Figs 4.11 and 4.12.

In the case of diffusion-controlled combustion, if the combusting char is a shrinking sphere (with decreasing size and constant density), we have

$$K_c = \frac{d(d_p^2)}{dt} = 4D\rho\,\mathrm{Nu}^* Y_{O_2,\infty}/(\rho_c\beta) \approx \mathrm{const}$$

$$d_{p0}^2 - d_p^2 = K_c t \tag{4.77}$$

$$\tau_b = d_{p0}^2/K_c = d_{p0}^2 \rho_c \beta/(4D\rho\,\mathrm{Nu}^* Y_{O_2,\infty}) \tag{4.78}$$

For $\mathrm{Nu}^* = 2$, the total char combustion rate is

$$G_{O_2} = - 2\pi d_p D\rho Y_{O_2,\infty} \tag{4.79}$$

This is the earliest diffusion theory of carbon combustion proposed by Nusselt.

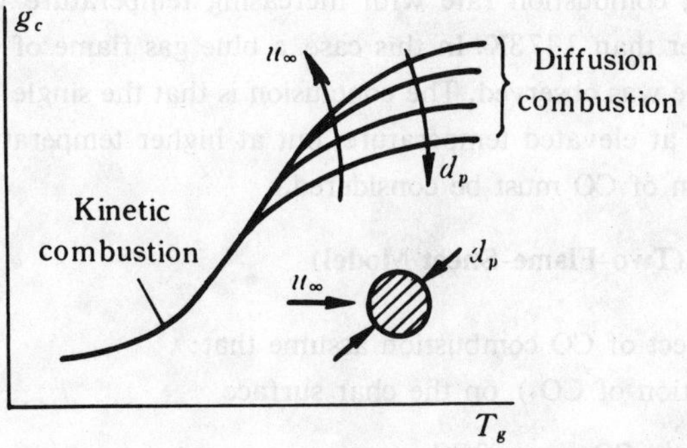

Fig. 4. 11

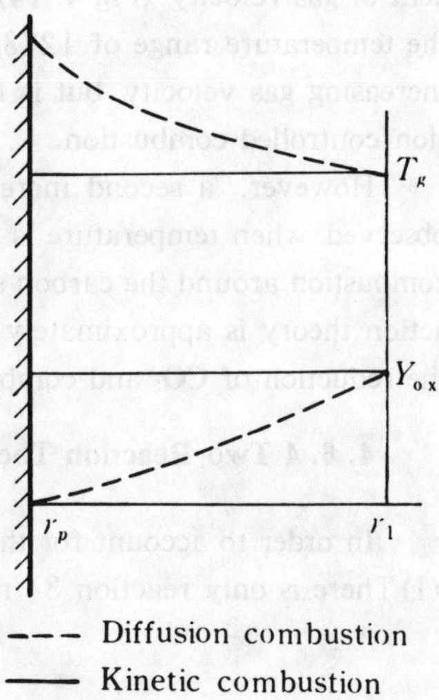

Fig. 4. 12

4. 6. 3 Some Experimental Results

The experimental results for combustion of electrode-carbon sphere of $d_p = 25$ mm and anthracite-char sphere of 15 mm, obtained by Hottel et al. and Khitrin et al. are shown in Figs 4. 13 and 4. 14. It can be seen that the single-reaction theory is qualitatively true in comparison with the experiments. As the temperature is lower than

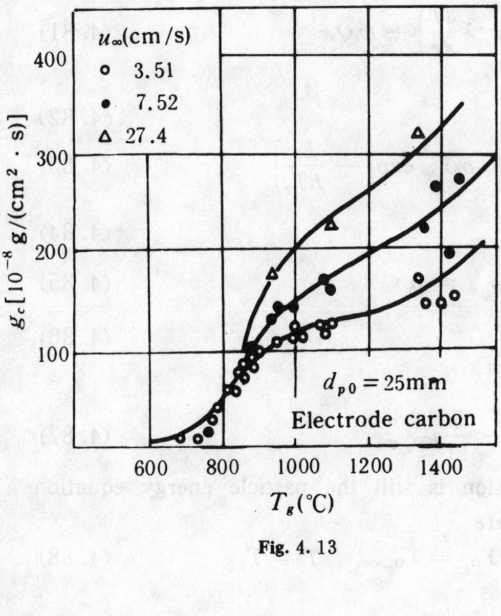

Fig. 4. 13

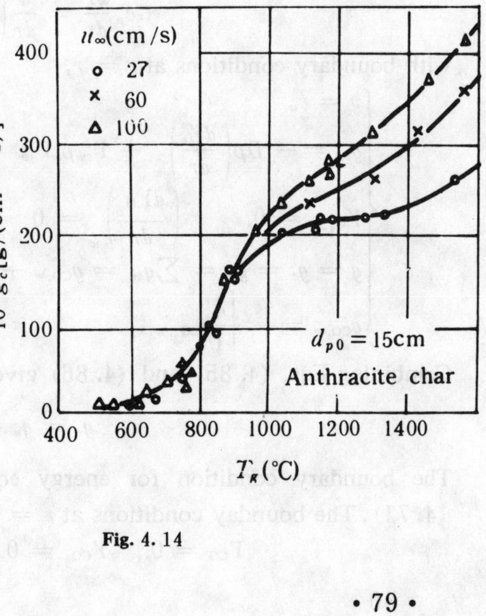

Fig. 4. 14

1173K, the combustion rate increases with the increase of temperature, but is independent of gas velocity (Fig 4. 14). This is the regime of kinetics-controlled combustion. In the temperature range of $1273K < T_g < 1473K$, the combustion rate is increasing with increasing gas velocity, but is independent of temperature. This is the regime of diffusion-controlled combustion.

However, a second increasing combustion rate with increasing temperature was observed, when temperature is higher than 1473K. In this case, a blue gas flame of CO combustion around the carbon sphere was observed. The conclusion is that the single-reaction theory is approximately true at elevated temperature, but at higher temperature the reduction of CO_2 and combustion of CO must be considered.

4. 6. 4 Two-Reaction Theory (Two-Flame-Sheet Model)

In order to account for the effect of CO combustion assume that:

(1) There is only reaction 3 (reduction of CO_2) on the char surface

$$\underset{12}{C} + \underset{44}{CO_2} \longrightarrow \underset{56}{2CO}$$

(2) The surface combustion is diffusion-kinetics controlled combution.

(3) There is diffusion-controlled CO combustion in the gas layer (stagnant film) around the particle

$$\underset{28}{CO} + \underset{16}{0.5O_2} \longrightarrow \underset{44}{CO_2}$$

(4) $\lambda = \text{const}$, $D\rho = \text{const}$, and $Le = 1$.

(5) No radiation.

The diffusion and energy equations in the gas layer are:

$$G\frac{dY_s}{dr} = \frac{d}{dr}\left(4\pi r^2 D\rho\frac{dY_s}{dr}\right) - w_s \qquad (s = CO, O_2, CO_2) \tag{4.80}$$

$$Gc_p\frac{dT}{dr} = \frac{d}{dr}\left(4\pi r^2\lambda\frac{dT}{dr}\right) + w_sQ_s \tag{4.81}$$

with boundary conditions at $r = r_p$

$$\begin{cases} v = v_w & (4.82) \\ g_{sw} = -D\rho\left(\dfrac{dY_s}{dr}\right)_w + Y_{sw}\rho_w V_{pw} = B_{c3}\rho_w Y_{sw}\exp\left(-\dfrac{E_3}{RT_p}\right) & (4.83) \\ Y_{O_2,w} = 0, \quad \left(\dfrac{dY_{O_2}}{dr}\right)_w = 0 & (4.84) \\ g = g_c = g_w = \sum g_{sw} = g_{CO,w} + g_{CO_2,w} = \rho_w v_w & (4.85) \\ g_{CO,w} = -\dfrac{14}{11}g_{CO_2,w} & (4.86) \end{cases}$$

Combining Eqs. (4. 85) and (4. 86) gives

$$g_c = g_w = -\frac{3}{11}g_{CO_2,w} \tag{4.87}$$

The boundary condition for energy equation is still the particle energy equation (4. 71). The bounday conditions at $r = r_1$ are

$$Y_{CO} = 0, \quad Y_{CO_2} = 0, \quad Y_{O_2} = Y_{O_2,\infty}, \quad T = T_g \tag{4.88}$$

By using the Zeldovich transformation

$$Y = Y_{O_2} + \frac{4}{11}Y_{CO_2}$$

$$Z = c_pT + Y_{O_2}Q_{O_2,4}$$

Equations (4. 80) and (4. 81) can be transformed into

$$G\frac{dY}{dr} = \frac{d}{dr}\left(4\pi r^2 D\rho\frac{dY}{dr}\right) \tag{4.89}$$

$$G\frac{dZ}{dr} = \frac{d}{dr}\left(4\pi r^2\frac{\lambda}{c_p}\frac{dZ}{dr}\right) \tag{4.90}$$

Integrating Eqs. (4. 89) and (4. 90) from r_p to r gives

$$G(Y - Y_w) = 4\pi r^2 D\rho\left(\frac{dY}{dr}\right) - 4\pi r_p^2 D\rho\left(\frac{dY}{dr}\right)_w$$

$$G(Z - Z_w) = 4\pi r^2\frac{\lambda}{c_p}\left(\frac{dZ}{dr}\right) - 4\pi r_p^2\frac{\lambda}{c_p}\left(\frac{dZ}{dr}\right)_w$$

Using boundary conditions, we have

$$-D\rho\left(\frac{dY}{dr}\right)_w = -D\rho\left[\left(\frac{dY_{O_2}}{dr}\right)_w + \frac{4}{11}\left(\frac{dY_{CO_2}}{dr}\right)_w\right]$$

$$= -\frac{4}{11}D\rho\left(\frac{dY_{CO_2}}{dr}\right)_w$$

$$= \frac{4}{11}(g_{CO_2,w} - Y_{CO_2,w}\,g_w)$$

$$= \frac{4}{11}\left(-\frac{11}{3}g_w - Y_{CO_2,w}\,g_w\right)$$

$$= g_w\left(-\frac{4}{3} - \frac{4}{11}Y_{CO_2,w}\right)$$

$$Y - Y_w = Y - \left(Y_{O_2,w} + \frac{4}{11}Y_{CO_2,w}\right) = Y - \frac{4}{11}Y_{CO_2,w}$$

and therefore

$$G\left(Y + \frac{4}{3}\right) = 4\pi r^2 D\rho\left(\frac{dY}{dr}\right)$$

Integration from r_p to r_1 gives

$$G_c = G = \pi d_p D\rho\,\mathrm{Nu}^*\ln\left(\frac{Y_\infty + 4/3}{Y_w + 4/3}\right)$$

or

$$G = \pi d_p D\rho\,\mathrm{Nu}^*\ln\left(1 + \frac{Y_{O_2,\infty} - \frac{4}{11}Y_{CO_2,w}}{\frac{4}{3} + \frac{4}{11}Y_{CO_2,w}}\right) \tag{4.91}$$

Using the boundary condition for energy equation, we have

$$\frac{\lambda}{c_p}\left(\frac{dZ}{dr}\right)_w = \lambda\left(\frac{dT}{dr}\right)_w + Q_{O_2,4}D\rho\left(\frac{dY_{O_2}}{dr}\right)_w = \lambda\left(\frac{dT}{dr}\right)_w = \frac{Gq_w}{\pi d_p^{\ 2}}$$

and therefore

$$G[(Z - Z_w) + q_w] = 4\pi r^2\frac{\lambda}{c_p}\frac{dZ}{dr}$$

Integration from r_p to r_1 gives

$$G = \pi d_p \text{Nu}^* \frac{\lambda}{c_p} \ln\left[1 + \frac{c_p(T_g - T_p) + Y_{O_2,\infty} Q_{O_2,4}}{q_w}\right] \tag{4.92}$$

Besides, we still have surface kinetics

$$G = G_c = \pi d_p^2 B_{c3} \rho_w Y_{CO_2,w} \exp\left(-\frac{E_3}{RT_p}\right) \tag{4.93}$$

and particle energy equation

$$\frac{\pi d_p^3}{6} \rho_c c_c \frac{dT_p}{dt} = \pi d_p^2 \varepsilon \sigma (T_g^4 - T_p^4) + G q_w - G_c Q_{c3} \tag{4.94}$$

Equations (4.91), (4.92), (4.93) and (4.94) will determine four unkowns G_c, $Y_{CO_2,w}$, T_P and q_w. It can be seen that the char combustion rate depends on diffusion and kinetics of surface reaction, but is independent of kinetics of CO combustion. The difference between the first model and this second model is: (1) the surface concentration affecting the combustion rate is $11/4\, Y_{CO_2,w}$ instead of $Y_{O_2,w}$; (2) the kinetics affecting the combustion rate is reaction 3 instead of reaction 1 or 2; (3) the particle temperature T_p depends on Q_{c3}, $Q_{O_2,4}$ instead of Q_{c1} or Q_{c2}. In the case of diffusion combustion at the surface, we have

$$G = \pi d_p \text{Nu}^* D\rho \ln\left(1 + \frac{3}{4} Y_{O_2,\infty}\right)$$

This result is nearly the same as that obtained from the first model except that T_p is different.

4.6.5 Multi-reaction and Single-Flame-Surface Model

Assume no CO combustion and simultaneous reactions 1, 2 and 3 at the surface, and λ=const, $D\rho$ =const. In this case, we can derive

$$G_s = -4\pi r^2 D\rho \left(\frac{dY_s}{dr}\right) + Y_s G = \text{const} \quad (s = CO, CO_2, O_2)$$

The solution is

$$G = \pi d_p D\rho \text{ Nu}^* \ln\left(\frac{G_s/G - Y_{s\infty}}{G_s/G - Y_{sw}}\right) \tag{4.95}$$

Then we have the surface kinetics

$$G_S = \pi d_p^2 \rho Y_{sw} \sum_l B_l \exp\left(-\frac{E_l}{RT_p}\right) \tag{4.96}$$

the solution of gas energy equation

$$G = \pi d_p \text{ Nu}^* \frac{\lambda}{c_p} \ln\left[1 + \frac{c_p(T_g - T_p)}{q_w}\right], \qquad q_w = \frac{\pi d_p^2 \lambda \left(\frac{dT}{dr}\right)_w}{G} \tag{4.97}$$

and the particle energy equation

$$\frac{\pi d_p^3}{6} \rho_c c_c \frac{dT_p}{dt} = \pi d_p^2 \varepsilon \sigma (T_g^4 - T_p^4) + G q_w + \sum_l G_{cl} Q_{cl} \tag{4.98}$$

Besides, we have the relation $\sum_s G_s = G$ and the stoichiometric relations.

4.7 COAL-PARTICLE COMBUSTION

The basic assumptions are:

(1) Simultaneous moisture evaporation, devolatilization and char combustion;

(2) Simultaneous three char reactions on the surface with thermal radiation;

(3) No gas-phase (CO or volatile) reaction in the gas layer (the stagnant film) around the particle.

We can write the diffusion and energy equations in the gas layer as:

$$G_s = -4\pi r^2 D\rho \frac{dY_s}{dr} + GY_s = \text{const}$$

$$4\pi r^2 \lambda \frac{dT}{dr} = G[c_p(T - T_w) + q_w]$$

with boundary conditions:

$$G_s = G_{sw} = \pi d_p^2 \rho Y_{sw} \sum B_l \exp\left(-\frac{E_l}{RT_p}\right)$$

$$G = G_c + G_v + G_w \quad (\dot{m} = \dot{m}_h + \dot{m}_v + \dot{m}_w)$$

$$G_c = -\frac{3}{4} G_{O_2} - \frac{3}{11} G_{CO_2}$$

$$m_p c_c \frac{dT_p}{dt} = \pi d_p^2 \varepsilon\sigma (T_g^4 - T_p^4) + Gq_w - G_w L_w - G_v \Delta h_v + \sum_l G_{cl} Q_{cl}$$

Their solution gives

$$G = \pi d_p \text{Nu}^* D\rho \ln\left[\frac{G_s/G - Y_{s\infty}}{G_s/G - Y_{sw}}\right] \tag{4.99}$$

$$G = \pi d_p \text{Nu}^* \frac{\lambda}{c_p} \ln[1 + c_p(T_g - T_p)/q_w] \tag{4.100}$$

Finally, we have ten equations to determine 10 unknowns; G, G_c, G_v, G_w, G_{O_2}, G_{CO_2}, $Y_{O_2,w}$, $Y_{CO_2,w}$, m_c, and T_p.

$$G_w = \pi d_p \text{Nu}^* D\rho \ln\left(\frac{1 - Y_{w\infty}}{1 - Y_{ww}}\right), \qquad Y_{ww} = B_w \exp\left(-\frac{E_w}{RT_p}\right) \tag{4.101}$$

$$G_v = m_c \left[\alpha_1 B_{v1} \exp\left(-\frac{E_{v1}}{RT_p}\right) + \alpha_2 B_{v2} \exp\left(-\frac{E_{v2}}{RT_p}\right)\right] \tag{4.102}$$

$$\dot{m}_c = -m_c \left[B_{v1} \exp\left(-\frac{E_{v1}}{RT_p}\right) + B_{v2} \exp\left(-\frac{E_{v2}}{RT_p}\right)\right] \tag{4.103}$$

$$G = \pi d_p \text{Nu}^* D\rho \ln\left[\frac{G_{CO_2}/G - Y_{CO_2,\infty}}{G_{CO_2}/G - Y_{CO_2,w}}\right] \tag{4.104}$$

$$G = \pi d_p \text{Nu}^* D\rho \ln\left[\frac{G_{O_2}/G - Y_{O_2,\infty}}{G_{O_2}/G - Y_{O_2,w}}\right] \tag{4.105}$$

$$G_{O_2} = -\pi d_p^2 \rho Y_{O_2,w} \left[B_1 \exp\left(-\frac{E_1}{RT_p}\right) + B_2 \exp\left(-\frac{E_2}{RT_p}\right)\right] \tag{4.106}$$

$$G_{CO_2} = \pi d_p^2 \rho \left[\frac{11}{8} Y_{O_2,w} B_1 \exp\left(-\frac{E_1}{RT_p}\right) - Y_{CO_2,w} B_3 \exp\left(-\frac{E_3}{RT_p}\right)\right] \tag{4.107}$$

$$G_c = -\frac{3}{4} G_{O_2} - \frac{3}{11} G_{CO_2} \tag{4.108}$$

$$G_c = G - G_v - G_w \tag{4.109}$$

$$m_p c_c \frac{dT_p}{dt} = \pi d_p^2 \varepsilon\sigma (T_g^4 - T_p^4) + Gq_w - G_w L_w - G_v \Delta h_v + |G_{O_2}^{(1)}| Q_{O_2,1} + |G_{O_2}^{(2)}| Q_{O_2,2} - |G_{CO_2}^{(3)}| Q_{CO_2}^{(3)} \tag{4.110}$$

where

$$|G_{O_2}{}^{(1)}| = \pi d_p{}^2 \rho Y_{O_2,w} B_1 \exp\left(-\frac{E_1}{RT_p}\right)$$

$$|G_{O_2}{}^{(2)}| = \pi d_p{}^2 \rho Y_{O_2,w} B_2 \exp\left(-\frac{E_2}{RT_p}\right)$$

$$|G_{O_2}{}^{(3)}| = \pi d_p{}^2 \rho Y_{CO_2,w} B_3 \exp\left(-\frac{E_3}{RT_p}\right)$$

and q_w is determined by Eq. (4. 100). Obviously Eqs. (4. 100) — (4. 110) can be solved only by numerical predictions. Figures 4. 15 and 4. 16 give the predicted mass loss and temperature rise of a Huainan bituminous coal particle of $d_p = 8$mm, which is in good agreement with experimental results.

Experimental results for combustion of 0. 75mm brown coal particles in air in the temperature range of 1473—1874K give the volatile ignition time, volatile combustion time and char combustion time:

$$\tau_i = 2.5 \times 10^{15}\, T_g{}^{-4} d_{p0} \qquad (\tau\text{: s, } d_p\text{: m})$$

$$\tau_v = 0.45 \times 10^6 d_{p0}^2$$

$$\tau_{ch} = 1.11 \times 10^9\, T_g{}^{-0.9} Y_{O_2,\infty}^{-1} d_{p0}^2$$

It has been found that the particle diameter kept almost unchanged during volatile combustion and particle diameter decreases during char combustion. For anthracite particles of 150 μm to 800 μm the ignition time and char combustion time are

$$\tau_i = 16.45 \times 10^{15}\, T_g{}^{-3.5} Y_{O_2,\infty}^{-0.15} d_{p0}^{1.2}$$

$$\tau_{ch} = 2.28 \times 10^6 (1 - 0.01A_c)\, T_g{}^{-0.9} Y_{O_2,\infty}^{-1} d_{p0}^2 \rho_c$$

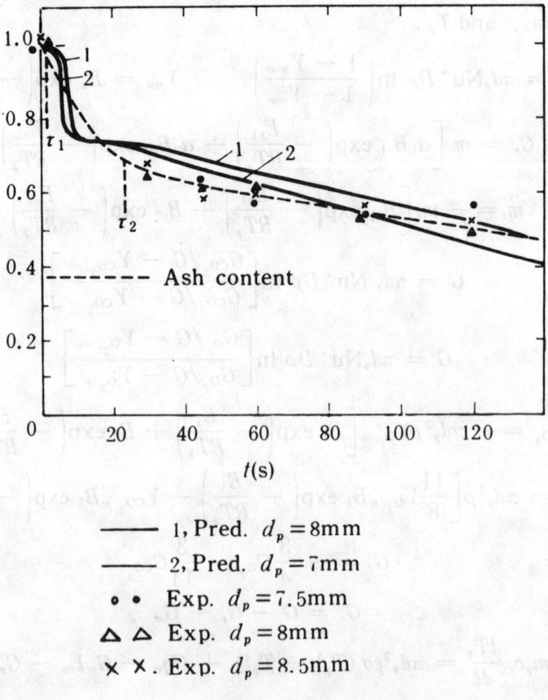

Fig. 4. 15

Where A_c and ρ_c are the ash content and char density respectively. It has also been found that the total combustion time of a coal particle depends not only on particle size, temperature, oxygen concentration, coal type, but also on the heating rate and the swelling behavior.

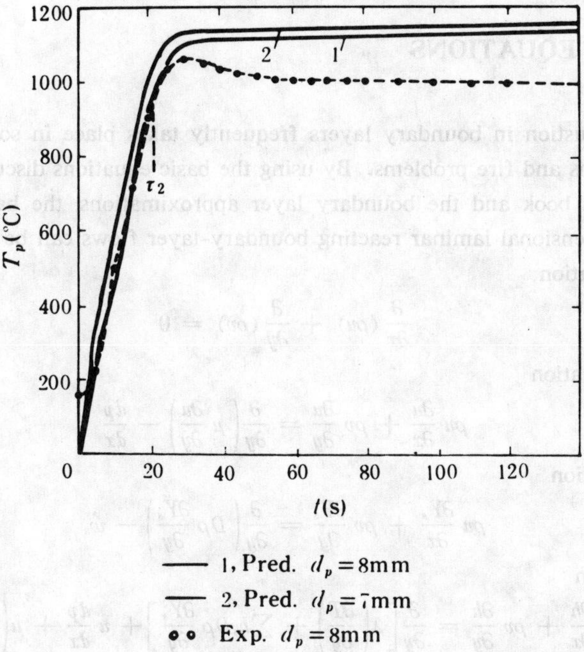

Fig. 4. 16

CHAPTER 5 COMBUSTION IN LAMINAR BOUNDARY LAYERS

5.1 BASIC EQUATIONS

The combustion in boundary layers frequently takes place in solid-fuel rockets, reentry problems and fire problems. By using the basic equations discussed in the First Chapter of this book and the boundary layer approximations, the basic equations of steady two-dimensional laminar reacting boundary-layer flows can be written as:

Continuity equation

$$\frac{\partial}{\partial x}(\rho u) + \frac{\partial}{\partial y}(\rho v) = 0 \tag{5.1}$$

Momentum equation

$$\rho u \frac{\partial u}{\partial x} + \rho v \frac{\partial u}{\partial y} = \frac{\partial}{\partial y}\left(\mu \frac{\partial u}{\partial y}\right) - \frac{dp}{dx} \tag{5.2}$$

Diffusion equation

$$\rho u \frac{\partial Y_s}{\partial x} + \rho v \frac{\partial Y_s}{\partial y} = \frac{\partial}{\partial y}\left(D\rho \frac{\partial Y_s}{\partial y}\right) - w_s \tag{5.3}$$

Energy equation

$$\rho u \frac{\partial h}{\partial x} + \rho v \frac{\partial h}{\partial y} = \frac{\partial}{\partial y}\left[\lambda\left(\frac{\partial T}{\partial y}\right) + \sum h_s D\rho \frac{\partial Y_s}{\partial y}\right] + u\frac{dp}{dx} + \mu\left(\frac{\partial u}{\partial y}\right)^2 \tag{5.4}$$

where the mixture enthalpy, including the chemical enthalpy, is

$$h = \sum Y_s h_s = \int c_p dT + \sum Y_s h_{0s}$$

The energy equation in the form of stagnant enthalpy is

$$\rho u \frac{\partial H}{\partial x} + \rho v \frac{\partial H}{\partial y} = \frac{\partial}{\partial y}\left[\lambda\left(\frac{\partial T}{\partial y}\right) + \sum h_s D\rho \frac{\partial Y_s}{\partial y}\right] + \frac{\partial}{\partial y}\left[\mu \frac{\partial}{\partial y}\left(\frac{u^2}{2}\right)\right] \tag{5.5}$$

For constant and equal specific heat for each species, we have

$$\rho u c_p \frac{\partial T}{\partial x} + \rho v c_p \frac{\partial T}{\partial y} = \frac{\partial}{\partial y}\left(\lambda \frac{\partial T}{\partial y}\right) + u\frac{dp}{dx} + \mu\left(\frac{\partial u}{\partial y}\right)^2 + w_s Q_s \tag{5.6}$$

and

$$\rho u \frac{\partial H}{\partial x} + \rho v \frac{\partial H}{\partial y} = \frac{\partial}{\partial y}\left[\frac{\mu}{\Pr}\left(\frac{\partial H}{\partial y}\right) + \mu\left(1 - \frac{1}{\Pr}\right)\frac{\partial}{\partial y}\left(\frac{u^2}{2}\right) + \sum h_s D\rho\left(1 - \frac{1}{\text{Le}}\right)\frac{\partial Y_s}{\partial y}\right] \tag{5.7}$$

In the case of laminar boundary layer over a flat plate and Pr=Le=Sc=1, we have

$$\begin{cases} \dfrac{\partial}{\partial x}(\rho u) + \dfrac{\partial}{\partial y}(\rho v) = 0 & (5.8) \\ \rho u \dfrac{\partial u}{\partial x} + \rho v \dfrac{\partial u}{\partial y} = \dfrac{\partial}{\partial y}\left(\mu \dfrac{\partial u}{\partial y} \right) & (5.9) \\ \rho u \dfrac{\partial H}{\partial x} + \rho v \dfrac{\partial H}{\partial y} = \dfrac{\partial}{\partial y}\left(\mu \dfrac{\partial H}{\partial y} \right) & (5.10) \\ \rho u \dfrac{\partial Y_s}{\partial x} + \rho v \dfrac{\partial Y_s}{\partial y} = \dfrac{\partial}{\partial y}\left(\mu \dfrac{\partial Y_s}{\partial y} \right) - w_s & (5.11) \end{cases}$$

It can be seen that there is a similarity between the velocity field and the stagnant enthalpy field

$$u/u_\infty = (H - H_w)/(H_\infty - H_w) \qquad (5.12)$$

5.2 LAMINAR BOUNDARY LAYER WITH SURFACE REACTION

Consider the laminar boundary layer over a flat carbon plate (Fig. 5.1) with surface reaction. Assume only reaction 1

$$C + O_2 \rightarrow CO_2$$

on the carbon surface and no CO gas reaction. The basic equations are :

$$\begin{cases} \dfrac{\partial}{\partial x}(\rho u) + \dfrac{\partial}{\partial y}(\rho v) = 0 & (5.1) \\ \rho u \dfrac{\partial u}{\partial x} + \rho v \dfrac{\partial u}{\partial y} = \dfrac{\partial}{\partial y}\left(\mu \dfrac{\partial u}{\partial y} \right) & (5.2) \\ \rho u \dfrac{\partial Y_{ox}}{\partial x} + \rho v \dfrac{\partial Y_{ox}}{\partial y} = \dfrac{\partial}{\partial y}\left(D\rho \dfrac{\partial Y_{ox}}{\partial y} \right) & (5.13) \\ \rho u \dfrac{\partial T_*}{\partial x} + \rho v \dfrac{\partial T_*}{\partial y} = \dfrac{\partial}{\partial y}\left(\dfrac{\mu}{\mathrm{Pr}} \dfrac{\partial T_*}{\partial y} \right) + \left(1 - \dfrac{1}{\mathrm{Pr}} \right) \dfrac{\partial}{\partial y}\left[\mu \dfrac{\partial}{\partial y}\left(\dfrac{u^2}{2c_p} \right) \right] & (5.14) \end{cases}$$

where the energy equation is written in the form of stagnant temperature, which is defined as

$$T_* = T + u^2/(2c_p) = T\left(1 + \frac{\gamma - 1}{2}\mathrm{Ma}^2 \right)$$

In the case of negligible Stefan flux, the boundary conditions are:

$$\begin{cases} \text{at} \quad y = 0; \quad u = v = 0 \\ \quad - g_{ox,w} = D_w \rho_w \left(\dfrac{\partial Y_{ox}}{\partial y} \right)_w = B\rho_w Y_{ox,w} \exp\left(- \dfrac{E}{RT_w} \right) \\ \quad \lambda_w \left(\dfrac{\partial T}{\partial y} \right)_w = g_{ox,w} Q_{ox} = - Q_{ox} D_w \rho_w \left(\dfrac{\partial Y_{ox}}{\partial y} \right)_w \\ \text{at} \quad y = \infty: \quad u = u_\infty \\ \quad T = T_\infty, \quad Y_{ox} = Y_{ox,\infty} \end{cases} \qquad (5.15)$$

After introducing the Howarth-Dorodnitzn transformation for compressible boundary-layer flows

$$\bar{x} = \int_0^x \frac{p}{p_\infty} dx = x, \qquad \bar{y} = \int_0^y \frac{\rho}{\rho_\infty} dy$$

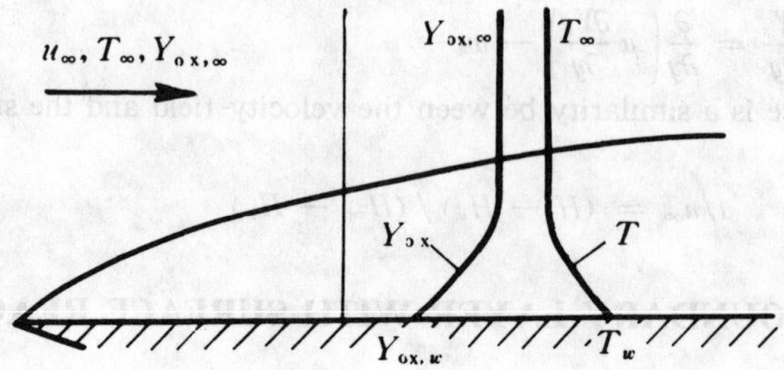

Fig. 5. 1

$$\bar{u} = u, \quad \bar{v} = \frac{\rho}{\rho_\infty} v + u \frac{\partial \bar{y}}{\partial x} \frac{p}{p_\infty} = \frac{\rho}{\rho_\infty} v + u \frac{\partial \bar{y}}{\partial x}$$

and assuming

$$\rho\mu = \text{const} \quad (\mu \sim T)$$

the basic equations in the $\bar{x}$-$\bar{y}$ plane of hypothetical incompressible flows are

$$\begin{cases} \dfrac{\partial \bar{u}}{\partial \bar{x}} + \dfrac{\partial \bar{v}}{\partial \bar{y}} = 0 & (5.16) \\ \bar{u}\dfrac{\partial \bar{u}}{\partial \bar{x}} + \bar{v}\dfrac{\partial \bar{u}}{\partial \bar{y}} = \nu_\infty \dfrac{\partial^2 \bar{u}}{\partial \bar{y}^2} & (5.17) \\ \bar{u}\dfrac{\partial Y_{ox}}{\partial \bar{x}} + \bar{v}\dfrac{\partial Y_{ox}}{\partial \bar{y}} = D_\infty \dfrac{\partial^2 Y_{ox}}{\partial \bar{y}^2} & (5.18) \\ \bar{u}\dfrac{\partial T_*}{\partial \bar{x}} + \bar{v}\dfrac{\partial T_*}{\partial \bar{y}} = \dfrac{\nu_\infty}{\text{Pr}} \dfrac{\partial^2 T_*}{\partial \bar{y}^2} + \left(1 - \dfrac{1}{\text{Pr}}\right) \nu_\infty \dfrac{\partial^2}{\partial \bar{y}^2}\left(\dfrac{\bar{u}^2}{2c_p}\right) & (5.19) \end{cases}$$

Hence, the continuity and momentum equations can be solved separately, not depending on the energy equation. This is a classical Blasius problem. By introducing the Blasius coordinate

$$\eta = \bar{y}\sqrt{\frac{u_\infty}{\nu_\infty \bar{x}}}$$

and the stream function

$$\psi = \sqrt{u_\infty \nu_\infty \bar{x}}\, f(\eta)$$

where ψ is defined by

$$\bar{u} = \frac{\partial \psi}{\partial \bar{y}}, \qquad \bar{v} = -\frac{\partial \psi}{\partial \bar{x}}$$

we have the Blasius equation

$$2f''' + ff'' = 0 \tag{5.20}$$

with boundary conditions

$$f(0) = f'(0) = 1, \qquad f'(\infty) = 1$$

Now, $f(\eta)$ is a known function, from which we can find $\bar{u} = u_\infty f(\eta)$. Assume that there are similar concentration and temperature profiles and define

$$Y_{\text{ox}}/Y_{\text{ox},\infty} = \varphi(\eta), \qquad T_*/T_{*,\infty} = \zeta(\eta)$$

Then, the diffusion and energy equations can be transformed into the following ordinary differential equations

$$2\varphi'' + \text{Sc}\varphi' f = 0 \tag{5.21}$$

$$2\zeta'' + \text{Pr}\zeta' f = \left(1 - \frac{1}{\text{Pr}}\right)(\gamma - 1)\text{Ma}_\infty^2 (f'^2)''/\left[2\left(1 + \frac{\gamma - 1}{2}\text{Ma}_\infty^2\right)\right] \tag{5.22}$$

with boundary conditions at $\eta = 0$

$$-\text{Le}Y_{\text{ox},\infty}Q_{\text{ox}}\varphi'(0)/\left[c_p T_\infty\left(1 + \frac{\gamma - 1}{2}Ma_\infty^2\right)\right] = \zeta'(0) \tag{5.23}$$

$$\omega\frac{\varphi'(0)}{\varphi(0)} = \left(1 + \frac{\gamma - 1}{2}\text{Ma}_\infty^2\right)\zeta(0)\exp\left[-\frac{E}{RT_\infty\zeta(0)\left(1 + \frac{\gamma - 1}{2}\text{Ma}_\infty^2\right)}\right] \tag{5.24}$$

where ω is the ratio of diffusivity over preexponential factor, multiplied by the local Reynolds number

$$\omega = \frac{D_w}{B}\sqrt{\frac{u_\infty}{v_\infty \bar{x}}} = \frac{D_w}{BL}\sqrt{\text{Re}_{\infty L}\frac{L}{\bar{x}}}$$

where $\text{Re}_{\infty L} = u_\infty L/v_\infty$, and L is a characteristic length. Obviously, ω is a function of $\bar{x}$. Only in case of

$$B = \sqrt{A/\bar{x}} = \sqrt{A/x}$$

$$\omega = \frac{D_w}{\sqrt{AL}}\sqrt{\text{Re}_{\infty L}} = D_w\sqrt{\frac{u_\infty}{v_\infty A}} \tag{5.25}$$

which is independent of the coordinate, we have the similar φ and ζ profiles, or φ and ζ are functions of only η . In this case, the general solution of diffusion and energy equations, Eqs. (5. 21) and (5. 22) is

$$\varphi(\eta) = \varphi(0) + A_1\int_0^\eta (f'')^{\text{Sc}}d\eta \tag{5.26}$$

$$\zeta(\eta) = \zeta(0) + A_2\int_0^\eta (f'')^{\text{Pr}}d\eta$$

$$+ \frac{\left(1 - \frac{1}{\text{Pr}}\right)(\gamma - 1)\text{Ma}_\infty^2}{2\left(1 + \frac{\gamma - 1}{2}\text{Ma}_\infty^2\right)}\int_0^\eta (f'')^{\text{Pr}}\left[\int_0^\eta (f'')^{-\text{Pr}}(f'^2)''d\eta\right]d\eta \tag{5.27}$$

The boundary conditions of diffusion and energy equationas at $\eta = \infty$ are

$$\zeta(\infty) = \varphi(\infty) = 1 \tag{5.28}$$

From Eqs. (5. 23), (5. 24), (5. 25), (5. 26), (5. 27), (5. 28) we can find $\varphi(\eta)$, $\zeta(\eta)$, $\varphi(0)$, $\zeta(0)$, $\varphi'(0)$, $\zeta'(0)$ and hence the combustion rate $g_{\text{ox},w}$, the oxygen surface concentration $Y_{\text{ox},w}$, and the surface temperature T_w . For the simplest case of Pr =Sc=Le=1, we have

$$\frac{\varphi - \varphi_w}{1 - \varphi_w} = \frac{\zeta - \zeta_w}{1 - \zeta_w} = f'(\eta)$$

or

$$\frac{Y_{ox} - Y_{ox,w}}{Y_{ox,\infty} - Y_{ox,w}} = \frac{T_* - T_w}{T_{*,\infty} - T_w} = f'(\eta) \qquad (5.29)$$

and

$$\varphi'(\eta) = (1 - \varphi_w) f''(\eta)$$
$$\varphi'(0) = (1 - \varphi_w) f''(0)$$
$$\zeta'(0) = (1 - \zeta_w) f''(0)$$

The combustion rate is determined by the following expressions:

$$\begin{cases} g_{ox,w} = - D_w \dfrac{\rho_w{}^2}{\rho_\infty} Y_{ox,\infty} \sqrt{\dfrac{u_\infty}{\nu_\infty x}} (1 - \varphi_w) f''(0) & (5.30) \\ - Y_{ox,\infty} Q_{ox} (1 - \varphi_w) \Big/ \left[c_p T_\infty \left(1 + \dfrac{\gamma - 1}{2} \mathrm{Ma}_\infty^2 \right) \right] = (1 - \zeta_w) & (5.31) \\ \omega \dfrac{(1 - \varphi_w) f''(0)}{\varphi_w} = \left(1 + \dfrac{\gamma - 1}{2} \mathrm{Ma}_\infty^2 \right) \zeta_w & \\ \cdot \exp \left[- \dfrac{E}{R T_\infty \zeta_w \left(1 + \dfrac{\gamma - 1}{2} \mathrm{Ma}_\infty^2 \right)} \right] & (5.32) \end{cases}$$

It can be seen that the combustion rate is proportional to $\sqrt{u_\infty}$, $1/\sqrt{x}$, D_w, and $Y_{ox,\infty}$. The factor $(1 - \varphi_w)$ expresses the effect of kinetics. From Eq. (5.31) we have

$$T_w = T_\infty + \frac{u_\infty^2}{2c_p} + \frac{Y_{ox,\infty} Q_{ox}}{c_p} \left(1 - \frac{Y_{ox,w}}{Y_{ox,\infty}} \right) \qquad (5.33)$$

Clearly, the surface temperature is higher than the gas temperature of oncoming flow due to the stagnation effect of high speed flow (the second term on the right-hand side) and the combustion effect (the third term on the right-hand side). In case of $\omega \ll 1$ or

$$D_w \ll B \sqrt{\nu_\infty \bar{x} / u_\infty}$$

we have $\varphi_w \approx 0$, $Y_{ox,w} \approx 0$, and

$$g_{ox,w} = - D_w \frac{\rho_w^2}{\rho_\infty} Y_{ox,\infty} \sqrt{\frac{u_\infty}{\nu_\infty x}} f''(0) \qquad (5.34)$$

$$T_w = T_\infty + \frac{u_\infty^2}{2c_p} + \frac{Y_{ox,\infty} Q_{ox}}{c_p} \qquad (5.35)$$

This is the case of diffusion-controlled combustion, and we have the highest surface temperature. For cases of finite reaction rate ($Y_{ox,w} \neq 0$), the temperature rise due to combustion is lower than that for the case of diffusion-controlled combustion. In the case of $\omega \gg 1$ or

$$D_w \gg B \sqrt{\nu_\infty \bar{x} / u_\infty}$$

we have $\varphi_w \approx 1$, $Y_{ox,w} \approx Y_{ox,\infty}$, and

$$T_w = T_\infty + \frac{u_\infty^2}{2c_p} \qquad (5.36)$$

This is the case of kinetics-controlled combustion, and we have the lowest surface temperature. The temperature rise is caused only by the stagnation effect, and the combustion effect is almost zero.

Using the stagnant film theory for Ma$\approx$0 and Le=1, we can obtain the following expressions for the combustion rate, the oxygen surface concentration and the surface temperature:

$$\begin{cases} g_{ox,w} = - f''(0)\overline{D\rho}\sqrt{\dfrac{u_\infty}{\nu_\infty x}}(Y_{ox,\infty} - Y_{ox,w}) & (5.37) \\ g_{ox,w} = - B\bar{\rho}Y_{ox,w}\exp\left(-\dfrac{E}{RT_w}\right) & (5.38) \\ c_p(T_w - T_\infty) = (Y_{ox,\infty} - Y_{ox,w})Q_{ox} & (5.39) \end{cases}$$

Obviously, there is no remarkable difference between the results obtained by the boundary-layer theory (Eqs. (5.30) — (5.32)) and the results obtained by the stagnant-film theory (Eqs. (5.37)-(5.39)), except some difference in the transport coefficients.

5.3 LAMINAR BOUNDARY LAYER WITH EVAPORATION AND GAS—PHASE COMBUSTION

For low Mach number two-dimensional laminar boundary layer over a flat plate with evaporation on the surface, the basic equations are

$$\begin{cases} \dfrac{\partial}{\partial x}(\rho u) + \dfrac{\partial}{\partial y}(\rho v) = 0 & (5.1) \\ \rho u\dfrac{\partial u}{\partial x} + \rho v\dfrac{\partial u}{\partial y} = \dfrac{\partial}{\partial y}\left(\mu\dfrac{\partial u}{\partial y}\right) & (5.9) \\ \rho u\dfrac{\partial Y_s}{\partial x} + \rho v\dfrac{\partial Y_s}{\partial y} = \dfrac{\partial}{\partial y}\left(D\rho\dfrac{\partial Y_s}{\partial y}\right) - w_s & (5.3) \\ \rho u c_p\dfrac{\partial T}{\partial x} + \rho v c_p\dfrac{\partial T}{\partial y} = \dfrac{\partial}{\partial y}\left(\lambda\dfrac{\partial T}{\partial y}\right) + w_s Q_s & (5.40) \end{cases}$$

with boundary conditions at the surface

at $\quad y = 0: \qquad u = 0,\ v = v_w$

$$g_{sw} = - D\rho\left(\frac{\partial Y_s}{\partial y}\right)_w + Y_{sw}\rho_w v_w = \alpha\rho_w v_w \qquad \begin{pmatrix} s = F,\ \alpha = 1 \\ s \neq F,\ \alpha = 0 \end{pmatrix}$$

$$\lambda_w\left(\frac{\partial T}{\partial y}\right)_w = \rho_w v_w q_e$$

For the simplest case of Pr=Sc=Le=1, using Zeldovich transformation

$$Y \equiv Y_{ox} - \beta Y_F, \qquad Z \equiv c_p T + Y_{ox}Q_{ox}$$

and

$$f_1 = \frac{Y - Y_W}{Y_\infty - Y_w}, \qquad f_2 = \frac{Z - Z_w}{Z_\infty - Z_w}$$

we have

$$\frac{u}{u_\infty} = f_1 = f_2 = \frac{Y - Y_w}{Y_\infty - Y_w} = \frac{Z - Z_w}{Z_\infty - Z_w} \qquad (5.41)$$

and hence

$$\left(\frac{\partial u}{\partial y}\right)_w = \frac{u_\infty}{Z_\infty - Z_w}\left(\frac{\partial Z}{\partial y}\right)_w$$

$$= \frac{u_\infty}{c_p(T_\infty - T_w) + (Y_{ox,\infty} - Y_{ox,w})Q_{ox}} \left[c_p \left(\frac{\partial T}{\partial y} \right)_w + Q_{ox} \left(\frac{\partial Y_{ox}}{\partial y} \right)_w \right]$$

For diffusion combustion in gas phase, we have $Y_{ox,w} = 0$, $g_{ox,w} = 0$ and

$$\left(\frac{\partial Y_{ox}}{\partial y} \right)_w = 0$$

$$\tau_w = \mu_w \left(\frac{\partial u}{\partial y} \right)_w = \left(\frac{\lambda}{c_p} \right)_w \frac{u_\infty}{c_p(T_m - T_w)} c_p \left(\frac{\partial T}{\partial y} \right)_w = \frac{u_\infty}{c_p(T_m - T_w)} \lambda_w \left(\frac{\partial T}{\partial y} \right)_w$$

$$= - \frac{u_\infty}{c_p(T_m - T_w)} q_w$$

Therefore, we obtain

$$q_w = - \left(\lambda \frac{\partial T}{\partial y} \right)_w = - \frac{c_p(T_m - T_w)}{u_\infty} \tau_w \tag{5.42}$$

or

$$\rho_w v_w = \frac{c_p(T_m - T_w)}{q_e} \frac{\tau_w}{u_\infty} \tag{5.43}$$

Eqs. (5.42) and (5.43) express the generalized Reynolds analogy in a laminar boundary layer with evaporation and combustion.

For a laminar boundary layer with surface evaporation and gas phase combustion (Fig. 5.2), we can introduce the Howwarth-Dorodnitzn-Blasius coordinate

$$\eta = \frac{1}{2} \int_0^y \frac{\rho}{\rho_\infty} dy \sqrt{\frac{u_\infty}{v_\infty x}}$$

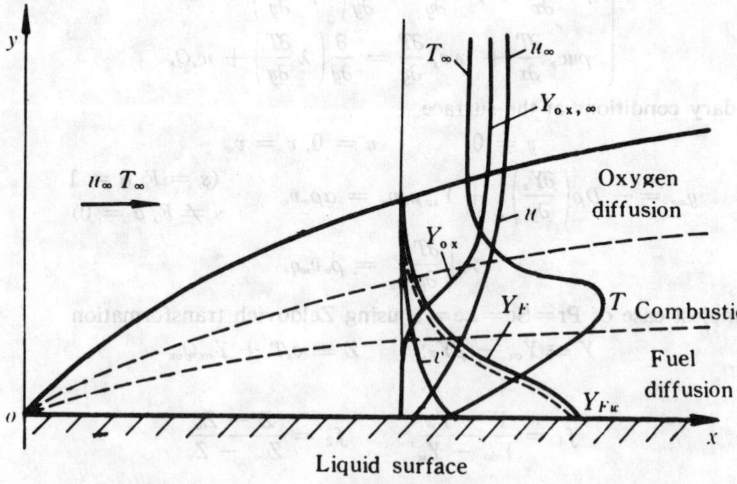

Fig. 5.2

and the stream function in the plane of hypothetical incompressible flows

$$\psi = \sqrt{u_\infty v_\infty x} f(\eta)$$

and assume

$$\rho\mu = \text{const}$$

Then, we have the momentum equation in the form similar to that of the Blasius equation

$$f''' + ff'' = 0 \tag{5. 44}$$

However, the boundary conditions are now different. They are:

at $y = 0$: $u = 0, v = v_w \neq 0$

at $y = \infty$: $u = u_\infty$

$$\text{or} \quad f'(0) = 0, \quad f'(\infty) = 2, \quad f(0) \neq 0 \tag{5. 45}$$

The condition of $v_w \neq 0$ or $f(0) \neq 0$ makes the solution of Eq. (5. 44) different from that of Blasius equation. v_w is related to q_w and g_{Fw}, and originally we need to simultaneously solve the momentum, diffusion and energy equations. However, this task was remarkably simplified by Emmons who used the above-described Reynolds analogy to find a new boundary condition for the momentum equation, and therefore solved the whole problem without using diffusion and energy equations.

The expression of Reynolds analogy, Eq. (5. 43) gives

$$\rho_w v_w = \frac{c_p (T_m - T_w)}{q_e} \frac{\tau_w}{u_\infty} = B \frac{\tau_w}{u_\infty}$$

or

$$\frac{\rho_w v_w}{\rho_\infty u_\infty} = B \frac{\tau_w}{\rho_\infty u_\infty^2}$$

The shear stress is

$$\tau_w = \mu_w \left(\frac{\partial u}{\partial y} \right)_w = \mu_w \frac{u_\infty}{2} f''(0) \left(\frac{\partial \eta}{\partial y} \right)_w = \frac{\mu_\infty u_\infty}{4} f''(0) \sqrt{\frac{u_\infty}{v_\infty x}}$$

Hence we have

$$\frac{\rho_w v_w}{\rho_\infty u_\infty} = \frac{B}{4} \frac{f''(0)}{\sqrt{\text{Re}_x}} \tag{5. 46}$$

where

$$\text{Re}_x = u_\infty x / v_\infty$$

On the other hand, using the expressions of HDB coordinate and stream function given before, we have

$$\frac{\rho v}{\rho_\infty u_\infty} = \frac{\eta f'(\eta) - f(\eta)}{2\sqrt{\text{Re}_x}} - \frac{f'(\eta)}{2} \frac{\partial}{\partial x} \int_0^y \frac{\rho}{\rho_\infty} dy$$

and hence

$$\frac{\rho_w v_w}{\rho_\infty u_\infty} = -\frac{f(0)}{2\sqrt{\text{Re}_x}} \tag{5. 47}$$

Comparing Eq. (5. 46) with Eq. (5. 47), we can find the third boundary condition for the momentum equation:

$$-\frac{f(0)}{f''(0)}=\frac{B}{2} \tag{5. 48}$$

which actually is the coupling condition of momentum equation with energy and diffusion equations. Obviously, $f(\eta)$, $f'(\eta)$, and $f''(\eta)$ are functions of B, so $f(\eta)$ is not the Blasius solution. Eq. (5. 44) together with its boundary conditions (5. 45) and (5. 48) was solved numerically by Emmons et al. to find $f(0)$ and $f''(0)$ as functions of B. On the basis of this solution, an approximate expression was given as

$$\frac{-f(0)}{2}\cong 0.18\ln(0.7+B) \tag{5. 49}$$

Substituting Eq. (5. 49) into Eq. (5. 47), the evaporation rate can be obtained as:

$$g_{Fw}=\rho_w v_w=0.18 D_\infty \rho_\infty \sqrt{\frac{u_\infty}{\nu_\infty x}}\ln(0.7+B) \tag{5. 50}$$

It is not difficult to prove that the evaporation rate given by the stagnant-film theory is

$$g_{Fw}=0.332\,\overline{D\rho}\sqrt{\frac{u_\infty}{\nu_\infty x}}\ln(1+B) \tag{5. 51}$$

So, the exact solution and the simplified solution give almost the same result except some difference in the transport coefficients and the constants. Because $f''(0)=-2f(0)/B$ decreases with the increase of B, the shear stress τ_w will decrease with the increase of B or with the increase of evaporation rate.

It can also be seen that this analytical method can be extended to more complicated cases, such as laminar boundary layer over a flat combusting carbon plate accounting for Stefan flux on the surface.

5. 4 IGNITION IN A LAMINAR BOUNDARY LAYER BY AN INERT HOT PLATE

In Chater 2 we have discussed the simplified solution of the ignition in a laminar boundary layer by an inert hot plate. Now we will discuss the exact solution. The ignition criterion is still zero temperature gradient at the surface

$$\left(\frac{\partial T}{\partial y}\right)_w=0$$

In a way similar to that used in the above-described examples, using the H-D-B coordinate

$$\eta=\frac{1}{2}\int_0^y\frac{\rho}{\rho_\infty}dy\sqrt{\frac{u_\infty}{\nu_\infty x}}$$

and the stream function

$$\psi = \sqrt{u_\infty v_\infty x} f(\eta)$$

the continuity and momentum equations can be transformed into the Blasius equation which can be separately solved, and $f(\eta)$, $f'(\eta)$ become known functions. Then, assuming Pr=Sc=Le=1 and using the generalized Reynolds analogy

$$\frac{Z - Z_w}{Z_\infty - Z_w} = \frac{u}{u_\infty} = \frac{1}{2} f'(\eta)$$

the combined concentration-temperature field can be obtained as:

$$\frac{c_p(T_w - T) - (Y_s - Y_{sw}) Q_s}{c_p(T_w - T_\infty) - (Y_{s\infty} - Y_{sw}) Q_s} = \frac{1}{2} f'(\eta) \qquad (5.52)$$

Introducing

$$\theta \equiv \frac{T_w - T}{T_w - T_\infty}, \qquad \chi = \frac{Q_s}{c_p(T_w - T_\infty)}$$

the concentration can be expressed as a function of temperature

$$Y_s = Y_{sw} + \frac{1}{\chi}\left\{\theta - [1 - \chi(Y_{s\infty} - Y_{sw})]\frac{f'(\eta)}{2}\right\} \qquad (5.53)$$

It is worthwhile to note that the concentration boundary condition at the surface give by Dooley

at $\qquad y = 0: \qquad Y_s = Y_{sw} = \text{const}$

is wrong, because for an inert plate this condition should be

$$\left(\frac{\partial Y_s}{\partial y}\right)_w = 0$$

which was used in Toong's solution. As far as the velocity field $f'(\eta)$ and concentration as a function of temperature have been found, the only thing need to be done is to solve the energy equation. The method used here is an asymptotic one. The basic idea is: before ignition the effect of reaction on temperature and concentration profiles is a small perturbation. The accumulation of deformation of temperature profile along the flow direction leads to a zero temperature gradient at the wall at a certain distance which is considered as the ignition distance.

The energy equation in the x, η coordinates is:

$$\frac{\partial^2 \theta}{\partial \eta^2} + f(\eta)\frac{\partial \theta}{\partial \eta} = 2x f'(\eta)\frac{\partial \theta}{\partial x} - \frac{4\chi w_s Q_s}{\rho u_\infty c_p (T_w - T_\infty)} \qquad (5.54)$$

In Eq. (5.54) the first and second terms on the left-hand side are conduction and convection terms in y direction respectively, and the first and second terms on the right-hand side are convection term in x direction and reaction term respectively. The boundary conditions are:

$$\theta(0, x) = 0, \qquad \theta(\infty, x) = 1 \qquad (5.55)$$

Assume that the effect of chemical reaction is a small perturbation, then Eq. (4.54) can be solved by using the asymptotic method. In the case of Pr=Sc=Le=1, setting the reaction term to be zero, the first-order approximate solution is

$$\theta^{(0)}(\eta) = u/u_\infty = \frac{1}{2} f'(\eta) \tag{5.56}$$

Substituting the lower-order approximation of $\theta(\eta)$ into the right-hand side of Eq. (5.54) as a known function of x, the higher-order approximation of $\theta(\eta)$ at each x location on the left-hand side can be calculated. The iteration is made till the zero temperature gradient at the wall at a certain distance x_i appears.

Introduce

$$M = 4Q_s/[c_p(T_w - T_\infty)]$$

$$S^{(n-1)}(\eta, x) = \frac{1}{f'(\eta)}\left[2\frac{u_\infty}{M}f'(\eta)\frac{\partial\theta^{(n-1)}}{\partial x} - \frac{w_s^{(n-1)}}{\rho}\right]$$

then the energy equation after n iterations is

$$\frac{\partial^2\theta^{(n)}}{\partial\eta^2} + f(\eta)\frac{\partial\theta^{(n)}}{\partial\eta} = \frac{x}{u_\infty}f''(\eta)MS^{(n-1)}(\eta, x) \tag{5.57}$$

If the right-hand side of Eq. (5.57) is considered as a known function of η, Eq. (5.57) will be a linear partial differential equation, and then its integral is

$$\theta^{(n)}(\eta, x) = \frac{1}{2}f'(\eta) - \frac{x}{u_\infty}M\left\{\frac{1}{2}f'(\eta)\int_0^\infty [f''(\eta)\int_0^\eta S^{(n-1)}(\eta, x)\,d\eta]d\eta - \int_0^\eta [f''(\eta)\int_0^\eta S^{(n-1)}(\eta, x)\,d\eta]d\eta\right\} \tag{5.58}$$

The gas density ρ and species concentration Y_s are determined by

$$\rho^{(n-1)} = p\{R[T_w - (T_w - T_\infty)\theta^{(n-1)}]\sum Y_s^{(n-1)}/\mathrm{Ma}_s\}^{-1} \tag{5.59}$$

$$Y_s^{(n-1)} = Y_{sw} + \frac{1}{\chi}\{\theta^{(n-1)} - [1 - \chi(Y_{s\infty} - Y_{s\infty})]f'(\eta)/2\} \tag{5.60}$$

The temperature gradient at the wall is

$$\left[\frac{\partial T^{(n)}}{\partial y}\right]_w = \left[\frac{\partial T^{(n)}}{\partial\eta}\right]_w\left(\frac{\partial\eta}{\partial y}\right)_w = -\frac{T_w - T_\infty}{2\sqrt{\frac{\nu_\infty x}{u_\infty}}}\left[\frac{\partial\theta^{(n)}}{\partial\eta}\right]_w$$

Taking the derivative of θ(η) for η from Eq. (5.58), we have

$$\left[\frac{\partial\theta^{(n)}}{\partial\eta}\right]_w = 0.664\left\{1 - \frac{x}{u_\infty}M\int_0^\infty [f''(\eta)\int_0^\eta S^{(n-1)}(\eta, x)\,d\eta]d\eta\right\} \tag{5.61}$$

Setting the ignition condition

$$\left(\frac{\partial\theta^{(n)}}{\partial\eta}\right)_w = 0$$

the expression of ignition delay can be obtained as:

$$\tau_i = x_i/u_\infty = \left\{M\int_0^\infty [f''(\eta)\int_0^\eta S^{(n-1)}(\eta, x)\,d\eta]d\eta\right\} \tag{5.62}$$

After asymptotic iterations τ_i should approches to a converged value.

For the dissociation reaction of aziethane $C_2H_6N_2$

$$C_2H_6N_2 \longrightarrow C_2H_6 + N_2$$

the calculated temperature profiles $\frac{T}{T_\infty}(\xi, \eta)$, concentration profiles $Y_F(\xi, \eta)$ and the ignition delay as a function of the surface temperature and the fuel/air ratio are shown in Fig. 5. 3, 5. 4, 5. 5 and 5. 6, where $\xi = x/x_i$. It can be seen that the results obtained by the exact solution are in qualitative agreement with those obtained by the simplified solution discussed in Chapter 2. For example, the ignition delay sharply decreases with the increase of surface temperature (Fig. 5. 5), and there is an optimum fuel/air ratio for minimum ignition delay.

Besides the above-discussed exact solution given by Dooley, Toong proposed a similar asymtotic method. The Karman—Pohlhausen integral method has also been used. In recent years, the ignition in laminar boundary layer by a hot plate has been studied by full numerical modeling which gives many detailed results of this phemomena.

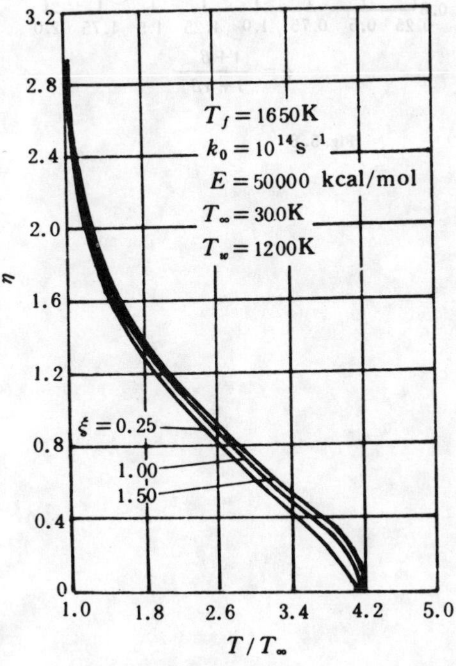

Fig. 5. 3

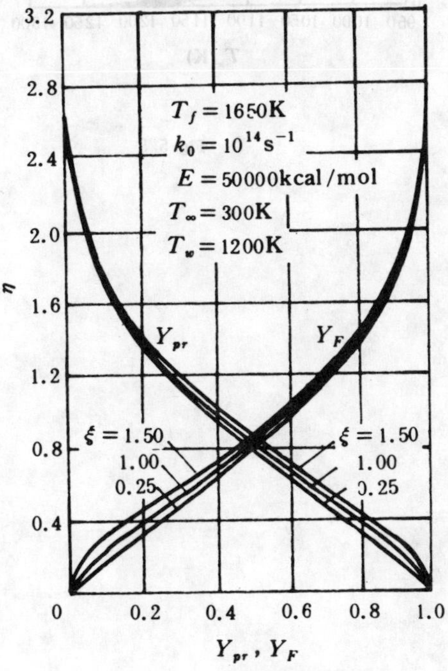

Fig. 5. 4

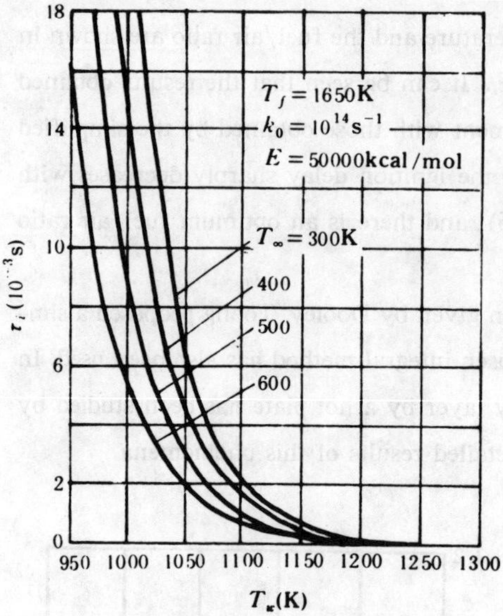

18
14
10
6
2
0
$T_f = 1650K$
$k_0 = 10^{14}s^{-1}$
$E = 50000kcal/mol$
$T_\infty = 300K$
400
500
600
$\tau_i\ (10^{-3}s)$
950 1000 1050 1100 1150 1200 1250 1300
$T_w(K)$

Fig. 5. 5

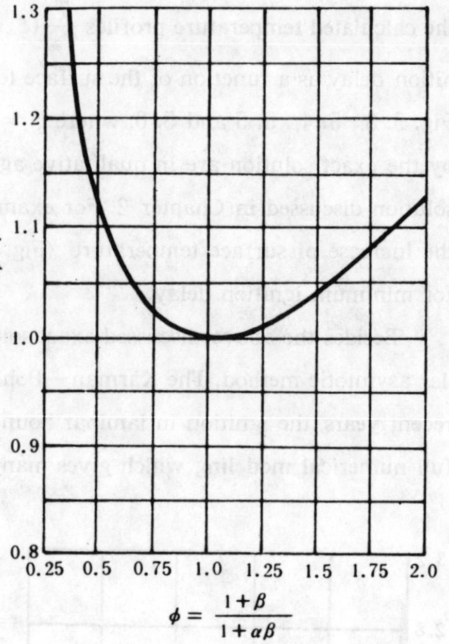

1.3
1.2
1.1
1.0
0.9
0.8
$\tau_i / \tau_{i,\phi=1}$
0.25 0.5 0.75 1.0 1.25 1.5 1.75 2.0
$\phi = \frac{1+\beta}{1+\alpha\beta}$

Fig. 5. 6

CHAPTER 6 GAS-PARTICLE FLOWS

6.1 BACKGROUND

Basically, there are two different approaches to treat gas-particle flows: one is to consider the fluid or gas as a continuum and the particles as a discrete system; the other is to consider both the fluid/gas and particles as coexisting and interpenetrating continua or the particles as a pseudo-fluid. Recently, combined treatment, such as continuum-trajectory approach to treat particle phase has also been studied for reacting gas-particle flows.

Earlier studies were limited to particle motion in a known flow field with neglected effect of particles on the fluid/gas flow field. One of the important features of the modern multiphase fluid dynamics is to thoroughly consider the mass, momentum and energy interaction between fluid/gas and particle phases. The classical multiphase fluid dynamics is based on the analytical studies, such as theory of boundary layers and asymptotic method in single-phase fluid dynamics.

6.2 BASIC PROPERTIES

6.2.1 Flow Regime

To identify the flow regime of gas-particle flows, the following characteristic times are defined as

Flow time (residence time) $\tau_f = L/v$

Relaxation time for diffusion $\tau_r = d_p^2\,\rho_p\,/\,(18\,\mu)$

Relaxation time for mean motion $\tau_{r1} = \tau_r\,(1 + \mathrm{Re}_p^{2/3}/6)$

Fluid fluctuation time $\tau_T = l/u' = k/\varepsilon$

Particle-particle impaction time $\tau_p = l_p/u'_p = (c\pi n_p r_p^2)^{-1}(u'_p)^{-1}$

where

$$\mathrm{Re}_p = |v - v_p|\,d_p/\nu$$

According to the magnitude of the similarity parameters obtained from the ratio of a certain characteristic time to the other, the following different flow regimes for limiting cases of gas-particle flows can be identified as:

$\tau_{r1}/\tau_f \ll 1$ — No-slip flow (equilibrium flow)

$\tau_{r1}/\tau_f \gg 1$ — Strong slip flow (frozen flow)

$\tau_r/\tau_f \gg 1$ — Diffusion-frozen flow

$\tau_r/\tau_f \ll 1$ — Diffusion-equilibrium flow
$\tau_{r1}/\tau_p \ll 1$ — Dilute suspension flow
$\tau_{r1}/\tau_p \gg 1$ — Dense suspension flow

It can be seen that

$$\tau_{r1}/\tau_p \sim \left(\frac{u_p' d_p}{\nu}\right) \Phi_p$$

where Φ_p is the particle-phase volume fraction.

6. 2. 2 Particle Size and Size Distribution

Frequently the particle size distribution can be expressed by the semiempirical Rosin-Rammler formula

$$R(d_k) = \exp[-(d_k/\bar{d})^n] \tag{6.1}$$

where $R(d_k)$ is the weight fraction of particles whose sizes are greater than d_k, n is the index of nonuniformness, $\bar{d}$ is a characteristic size. Both n and $\bar{d}$ are determined by experiments. The derivative of $R(d_k)$ is

$$\frac{dR}{d(d_k)} = \frac{n}{(\bar{d})^n}(d_k)^{(n-1)} \exp[-(d_k/\bar{d})^n] \tag{6.2}$$

which is called the differential size distribution, while $R(d_k)$ is called the integral size distribution.

The mean particle size can be defined as

radius averaged $$d_{10} = \sum n_k d_k / \sum n_k \tag{6.3}$$

surface averaged $$d_{20} = (\sum n_k d_k^2 / \sum n_k)^{\frac{1}{2}} \tag{6.4}$$

volume averaged $$d_{30} = (\sum n_k d_k^3 / \sum n_k)^{\frac{1}{3}} \tag{6.5}$$

Sauter diameter $$d_{32} = \sum n_k d_k^3 / (\sum n_k d_k^2) \tag{6.6}$$

Among them the Sauter diameter is most widely used in engineering.

The typical particle sizes are

Coal particles in fluidized bed	1—10mm
Spray	10—200μm
Pulverized coal	1—100μm
Soot	1—5μm

6. 2. 3 Apparent Density and Volume Fraction

There are differently defined densities in gas-particle flows.

The expression related to these densities is

$$\rho_m = \rho + \rho_p = \rho + \sum \rho_k = \rho + (\sum n_k \pi d_k^3/6)\,\bar{\rho}_p \tag{6.7}$$

where

ρ_m — mixture density
ρ — fluid/gas apparent density
ρ_p, ρ_k — particle apparent density
$\bar{\rho}_p$ — particle material density

The volume fraction of particle and gas phase is defined as

$$\begin{cases} \Phi_p = \rho_p/\bar{\rho}_p \\ \Phi = 1 - \Phi_p = 1 - \rho_p/\bar{\rho}_p \end{cases} \tag{6.8}$$

For dilute gas-particle flows we have

$$\rho = \bar{\rho}(1 - \rho_p / \bar{\rho}_p) \approx \bar{\rho}$$

where $\bar{\rho}$ is the gas material density. Obviously, the gas apparent density is nearly equal to the gas material density in dilute gas-particle flows.

The mass flux ratio, or the ratio of particle mass flux over the gas mass flux is defined as

$$\rho_{p0}u_{p0} / (\rho_0 u_0)$$

which is also called mass loading. In case of equality of initial gas and particle velocity the mass loading will be equal to the apparent density ratio ρ_{p0}/ρ_0. For example, the typical value for spray or pulverized-coal flame is

$$\rho_p/\rho \approx 1/15 = \rho_p/[\bar{\rho}/(1 - \rho_p/\bar{\rho}_p)] = \frac{\bar{\rho}_p}{\bar{\rho}}\left(\frac{\Phi_p}{1 - \Phi_p}\right) \approx 1000 \frac{\Phi_p}{1 - \Phi_p}$$

or

$$\Phi_p < 0.01\%$$

Thus the liquid spray flame and the pulverized-coal flame are dilute gas-particle flows. Other examples are

Pneumatic conveying $\Phi_p \approx 0.1\%$ (mass loading ≈ 1)

Fluidized bed
Flow in gun barrels } $\Phi_p = 0$—1

In case of $\Phi_p = 0.1\%$, we have $1 = 1000\, n\pi d_p^3/6$, or the mean distance between particles is

$$\Delta \approx n^{-\frac{1}{3}} = \left(\frac{1000\pi}{6}\right)^{\frac{1}{3}} d_p = 8.1 d_p$$

In flames we have $\Delta > 20 d_p$

6.2.4 Various Model Approximations

The basic problem in describing gas-particle flows is to model the particle phase. Various particle-phase models are shown in Tab. 6.1.

Table 6.1 Particle-phase Models

Model	Approach	Effect of particles on gas phase	Phase slip	Coordinate	Particle transport properties
Single-particle-dynamics model (SPD)	Discrete system	Neglected	Yes	Lagrangian	No (Diffusion frozen)
Small-slip model (SS)	Continuum	Neglected	Yes (Slip= Diffusion)	Eulerian	Yes (Diffusion =slip)

(Continued)

No-slip model (NS)	Continuum	Partially considered	No	Eulerian	Yes (Diffusion equilibrium)
Trajectory model (PT)	Discrete system	Considered	Yes	Lagrangian	No (determ. traj.) or Yes (stoch. traj.)
Pseudofluid or multifluid model (MF)	Continuum	Considered	Yes	Eulerian	Yes

The earliest model is the single-particle-dynamics model, in which the particle trajectory of mean motion or convection in a known flow field (frequently uniform velocity and temperature field) and the particle velocity and temperature change along the trajectory are considered, neglecting the effect of particles on the flow field. This is an oversimplified model. In the meantime, Tchen, Hinze, and others studied particle behavior in a fluctuating flow field—its turbulent diffusion. In the late 1960's to early 1970's, the single-fluid model or no-slip model was proposed. This model is a dynamically and thermally equilibrium model, in which the particle velocity and temperature are assumed to be equal to those of gas phase everywhere, and particle diffusion is assumed to be equal to gas species diffusion. This is another kind of simplified model. Also in the late 1960's S. L. Soo proposed a pseudo-fluid model or small-slip model of particle phase based on a continuum concept, and the slip between fluid and particles was taken into account. However, the slip was considered as a result of particle diffusion, so this model can be called a "small slip" model. More recently developed models are those which accounts for slip, mass, momentum and energy interaction between fluid and particle phases, in particular, the effect of particles on fluid/gas flows. These models are particle trajectory model (including deterministic trajectory model and stochastic trajectory model) and pseudo-fluid model with simultaneous phase slip and particle diffusion.

Actually all of different models are an approximation to the real process from different point of view. The feasibility of these models depends on the particle volume fraction, relaxation time, flow time and particle-particle impaction time. The range of applicability of various models can be illustratively shown in Tab. 6. 2.

Table 6. 2 Applicability of Various Particle-phase Models

Model	n_p (Φ_p)	τ_{r1}/τ_f	τ_r/τ_f	τ_{r1}/τ_p
Frozen flow		very large		
SPD	very small	very large	very large	very small

(Continued)

SS	small	small	moderate	very small
NS	no limit	very small	very small	no limit
PT	moderate	moderate	very large	very small
MF	moderate or large	moderate	moderate	no limit

6. 2. 5 Particle Drag, Heat and Mass Transfer and Reaction

In analyzing gas-particle flows the particle drag, heat and mass transfer and reaction rate are important basic properties. The particle drag for different range of Reynolds number of relative motion can be given as:

Newton's formula

$$c_d = 0.44 \qquad (\mathrm{Re}_p > 1000) \qquad (6.9)$$

Wallis-Kliachko's formula

$$c_d = (1 + \mathrm{Re}_p^{2/3}/6)\,24/\mathrm{Re}_p \qquad (1 < \mathrm{Re}_p < 1000) \qquad (6.10)$$

Stokes' formula

$$c_d = 24/\mathrm{Re}_p \qquad (\mathrm{Re}_p < 1) \qquad (6.11)$$

where

$$\mathrm{Re}_p = |v - v_p|\,d_p/\nu$$

In case of higher particle temperature than gas temperature, the particle drag will be higher than that for isothermal case. In this case the gas viscosity in Re_p is determined by the so-called 1/3 Law, that is

$$\nu = \frac{1}{3}\nu_p + \frac{2}{3}\nu_g \qquad (6.12)$$

where ν_p and ν_g are gas viscosity coefficients at temperature T_p and T_g respectively, and T_p and T_g are particle and gas temperature respectively. The particle mass change due to evaporation, devolatilization or heterogeneous reaction will reduce the drag, using the expression

$$c_d = c_{d0} \ln(1 + B)/B \qquad (6.13)$$

where B is a dimensionless parameter related to the mass changing rate $\dot{m}$ by

$$\ln(1 + B) = \dot{m}/(\pi d_p\,\mathrm{Nu}\,D\rho) \qquad (6.14)$$

The particle heat and mass transfer can be described by Ranz-Marshell expressions

$$\begin{cases} \mathrm{Nu} = 2 + 0.6\,\mathrm{Re}_p^{0.5}\,\mathrm{Pr}^{0.33} \\ \mathrm{Sh} = 2 + 0.6\,\mathrm{Re}_p^{0.5}\,\mathrm{Sc}^{0.33} \end{cases} \qquad (6.15)$$

For particle evaporation during its preheating period, the mass, diameter and temperature change are described by

$$m_p c_l \frac{dT_p}{dt} = \pi d_p\,\mathrm{Nu}\,\lambda(T - T_p)\ln(1 + B)/B - \dot{m}\,q_e \qquad (6.16)$$

$$\dot{m} = \pi d_p D\rho\,\mathrm{Nu}\ln(1 + B) \qquad (6.17)$$

$$\frac{d(d_p^2)}{dt} = -\,4\mathrm{Nu}\,D\rho\ln(1 + B)/\overline{\rho_p} \qquad (6.18)$$

where

$$B = (Y_{Fw} - Y_{F\infty}) / (1 - Y_{Fw})$$

$$Y_{Fw} = B_w \exp(- E_w/RT_p)$$

For particle steady evaporation, the mass, diameter and temperature change are described by

$$\begin{cases} \dot{m} = \pi d_p \,\mathrm{Nu} \dfrac{\lambda}{c_p} \ln(1 + B) & (6.19) \\ \dfrac{d(d_p^2)}{dt} = \dfrac{4\mathrm{Nu}\,\lambda}{\rho_p c_p} \ln(1 + B) & (6.20) \\ B = c_p (T - T_p) / q_e = (Y_{Fw} - Y_{F\infty}) / (1 - Y_{Fw}) & (6.21) \\ Y_{Fw} = B_w \exp(- E_w/RT_p) \end{cases}$$

In case of coal-particle simultaneous moisture evaporation, devolatilization and char combustion, the mass changing rate and temperature change based on a simple pyrolyzation and a single reaction model are described by

$$\dot{m} = \dot{m}_w + \dot{m}_v + \dot{m}_h \tag{6.22}$$

$$\begin{cases} \dot{m}_w = \pi d_p \,\mathrm{Nu}\, D\rho \ln[1 + (Y_{ww} - Y_{w\infty}) / (1 - Y_{ww})] & (6.23) \\ Y_{ww} = B_w \exp(- E_w/RT_p) & (6.24) \end{cases}$$

$$\begin{cases} \dot{m}_v = m_c \alpha B_v \exp(- E_v/RT_p) & (6.25) \\ \dot{m}_v = - m_c B_v \exp(- E_v/RT_p) & (6.26) \end{cases}$$

$$\begin{cases} \dot{m}_h = \pi d_p \,\mathrm{Nu}\, D\rho \ln[1 + (Y_{\mathrm{ox},\infty} - Y_{\mathrm{ox},w}) / (\beta + Y_{\mathrm{ox},w})] & (6.27) \\ \dot{m}_h = \pi d_p \,\mathrm{Nu}\, (\lambda/c_p) \ln[1 + c_p (T - T_p) / q_w] & (6.28) \\ \dot{m}_h = \pi d_p^2 \beta B Y_{\mathrm{ox},w} \rho \exp(- E/RT_p) & (6.29) \\ \dfrac{\pi d_p^3}{6} \rho_p c \dfrac{dT_p}{dt} = \pi d_p^2 \sigma \varepsilon (T^4 - T_p^4) - \dot{m}\, q_w + \dot{m}_h Q_c & (6.30) \end{cases}$$

6.2.6 Statistical Conservation Equation of a Particle Group

Generally speaking, for a highly dispersed system, each particle in a particle cloud, may have its own size d_p, velocity v_p, temperature T_p, material density ρ_p, different from those of other particles. Particles of the same size may have different velocity and different temperature. Hence, the most general description is to treat the particle cloud as a statistical group. The particle number density in the range of $v_p \longrightarrow v_p + dv_p$, $d_p \longrightarrow d_p + d(d_p)$, $T_p \longrightarrow T_p + dT_p$, $\overline{\rho_p} \longrightarrow \overline{\rho_p} + d\overline{\rho_p}$ can be defined by

$$\varphi(v_p, T_p, \overline{\rho_p}, d_p, x, t)\, dv_p dT_p d\overline{\rho_p} d(d_p)\, dx dt$$

where φ is a probability density distribution function. According to the Louville Theorem in statistical mechanics, for a dilute suspension (without particle-particle interaction), we have

$$\frac{\partial \varphi}{\partial t} + \frac{\partial \varphi}{\partial v_p}\frac{dv_p}{dt} + \frac{\partial \varphi}{\partial d_p}\frac{d(d_p)}{dt} + \frac{\partial \varphi}{\partial \overline{\rho_p}}\frac{d\overline{\rho_p}}{dt} + \frac{\partial \varphi}{\partial T_p}\frac{dT_p}{dt} + \frac{\partial \varphi}{\partial x}\frac{dx}{dt} = 0 \tag{6.31}$$

This is an equation similar to a sourceless Boltzmann equation or a Vlasov equation. Owing to zero volume changing rate in the phase space, Eq. (6.31) can also be expressed as

$$\frac{\partial \varphi}{\partial t} + \frac{\partial}{\partial v_p}(\varphi \dot{v}_p) + \frac{\partial}{\partial d_p}(\varphi \dot{d}_p) + \frac{\partial}{\partial \rho_p}(\varphi \dot{\rho}_p) + \frac{\partial}{\partial T_p}(\varphi \dot{T}_p) + \frac{\partial}{\partial x}(\varphi \dot{x}) = 0 \tag{6.32}$$

Up to now, in practical flow field, Eq. (6. 32) is difficult to be solved. Therefore, in many practical problems, the continuous size distribution is discretized to finite size groups and it is assumed that each size group has the same velocity, temperature and material density.

6. 3 BASIC EQUATIONS

6. 3. 1 The Control Volume of a Multiphase System

The general description of a multiphase flow or two-phase flow system is to consider this system as a multifluid mixture in which particles and fluid occupy the same space macroscopically (but occupy different space microscopically), interpenetrating into each other and each phase having its own size, velocity and temperature. Consider a control volume of the multiphase mixture dV with its external surface dA (Fig. 6. 1). Assume that the total number of particle phases in this control volume is k and each particle phase occupies a volume dV_k with its surface (the interface of k -th particle phase with the fluid) dA_k. The ratio $dV_k/dV = \Phi_k$ is just the local volume fraction of k- th particle phase. In order to describe the particulate suspension by using the multifluid concept in Eulerian coordinates, the condition having to be satisfied is that the size of the control volume l must be much smaller than the geometrical size of the flow field L, and in the meantime much greater than the particle size d_k. This is to require

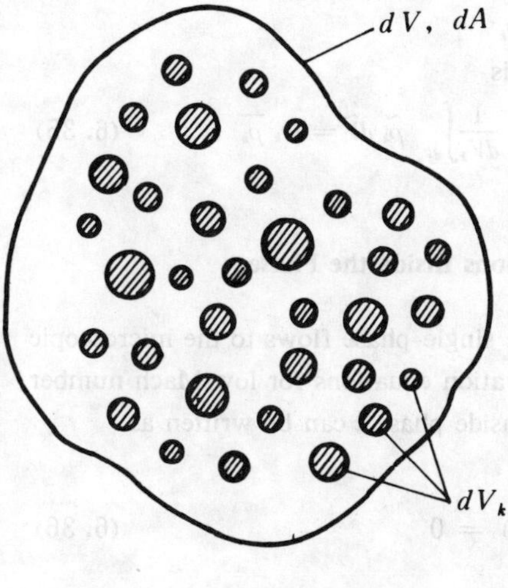

Fig. 6. 1

$$d_k \ll l \ll L \tag{6.33}$$

Only under this condition the control volume can behave like an elementary volume of the macroscopic flow field, and the local values of each variable in the control volume express the macroscopic properties of the flow field, but not the detailed flow field around each particle. For example, in a cyclone separator, L is the order of 0. 5—1m, d_k is the order of 100μm, we may choose l to be 0. 01m. For a large-size cyclone furnace, L is the order of 10m, d_k is the order of 1mm, then l can be given as the order of 0. 1m.

6. 3. 2 The Concept of Volume Averaging

As we have already indicated, if the "microscopic" flow field inside a control volume is to be considered, then different phases will occupy different volume. However, if the "macroscopic" flow field is to be considered, then different phases are treated as if they occupy the same space and interpenetrate into each other, each having its own velocity, temperature and volume fraction. The basic equations derived for single-phase flows are feasible only inside each phase for describing the "microscopic" flows. Therefore, for describing the "macroscopic" flow field, we have to use the concept of volume averaging. That is, each variable expressing the behavior of the "macroscopic" flow field is the volume-averaged value in the control volume dV .

Denoting the "microscopic" "real" value inside the phase by the symbol "$\sim$" and the volume-averaged value in the control volume dV by the symbol "$\langle \quad \rangle$", and using φ_k to express the generalized variable of k- th phase ($k = 1, 2, 3, \cdots f$; 1, 2, 3, ⋯denote particle phases and f denotes the fluid phase), the volume-averaged value is defined as:

$$\varphi_k = \langle \widetilde{\varphi}_k \rangle = \frac{1}{dV}\int_{dV_k} \widetilde{\varphi}_k \, d\widetilde{V} \tag{6. 34}$$

Where φ_k may be a scalor, a vector or a tensor, and

$$dV = \sum dV_k \qquad (k = 1, 2, 3\cdots, f)$$

$$dV_k / dV = \Phi_k$$

For example, the apparent density of k -th phase is

$$\rho_k = \langle \widetilde{\rho}_k \rangle = \frac{1}{dV}\int_{dV_k} \widetilde{\rho}_k \, d\widetilde{V} = \frac{dV_k}{dV}\frac{1}{dV_k}\int_{dV_k} \widetilde{\rho}_k \, d\widetilde{V} = \Phi_k \, \overline{\rho_k} \tag{6. 35}$$

where $\overline{\rho_k}$ is the material density of k- th phase.

6. 3. 3 "Microscopic" Conservation Equations Inside the Phase

Applying the basic conservation laws for the single-phase flows to the microscopic volume dV_k of k- th phase, the microscopic conservation equations for low Mach number flows with constant specific heat of each phase inside phase k can be written as

continuity equation

$$\frac{\partial \widetilde{\rho}_k}{\partial t} + \frac{\partial}{\partial x_j}(\widetilde{\rho}_k \, \widetilde{v}_{kj}) = 0 \tag{6. 36}$$

momentum equation

$$\frac{\partial}{\partial t}(\widetilde{\rho}_k \, \widetilde{v}_{ki}) + \frac{\partial}{\partial x_j}(\widetilde{\rho}_k \, \widetilde{v}_{kj} \widetilde{v}_{ki}) = \frac{\partial \widetilde{p}_k}{\partial x_i} + \frac{\partial}{\partial x_j}(\widetilde{\tau}_{k,\, ji}) + \widetilde{\rho}_k \, g_i \tag{6. 37}$$

energy equation

$$\frac{\partial}{\partial t}(\widetilde{\rho}_k \, c_k \widetilde{T}_k) + \frac{\partial}{\partial x_j}(\widetilde{\rho}_k \, \widetilde{v}_{kj} c_k \widetilde{T}_k) = \frac{\partial}{\partial x_j}(\lambda_k \frac{\partial \widetilde{T}_k}{\partial x_j}) + \widetilde{q}_c - \widetilde{q}_r \tag{6. 38}$$

species equation

$$\frac{\partial}{\partial t}(\widetilde{\rho}_k \, \widetilde{Y}_{ks}) + \frac{\partial}{\partial x_j}(\widetilde{\rho}_k \, \widetilde{v}_{kj} \widetilde{Y}_{ks}) = \frac{\partial}{\partial x_j}(D_k \, \widetilde{\rho}_k \frac{\partial \widetilde{Y}_{ks}}{\partial x_j}) - \widetilde{w}_{ks} \tag{6. 39}$$

where $\widetilde{q}_c$, $\widetilde{q}_r$ are heat of reaction and heat radiation respectively, $\widetilde{Y}_{ks}$ is the mass fraction of species s inside k- th phase, and $\widetilde{w}_{ks}$ is the reaction rate of species s inside k -th phase.

6. 3. 4 Volume-averaged Conservation Equations for Laminar Multiphase Flows

For a practical multiphase flow system what we need to know is the macroscopic flow behavior, but not the detailed microscopic flow field, such as the flow field around each particle, therefore, we need to derive the volume-averaged equations. By using the concept of volume averaging, the relationship between the volume-averaged value of derivative of scalor φ_k or tensor (vector) φ_{kj} and the derivative of volume-averaged value can be given as:

$$\left\langle \frac{\partial \widetilde{\varphi}_k}{\partial t} \right\rangle = \frac{\partial \langle \widetilde{\varphi}_k \rangle}{\partial t} - \frac{1}{dV}\int_{dA_k} \widetilde{\varphi}_k \, \widetilde{v}_{sj} \cdot n_{kj} \cdot d\widetilde{A} \tag{6. 40}$$

$$\left\langle \frac{\partial \widetilde{\varphi}_k}{\partial x_i} \right\rangle = \frac{\partial \langle \widetilde{\varphi}_k \rangle}{\partial x_i} + \frac{1}{dV}\int_{dA_k} \widetilde{\varphi}_k \, n_{ki} \cdot d\widetilde{A} \tag{6. 41}$$

$$\left\langle \frac{\partial \widetilde{\varphi}_{kj}}{\partial x_j} \right\rangle = \frac{\partial \langle \widetilde{\varphi}_{kj} \rangle}{\partial x_j} + \frac{1}{dV}\int_{dA_k} \widetilde{\varphi}_{kj} \cdot n_{kj} \cdot d\widetilde{A} \tag{6. 42}$$

where $\widetilde{v}_{sj}$ is the displacement velocity at the interface, n_{ki} or n_{kj} is the unit vector in the direction externally normal to the interface.

On the basis of Eqs. (6. 40)—(6. 42), putting $\widetilde{\varphi}_k = \widetilde{\rho}_k$, and taking the volume-averaged value of each term in the continuity equation (6. 36), we have

$$\left\langle \frac{\partial \widetilde{\rho}_k}{\partial t} \right\rangle = \frac{\partial \langle \widetilde{\rho}_k \rangle}{\partial t} - \frac{1}{dV}\int_{dA_k} \widetilde{\rho}_k \, \widetilde{v}_{sj} \cdot n_{kj} \cdot d\widetilde{A}$$

$$\left\langle \frac{\partial}{\partial x_j} \rho_k \cdot v_{kj} \right\rangle = \frac{\partial}{\partial x_j} \langle \rho_k v_{kj} \rangle + \frac{1}{dV}\int_{dA_k} \rho_k \, \widetilde{v}_{kj} \cdot n_{kj} \cdot d\widetilde{A}$$

and hence

$$\frac{\partial \langle \widetilde{\rho}_k \rangle}{\partial t} + \frac{\partial}{\partial x_j} \langle \widetilde{\rho}_k \cdot \widetilde{v}_{kj} \rangle = -\frac{1}{dV}\int_{dA_k} \widetilde{\rho}_k \, (\widetilde{v}_{kj} - \widetilde{v}_{sj}) \cdot n_{kj} \cdot d\widetilde{A}$$

It should be pointed out that $\widetilde{v}_{sj}$ is the displacement velocity at the interface due to phase change, and $\widetilde{v}_{kj} - \widetilde{v}_{sj}$ is the relative velocity of phase k with reference to the interface, that is the velocity of Stefan flux. Therefore, we have

$$-\frac{1}{dV}\int_{dA_k} \widetilde{\rho}_k \, (\widetilde{v}_{kj} - \widetilde{v}_{sj}) \cdot n_{kj} \cdot d\widetilde{A} = \frac{1}{dV}\int_{dV_k} \widetilde{S}_k \, d\widetilde{V} = \langle \widetilde{S}_k \rangle = S_k$$

where S_k is the volume-averaged mass source in the unit volume of multiphase mixture. Dropping the symbols "$\sim$" and "$\langle \quad \rangle$", the volume-averaged continuity equation of phase k in the mixture can be written as:

$$\frac{\partial \rho_k}{\partial t} + \frac{\partial}{\partial x_j} (\rho_k v_{kj}) = S_k \qquad (k = 1, 2, 3, \cdots) \tag{6. 43}$$

For the fluid/gas phase, we have

$$\frac{\partial \rho}{\partial t} + \frac{\partial}{\partial x_j} (\rho \, v_j) = S \tag{6. 44}$$

and

$$S = -\sum_k S_k = -\sum_k n_k \dot{m}_k \qquad (k = 1, 2, 3, \cdots)$$

$$\dot{m}_k = \frac{dm_k}{dt}, \quad \rho_k = n_k m_k = n_k \frac{\pi d_k^3}{6} \bar{\rho}_k$$

where n_k and m_k are number density and single-particle mass of k -th particle phase respectively. Similarly, we can take the volume-averaged value of each term in the momentum equation (6. 37). Owing to

$$\left\langle \frac{\partial}{\partial t} \widetilde{\rho}_k \widetilde{v}_{ki} \right\rangle = \frac{\partial}{\partial t} \langle \widetilde{\rho}_k \widetilde{v}_{ki} \rangle - \frac{1}{dV} \int_{dA_k} \widetilde{\rho}_k \widetilde{v}_{kj} \widetilde{v}_{sj} \bullet n_{kj} \bullet d\widetilde{A}$$

$$\left\langle \frac{\widetilde{\partial}}{\partial x_j} (\widetilde{\rho}_k \widetilde{v}_{kj} \widetilde{v}_{ki}) \right\rangle = \frac{\partial}{\partial x_j} \langle \widetilde{\rho}_k \widetilde{v}_{kj} \widetilde{v}_{ki} \rangle + \frac{1}{dV} \int_{dA_k} \widetilde{\rho}_k \widetilde{v}_{ki} \widetilde{v}_{kj} \bullet n_{kj} \bullet d\widetilde{A}$$

$$-\left\langle \frac{\widetilde{\partial}}{\partial x_i} \widetilde{p}_k \right\rangle = -\frac{\partial \langle \widetilde{p}_k \rangle}{\partial x_i} - \frac{1}{dV} \int_{dA_k} \widetilde{p}_k \delta_{ij} \bullet n_{kj} \bullet d\widetilde{A}$$

$$\left\langle \frac{\widetilde{\partial}}{\partial x_j} \widetilde{\tau}_{k,ji} \right\rangle = \frac{\partial}{\partial x_j} \langle \widetilde{\tau}_{k,ji} \rangle + \frac{1}{dV} \int_{dA_k} \widetilde{\tau}_{k,ji} \bullet n_{kj} \bullet d\widetilde{A}$$

$$\langle \widetilde{\rho}_k g_i \rangle = \rho_k g_i, \qquad \langle \widetilde{p}_k \rangle = \Phi_k p$$

we have

$$\frac{\partial}{\partial t} \langle \widetilde{\rho}_k \widetilde{v}_{kj} \rangle + \frac{\partial}{\partial x_j} \langle \widetilde{\rho}_k \widetilde{v}_{kj} \widetilde{v}_{ki} \rangle = \frac{\partial}{\partial x_i} (\Phi_k p) + \frac{\partial}{\partial x_j} \langle \widetilde{\tau}_{k,ji} \rangle + \rho_k g_i$$
$$+ \frac{1}{dV} \int_{dA_k} (-\widetilde{p}_k \delta_{ij} + \widetilde{\tau}_{k,ji}) \bullet n_{kj} \bullet d\widetilde{A} - \frac{1}{dV} \int_{dA_k} \widetilde{\rho}_k \widetilde{v}_{ki} (\widetilde{v}_{kj} - \widetilde{v}_{sj}) \bullet n_{kj} \bullet d\widetilde{A}$$

The fourth term on the right-hand side of this equation is the sum of pressure force and viscous force at the interface, which can be expressed by

$$\frac{1}{dV} \int_{dA_k} (-p_k \delta_{ij} + \widetilde{\tau}_{k,ji}) n_{kj} \bullet d\widetilde{A} = \frac{1}{dV} \int_{dV_k} \widetilde{F}_i \, d\widetilde{V} = F_i$$

For a particulate suspension, F_i is the sum of volume-averaged particle drag force and lifting force per unit volume in the control volume dV, namely

$$F_i = F_{di} + F_{Mi}$$

$$F_{di} = \sum_k n_k \frac{\pi d_k^2}{4} c_{dk} \frac{\rho}{2} |v_k - v| (v_{ki} - v_i) = \sum_k \frac{\rho_k}{\tau_{rk}} (v_{ki} - v_i)$$
$$(k = 1, 2, 3, \cdots, k \neq f)$$

where

$$\tau_{rk} = \frac{d_k^2 \rho_p}{18\mu} \left(1 + \frac{\mathrm{Re}_k^{2/3}}{6}\right)^{-1}, \quad \mathrm{Re}_k = |v_k \quad v| d_k / \nu$$

and F_{Mi} is the Magnus force which will be discussed later in this chapter. The last term on the right-hand side of above-derived equation is the momentum source term due to phase change, which can be expressed as (for the fluid/gas phase):

$$-\frac{1}{dV} \int_{dA_k} \widetilde{\rho}_k \widetilde{v}_{ki} (\widetilde{v}_{kj} - \widetilde{v}_{sj}) \bullet n_{kj} \bullet d\widetilde{A} = \langle \widetilde{v}_i \widetilde{S} \rangle = v_i S$$

Therefore, after dropping the symbols "$\sim$" and "$\langle \quad \rangle$", the volume-averaged momentum equation of fluid/gas phase can be written as:

$$\frac{\partial}{\partial t} (\rho v_i) + \frac{\partial}{\partial x_j} (\rho v_j v_i) = -\frac{\partial p}{\partial x_i} + \frac{\partial}{\partial x_j} \left[\mu \left(\frac{\partial v_j}{\partial x_i} + \frac{\partial v_i}{\partial x_j} \right) - \frac{2}{3} \left(\mu \frac{\partial v_j}{\partial x_j} \right) \delta_{ij} \right]$$
$$+ \Delta\rho g_i + \sum_k \frac{\rho_k}{\tau_{rk}} (v_{ki} - v_i) + F_{Mi} + v_i S \qquad (6.45)$$

where $\Delta\rho g_i$ is the gravitational force with buoyancy effect. The last three terms on the right-hand side of Eq. (6. 45) are source terms due to interaction between phases. For particle phases in dilute suspensions, there is no particle-particle impaction, and hence no particle pressure and particle viscosity, or $\langle \widetilde{p}_k \rangle = 0$ and $\langle \widetilde{\tau}_{k,ji} \rangle = 0$, the final form of volume-averaged momentum equation of particle phase is

$$\frac{\partial}{\partial t}(\rho_k v_{ki}) + \frac{\partial}{\partial x_j}(\rho_k v_{kj} v_{ki}) = \rho_k g_i + \frac{\rho_k}{\tau_{rk}}(v_i - v_{ki}) + F_{k,Mi} + v_i S_k \tag{6.46}$$

The last three terms on the right-hand side of Eq. (6.46) are also particle-fluid/gas interaction terms, and we have

$$S = -\sum S_k, \quad F_{Mi} = -\sum_k F_{k,Mi}$$

In a similar way the volume-averaged energy equations of gas and particle phases and the species equations of gas phase can be derived. To summarize, the volume-averaged conservation equations of laminar multiphase flows in Eulerian coordinates can be obtained as follows:

fluid/gas phase continuity

$$\frac{\partial \rho}{\partial t} + \frac{\partial}{\partial x_j}(\rho v_j) = S \tag{6.44}$$

k-th particle phase continuity

$$\frac{\partial \rho_k}{\partial t} + \frac{\partial}{\partial x_j}(\rho_k v_{kj}) = S_k \tag{6.43}$$

mixture continuity

$$\frac{\partial}{\partial t}(\rho + \sum \rho_k) + \frac{\partial}{\partial x_j}(\rho v_j + \sum \rho_k v_{kj}) = 0 \tag{4.47}$$

fluid/gas phase momentum

$$\frac{\partial}{\partial t}(\rho v_i) + \frac{\partial}{\partial x_j}(\rho v_j v_i) = -\frac{\partial p}{\partial x_i} + \frac{\partial \tau_{ji}}{\partial x_j} + \Delta \rho g_i + \sum_k \frac{\rho_k}{\tau_{rk}}(v_{ki} - v_i) + v_i S + F_{Mi} \tag{6.45}$$

k-th particle phase momentum

$$\frac{\partial}{\partial t}(\rho_k v_{ki}) + \frac{\partial}{\partial x_j}(\rho_k v_{kj} v_{ki}) = \rho_k g_i + \frac{\rho_k}{\tau_{rk}}(v_i - v_{ki}) + v_i S_k + F_{k,Mi} \tag{6.46}$$

mixture momentum

$$\frac{\partial}{\partial t}(\rho v_i + \sum \rho_k v_{ki}) + \frac{\partial}{\partial x_j}(\rho v_j v_i + \sum_k \rho_k v_{kj} v_{ki}) = -\frac{\partial p}{\partial x_i} + \frac{\partial \tau_{ji}}{\partial x_j} + \rho g_i + \sum_k \rho_k g_i \tag{6.48}$$

fluid/gas phase energy

$$\frac{\partial}{\partial t}(\rho c_p T) + \frac{\partial}{\partial x_j}(\rho v_j c_p T) = \frac{\partial}{\partial x_j}\left(\lambda \frac{\partial T}{\partial x_j}\right) + w_s Q_s - q_r + \sum n_k Q_k + c_p T S \tag{6.49}$$

k-th particle phase energy

$$\frac{\partial}{\partial t}(\rho_k c_k T_k) + \frac{\partial}{\partial x_j}(\rho_k v_{kj} c_k T_k) = n_k (Q_h - Q_k - Q_{rk}) + c_p T S_k \tag{6.50}$$

mixture energy

$$\frac{\partial}{\partial t}(\rho c_p T + \sum_k \rho_k c_k T_k) + \frac{\partial}{\partial x_j}(\rho v_j c_p T + \sum_k \rho_k v_{kj} c_k T_k)$$

$$= \frac{\partial}{\partial x_j}(\lambda \frac{\partial T}{\partial x_j}) + w_s Q_s - q_r + \sum n_k Q_h - \sum n_k Q_{rk} \tag{6.51}$$

fluid/gas phase species

$$\frac{\partial}{\partial t}(\rho Y_s) + \frac{\partial}{\partial x_j}(\rho v_j Y_s) = \frac{\partial}{\partial x_j}\left(D\rho \frac{\partial Y_s}{\partial x_j} \right) - w_s + \alpha_s S \tag{6.52}$$

where Q_k is the convective heat transfer between particles and fluid/gas, B has been already defined before, $c_p TS$ and $c_p TS_k$ are energy source terms due to phase change for fluid phase and particle phase respectively, q_r is the radiative heat transfer for fluid phase and Q_{rk} is the radiative heat transfer for k-th particle phase, w_s is the reaction rate of s- species in fluid phase, $w_s Q_s$ is the heat release per unit volume of gas-phase reaction, Q_h is the heat released by heterogeneous reaction on the particle surface (including evaporation or condensation), α_s is the fraction of contribution of s-species in phase change.

The above-derived conservation equations are written in Eulerian coordinates. However, the particle phase conservation equations can also be written in Lagrangian coordinates.

Putting $\rho_k = n_k m_k$ into the particle-phase continuity equation (6. 43), we have

$$\frac{\partial}{\partial t}(n_k m_k) + \frac{\partial}{\partial x_j}(n_k m_k v_{kj}) = \left[\frac{\partial n_k}{\partial t} + \frac{\partial}{\partial x_j}(n_k v_{kj}) \right] m_k$$

$$+ n_k \left(\frac{\partial m_k}{\partial t} + v_{kj} \frac{\partial m_k}{\partial x_j} \right) = m_k \left[\frac{\partial n_k}{\partial t} + \frac{\partial}{\partial x_j}(n_k v_{kj}) \right] + n_k \dot{m}_k = S_k$$

or

$$\frac{\partial n_k}{\partial t} + \frac{\partial}{\partial x_j}(n_k v_{kj}) = 0 \tag{6.53}$$

and

$$\frac{\partial}{\partial x_j}(n_k v_{kj}) = 0 \tag{6.54}$$

for steady flows, which means that the divergence of particle mass flux is zero. For a particle stream tube with two cross sections A_1 and A_2, using the Gauss-Ostragradsky Theorem, we have

$$\int_{A_1} n_k v_{kn} dA = \int_{A_2} n_k v_{kn} dA$$

or

$$N_k = \int_A n_k v_{kn} dA = \text{const} \tag{6.55}$$

where v_{kn} is the particle velocity component perpendicular to the stream-tube cross section or the modulus of particle velocity $|v_k|$, N_k is the total particle number flux. Eq. (6. 55) implies that the total particle number flux keeps constant along the stream tube or computationally a particle trajectory.

By using the particle continuity equation (6. 43), the particle momentum equation (6. 46) can be transformed into

$$\rho_k \frac{\partial v_{ki}}{\partial t} + \rho_k v_{kj} \frac{\partial v_{ki}}{\partial x_j} = \rho_k \frac{dv_{ki}}{dt_k}$$

$$= \rho_k g_i + \frac{\rho_k}{\tau_{rk}} (v_i - v_{ki}) + F_{k,Mi} + S_k (v_i - v_{ki})$$

Dividing each term by $n_k m_k$, the particle momentum equation in Lagrangian coordinate can be obtained as:

$$\frac{dv_{ki}}{dt_k} = (v_i - v_{ki}) / \tau_{rk} + g_i + \frac{F_{k,Mi}}{n_k m_k} + (v_i - v_{ki}) \dot{m}_k / m_k \tag{6.56}$$

In a similar way we can derive the particle energy equation in Lagrangian coordinate. Therefore, the particle-phase conservation equations in Lagrangian coordinate are:

$$\begin{cases} \int_A n_k v_{kn} dA = N_k = \text{const} & (6.55) \\ \dfrac{dv_{ki}}{dt_k} = (v_i - v_{ki}) / \tau_{rk} + g_i + \dfrac{F_{k,Mi}}{n_k m_k} + (v_i - v_{ki}) \dot{m}_k / m_k & (6.56) \\ \dfrac{dT_k}{dt_k} = [Q_h - Q_k - Q_{rk} + \dot{m}_k (c_p T - c_k T_k)] / (m_k c_k) & (6.57) \end{cases}$$

6.4 SINGLE PARTICLE DYNAMICS

The simplest way to study gas-particle flows is the single particle dynamics, in which the effect of particles on fluid flow is neglected and separately moved single particles without fluctuation in known flow field are considered. This is an earliest model. However, for studying turbulent gas-particle flows, the results obtained by the analysis of single particle dynamics can be considered as basic phenomena in practical gas-particle flows.

6.4.1 Equation of Motion for a Single Particle

On the basis of Eq. (6.56), neglecting the lifting force and the momentum exchange due to phase change, the equation of motion for a single particle can be obtained as:

$$\frac{dv_{ki}}{dt_k} = (v_i - v_{ki}) / \tau_{rk} + g_i \tag{6.58}$$

6.4.2 Motion of a Single Particle in a Uniform Flow Field

Assume that a particle with initial velocity v_{p0} moves in a uniform flow field (Fig. 6.2), and the particle drag obeys the Stokes Law. Neglecting the gravitational force, the particle momentum equation in x direction is

$$\frac{\pi d_p^3}{6} \bar{\rho}_p \frac{du_p}{dt} = c_d \frac{\pi d_p^2}{4} \frac{\rho}{2} |\boldsymbol{u}_\infty - \boldsymbol{v}_p| (u_\infty - u_p) \tag{6.59}$$

Substituting

$$c_d = 24/\text{Re}_p = 24\nu / (|\boldsymbol{u}_\infty - \boldsymbol{v}_p| d_p)$$

into Eq. (6.59), we have

$$\frac{du_p}{dt} = (u_\infty - u_p) / \tau_r \tag{6. 60}$$

where

$$\tau_r = d_p^2 \rho_p / (18\mu)$$

Integrating Eq. (6. 60) with the initial condition of $u_p = u_{p0}$ at $t = 0$ gives

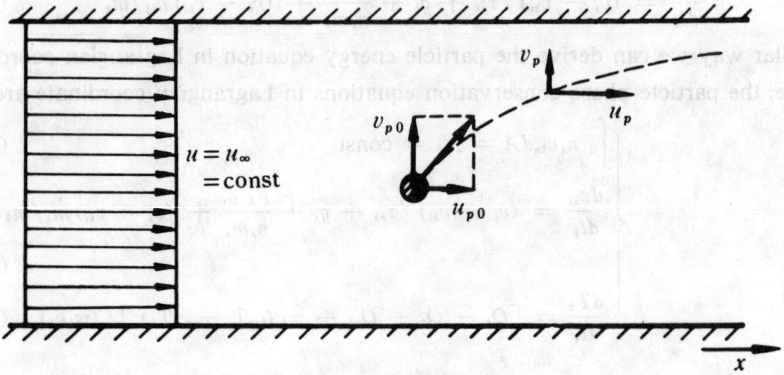

Fig. 6. 2

$$u_p = u_\infty - (u_\infty - u_{p0}) \exp(-t/\tau_r) \tag{6. 61}$$

Similarly, the transverse particle velocity can be obtained as

$$v_p = v_{p0} \exp(-t/\tau_r) \tag{6. 62}$$

Integrating Eqs. (6. 61) and (6. 62) with respect to t gives the particle trajectory equations

$$\begin{cases} x_p = u_\infty t - (u_\infty - u_{p0}) \tau_r (1 - e^{-t/\tau_r}) & (6. 63) \\ y_p = v_{p0} \tau_r (1 - e^{-t/\tau_r}) & (6. 64) \end{cases}$$

Similar equations can be found for non-Stokes particle drag. Eqs. (6. 61) — (6. 64) point out that: the particle longitudinal velocity approaches to the fluid velocity, the particle transverse velocity approaches to zero, and the particle trajectory approaches to its asymptote $y = v_{p0}\tau_r$. When $t = \tau_r$, $v_p = v_{p0}/e$. Hence the physical meaning of the relaxation time τ_r is the time needed for the reduction of particle-fluid velocity slip to $1/e$ times of its initial value. The smaller τ_r, the easier particle tracks the fluid.

6. 4. 3 Gravitational Deposition of a Particle

Assume that only gravitational force and drag force are exerted on an initially motionless particle with a Stokesian drag law, the equation of motion is

$$\frac{dv_p}{dt} + \frac{v_p}{\tau_r} - g = 0 \tag{6. 65}$$

with the initial condition

$$v_{p0} = 0 \quad \text{at} \quad t = 0$$

The solution is

$$v_p = \tau_r g (1 - e^{-t/\tau_r}) \tag{6. 66}$$

When $t \to \infty$, v_p approaches $\tau_r g = \text{const} = v_{pr}$, the particle acceleration will be zero, and there will be an equilibrium between the drag force and the gravitational force. v_{pr} is called terminal velocity.

6. 4. 4 Particle Behavior in Nonuniform Flow Field

(1) Magnus Force

As a nonspherical particle moves in the fluid, in particular after its impaction on the wall, it may rotate, causing a lifting force perpendicular to the direction of relative velocity, called Magnus force, whose magnitude is

$$F_M = \frac{\pi d_p^3}{6} \bar{\rho} |v - v_p| |\omega_p - \Omega| \tag{6.67}$$

where ω_p is the angular velocity of particle rotation, Ω is the half of fluid vorticity. It has been estimated that the ratio of F_M to the drag force is 0. 03 for $d_p = 1\mu\text{m}$ and 3 for $d_p = 10\ \mu\text{m}$. However, recent laser holography studies have shown that in most regions of the flow field the particle does not rotate. Therefore, except in the region adjacent to the wall, the Magnus force is not important.

(2) Saffman Force

If the particle is sufficiently large and there is large velocity gradient (for example, near the wall) in the flow field, there will be a particle lifting force called Saffman force, whose magnitude is

$$F_s = 1.6 (\mu \bar{\rho})^{\frac{1}{2}} d_p^2 |v - v_p| \left| \frac{\partial v}{\partial y} \right|^{\frac{1}{2}} \tag{6.68}$$

The ratio of Saffman force to Magnus force

$$0.17 d_p \left| \frac{\partial v}{\partial y} \right|^{\frac{1}{2}} \nu^{-\frac{1}{2}}$$

is much greater than unity, hence the Saffman force may play an important role, in particular in the regions of large velocity gradient.

(3) Particle Thermophoresis, Electrophoresis and Photophoresis

Tiny particles ($d_p < 1\mu\text{m}$) may move under the effect of the so-called "thermophoresis" or "electrophoresis" or "photophoresis", caused by large temperature gradient, electric field gradient and nonuniform radiation. The forces of thermophoresis and electrophoresis can be estimated by

$$F_{Tj} = -4.5 \nu^2 (\rho / T) d_p [\lambda / (2\lambda + \lambda_p)] \frac{\partial T}{\partial x_j} \tag{6.69}$$

$$F_E = (\pi/6) \rho_p d_p^3 q E \tag{6.70}$$

where λ and λ_p are gas and particle thermoconductivities, E and q are electric field strength and particle electric charge respectively. All these forces are significant merely for submicron or ultrafine particles.

6.4.5 Generalized Equation of Particle Motion

Equation (6.58) is only the simplest equation of particle motion. S. M. Tchen, using the method of intuitive superposition of various possible forces, proposed a generalized equation of particle motion with Stokesian drag as:

$$m_p \frac{dv_{pi}}{dt_p} = F_{di} + F_{vmi} + F_{pi} + F_{Bi} = 3\pi d_p \mu (v_i - v_{pi})$$

$$+ 0.5 \frac{\pi d_p^3}{6} \rho \frac{d}{dt_p}(v_i - v_{pi}) + \frac{\pi d_p^3}{6} \rho \frac{dv_i}{dt}$$

$$+ 1.5 (\pi \rho \mu)^{\frac{1}{2}} d_p^2 \int_{-\infty}^{t} \frac{d}{d\tau}(v_i - v_{pi}) \cdot (\tau - t)^{-\frac{1}{2}} \cdot d\tau \qquad (6.71)$$

Where F_{di}, F_{vmi}, F_{pi}, F_{Bi} are drag force, virtual mass force, pressure gradient force and Basset force (due to unsteady flow) respectively. After the discussion of the forces in nonuniform flow field, the more generalized equation of particle motion can be expressed as:

$$m_p \frac{dv_{pi}}{dt_p} = F_{di} + F_{vmi} + F_{pi} + F_{Bi} + F_{Mi} + F_{Si} + F_{Ti} + F_{Ei} + \cdots \qquad (6.72)$$

It should be noticed that in most cases forces other than the drag force are of minor importance, so the approximation made in Eq. (6.58) is still valid.

6.5 ANALYTICAL SOLUTIONS

6.5.1 Small-slip Continuum Model

From the 1950's to 1960's F. Marble and S. L. Soo proposed the continuum model or pseudo-fluid model of particle phase, in which either the effect of particle phase on fluid flow is considered as a small perturbation, or this effect is fully neglected. In this model it is assumed that the particle movement is caused purely by fluid flow, and the velocity slip between gas and particle phases is a small quantity relative to the mean fluid flow. The slip is considered as a result of particle diffusion. Actually, this is an extension of the concept of multicomponent single-phase fluid to multiphase fluid mixture. For a multicomponent single-phase fluid we have (see Chapter 1)

$$\begin{cases} v_{si} = v_i + V_{si} \\ \rho_s v_{si} = \rho_s v_i + \rho_s V_{si} \\ \sum \rho_s v_{si} = \rho v_i \end{cases}$$

where v_i is the mean flow velocity of the mixture, V_{si} is diffusion drift velocity of s-species relative to the mean flow, which is determined by the Fick Law

$$-\rho_s V_{si} = -\rho_s (v_{si} - v_i) = D_s \rho \frac{\partial Y_s}{\partial x_i}$$

where ρ_s is the absolute concentration of s-species, Y_s is the mass fraction of s-species. Similarly, in the small-slip model it is assumed that

$$\begin{cases} v_{ki} = v_{mi} + V_{ki} \\ \rho_k v_{ki} = \rho_k (v_{mi} + V_{ki}) \\ \rho_m v_{mi} = \rho v_i + \sum \rho_k v_{ki} \end{cases} \tag{6.73}$$

where V_{ki} is the particle velocity slip relative to the mixture velocity, and it is equal to the diffusion drift velocity, that is

$$-\rho_k (v_{ki} - v_{mi}) = -\rho_k V_{ki} = D_k \rho_m \frac{\partial Y_k}{\partial x_i} \tag{6.74}$$

where Y_k is the mass fraction of k-th particle phase in the multiphase fluid mixture, which is defined by

$$Y_k = \rho_k / \rho_m$$

The continuity equation of s-species in a multicomponent fluid is

$$\frac{\partial \rho_s}{\partial t} + \frac{\partial}{\partial x_j}(\rho_s v_{sj}) = 0$$

It is identical to the diffusion equation

$$\frac{\partial \rho_s}{\partial t} + \frac{\partial}{\partial x_j}(\rho_s v_j) = \frac{\partial}{\partial x_j}\left(D\rho \frac{\partial Y_s}{\partial x_j} \right)$$

For the small-slip model, the continuity equation of k-th phase in a multiphase fluid mixture is

$$\frac{\partial \rho_k}{\partial t} + \frac{\partial}{\partial x_j}(\rho_k v_{kj}) = 0 \tag{6.75}$$

It is equivalent to the particle diffusion equation

$$\frac{\partial \rho_k}{\partial t} + \frac{\partial}{\partial x_j}(\rho_k v_{mj}) = \frac{\partial}{\partial x_j}\left(D_k \rho_m \frac{\partial Y_k}{\partial x_j} \right) \tag{6.76}$$

In the small-slip model, in some cases, the particle diffusion is even considered as a drift relative to the fluid flow, that is

$$-\rho_k (v_{ki} - v_i) = D_k \frac{\partial \rho_k}{\partial x_j} \tag{6.77}$$

In this case, the particle diffusion equation is

$$\frac{\partial \rho_k}{\partial t} + \frac{\partial}{\partial x_j}(\rho_k v_j) = \frac{\partial}{\partial x_j}\left(D_k \frac{\partial \rho_k}{\partial x_j} \right) \tag{6.78}$$

It should be noticed that the small-slip model is different from the full multifluid model, in which the particle diffusion originates from the correlation of particle mass flux fluctuation not related to the velocity slip. Actually the simplification made in the small-slip model is a simplest way to solve the closure problem of particle turbulence modeling. Later we will discuss this problem in more detail in Chapter 8.

6.5.2 Laminar Boundary Layer Gas-particle Flows over a Flat Plate

Assuming that the particle drag obeys Stokes Law, and there is small velocity slip between particle and gas phases, the basic equations of laminar boundary layer gas-particle flow over a flat plate (Fig. 6.3) proposed by F. Marble can be written as:

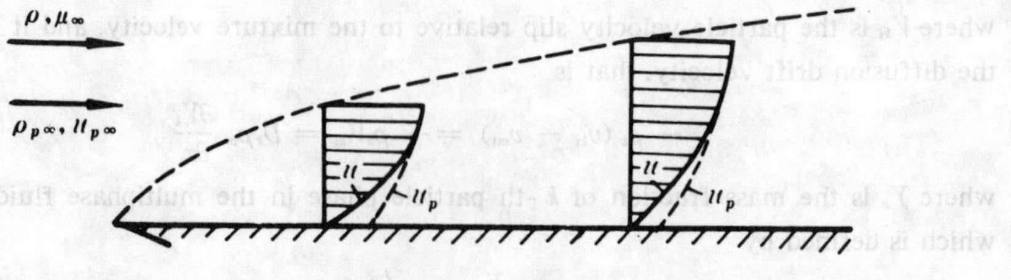

Fig. 6. 3

Gas continuity

$$\frac{\partial u}{\partial x} + \frac{\partial v}{\partial y} = 0 \tag{6.79}$$

Gas momentum

$$u\frac{\partial u}{\partial x} + v\frac{\partial u}{\partial y} = \nu\frac{\partial^2 u}{\partial y^2} + \frac{\rho_p}{\rho\tau_r}(u_p - u) \tag{6.80}$$

Particle continuity

$$\frac{\partial}{\partial x}(\rho_p u_p) + \frac{\partial}{\partial y}(\rho_p v_p) = 0 \tag{6.81}$$

Particle momemtum

$$u_p\frac{\partial u_p}{\partial x} + v_p\frac{\partial u_p}{\partial y} = (u - u_p)/\tau_r \tag{6.82}$$

$$u_p\frac{\partial v_p}{\partial x} + v_p\frac{\partial v_p}{\partial y} = (v - v_p)/\tau_r \tag{6.83}$$

Marble considered the far downstream region $x \gg \tau_r u_\infty$, where the velocity slip is very small, namely, $(u_p - u)/u_\infty \ll 1$, $v_p - v \approx 0$. Taking $(u_p - u)/u_\infty$ and $\rho_p/\rho_{p\infty}$ as dependent variables and $\kappa = \rho_{p\infty}/\rho$ as a parameter, by using the gas continuity and momentum equations, the particle continuity and momentum equations can be transformed into

$$\frac{\rho_p}{\rho_{p\infty}}\frac{\partial}{\partial x}\left(\frac{u_p - u}{u_\infty}\right) + \frac{u_p - u}{u_\infty}\frac{\partial}{\partial x}\left(\frac{\rho_p}{\rho_{p\infty}}\right) + \frac{u}{u_\infty}\frac{\partial}{\partial x}\left(\frac{\rho_p}{\rho_{p\infty}}\right) + \frac{u}{u_\infty}\frac{\partial}{\partial y}\left(\frac{\rho_p}{\rho_{p\infty}}\right) = 0 \tag{6.84}$$

$$u\frac{\partial}{\partial x}\left(\frac{u_p - u}{u_\infty}\right) + v\frac{\partial}{\partial y}\left(\frac{u_p - u}{u_\infty}\right) + (u_p - u)\frac{\partial}{\partial x}\left(\frac{u}{u_\infty}\right) + (u_p - u)\frac{\partial}{\partial x}\left(\frac{u_p - u}{u_\infty}\right)$$
$$+ \nu\frac{\partial^2}{\partial y^2}\left(\frac{u}{u_\infty}\right) = \left[1 + \kappa\frac{\rho_p}{\rho_{p\infty}}\right]\frac{u_\infty}{\tau_r u_\infty}\left(\frac{u_p - u}{u_\infty}\right) \tag{6.85}$$

Assume that the difference between single-phase flow velocity and gas-phase flow velocity in two-phase flows, the difference between gas-phase velocity and particle-phase velocity and the difference between the particle concentration at any location ρ_p and that of oncoming flow $\rho_{p\infty}$ are all small perturbation. Then the basic equations can be solved by using the asymptotic method. By using the small parameter $l_p/x = u_\infty\tau_r/x$, the gas velocity, particle velocity and particle concentration can be expressed by the following series

$$\psi(x,\eta) = \sqrt{\nu^* u_\infty x}\left[f^{(0)}(\eta) + \frac{l_p}{x}f^{(1)}(\eta) + (\frac{l_p}{x})^2 f^{(2)}(\eta) + \cdots\right]$$

$$(u_p - u)/u_\infty = \frac{l_p}{x}g^{(1)}(\eta) + (\frac{l_p}{x})^2 g^{(2)}(\eta) + \cdots$$

$$(\rho_p - \rho_{p\infty})/\rho_{p\infty} = \frac{l_p}{x}h^{(1)}(\eta) + (\frac{l_p}{x})^2 h^{(2)}(\eta) + \cdots$$

where

$$\nu^* = \nu/(1+\kappa),\ \eta = y\left(\frac{\nu^* x}{u_\infty}\right)^{-\frac{1}{2}} = y\left[\frac{u_\infty(1+\kappa)}{\nu x}\right]^{\frac{1}{2}}$$

The zero-order gas momentum equation or the equation of $f^{(0)}(\eta)$ is

$$\frac{d^3 f^{(0)}(\eta)}{d\eta^3} + \frac{1}{2}f^{(0)}\frac{d^2 f^{(0)}}{d\eta^2} = 0 \tag{6.86}$$

This is similar to the Blasius equation for single-phase boundary layers except the fluid density is modified by $\rho(1+\kappa) = \rho + \rho_{p\infty}$. Substituting the expressions of gas-velocity, particle-velocity and particle-concentration profiles into the transformed conservation equations, the first-order gas-velocity perturbation, particle velocity slip and particle concentration perturbation can be found by solving the following differential equations:

$$\frac{d^3 f^{(1)}}{d\eta^3} + \frac{1}{2}f^{(0)}\frac{d^2 f^{(1)}}{d\eta^2} + \frac{df^{(0)}}{d\eta}\frac{df^{(1)}}{d\eta} - \frac{1}{2}\frac{d^2 f^{(0)}}{d\eta^2}f^{(1)}$$
$$= -\frac{\kappa}{1+\kappa}\left(\frac{1}{2}f^{(0)}\frac{d^2 f^{(0)}}{d\eta^2}\right)\left[h^{(1)} + \frac{3}{2}\frac{df^{(0)}}{d\eta} + \frac{\eta}{2}\frac{d^2 f^{(0)}}{d\eta^2} - \frac{1}{4}(f^{(0)})^2\right] \tag{6.87}$$

$$g^{(1)}(\eta) = \frac{1}{2}f^{(0)}\frac{d^2 f^{(0)}}{d\eta^2} \tag{6.88}$$

$$\frac{1}{2}f^{(0)}\frac{dh^{(1)}}{d\eta} + h^{(1)}\frac{df^{(0)}}{d\eta}$$
$$= -g^{(1)} - \frac{\eta}{2}\frac{dg^{(1)}}{d\eta} - \left[f^{(0)}(1 - \frac{\eta}{4}f^{(0)}) + \frac{\eta}{2}\frac{df^{(0)}}{d\eta}\right]\frac{1}{2}\frac{d^2 f^{(0)}}{d\eta^2} \tag{6.89}$$

The solution procedure is: (1) solving Eq. (6. 88) to find $g^{(1)}(\eta)$ based on the known function $f^{(0)}(\eta)$; (2) solving Eq. (6. 89) to find $h^{(1)}(\eta)$ based on $f^{(0)}(\eta)$ and $g^{(1)}(\eta)$; and (3) solving Eq. (6. 87) to find $f^{(1)}(\eta)$ based on $f^{(0)}(\eta)$ and $h^{(1)}(\eta)$. The second-order approximation functions $g^{(2)}(\eta)$, $h^{(2)}(\eta)$, and $f^{(2)}(\eta)$ can be found in a similar way. The gas-phase velocity perturbation (relative to single-phase velocity), particle velocity slip (relative to gas velocity) and particle concentration perturbation (relative to the concentration of oncoming flow) are shown in Fig. 6. 4, based on the solution of first-order approximation. The results indicate that under the effect of particle inertia the gas-phase velocity in two-phase flows is larger than the single-phase velocity, and the particle velocity is larger than the gas velocity, in particular at the locations between the wall and the outer edge of the boundary layer. The particle concentration at the outer edge of the boundary layer is larger than that of the oncoming flow, but is lower at some distance from the wall. However, the analytical solution made by S. L Soo and the LDV measurements made by S. L. Lee have shown that in most regions, especially in the far downstream region the particle velocity does not exceed, but lags behind the gas velocity, and no particle concentration peak was observed in experiments. S. L. Lee introduced the Saffman force into boundary-layer equation, which makes the predicted velocity profiles in better agreement with experiments, but the predicted particle concentration profiles are still not in agreement with experiments.

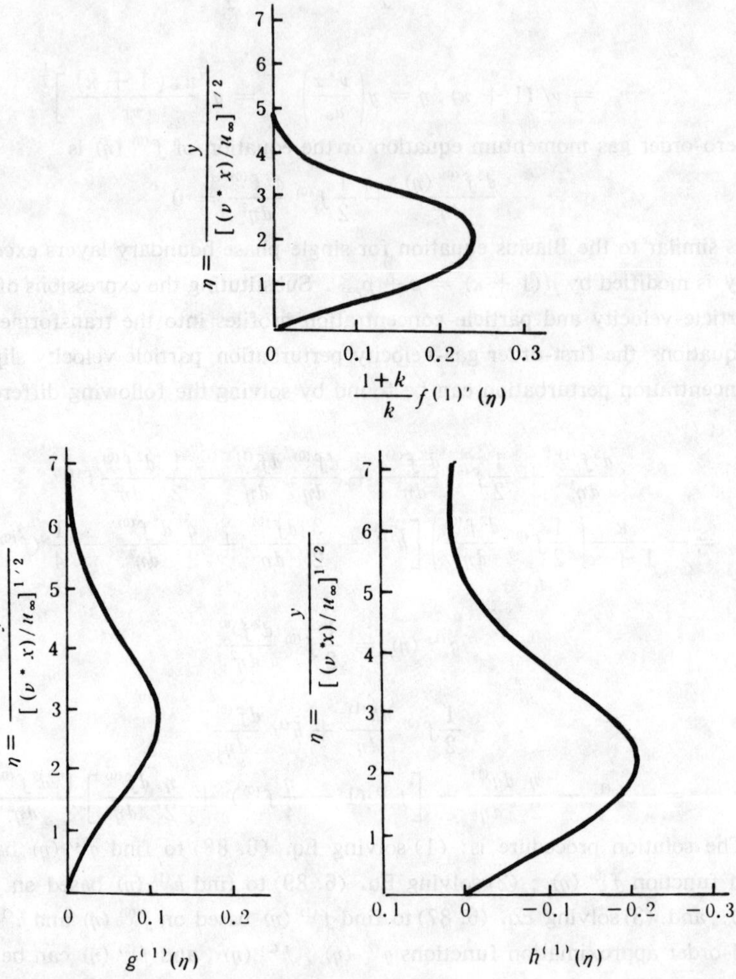

Fig. 6. 4

6. 5. 3 Fully-developed Gas-particle Pipe Flows

S. L. Soo first pointed out that fully-developed pipe flows are simplest to be analyzed. The particle diffusion equation based on the small-slip model given by Soo is

$$D_p \frac{\partial \rho_p}{\partial x_j} = - \rho_p (v_{pj} - v_j) \tag{6. 90}$$

Neglecting the convective and viscous terms, the particle momentum equation is simplified to an expression of equilibrium between the drag force and the gravitational force as

$$\rho_p (v_j - v_{pj}) / \tau_r = - \rho_p g_j (1 - \bar{\rho}/\bar{\rho}_p)$$

or

$$(v_{pj} - v_j) / \tau_r = g_j (1 - \bar{\rho}/\bar{\rho}_p) \qquad (6.91)$$

The fluid flow field is considered as a well-known 1/7 law of single-phase fully-developed flows (Fig. 6. 5), that is, for $y_s < y < R$, we have

$$u/u_0 = (y/R)^{1/7}$$

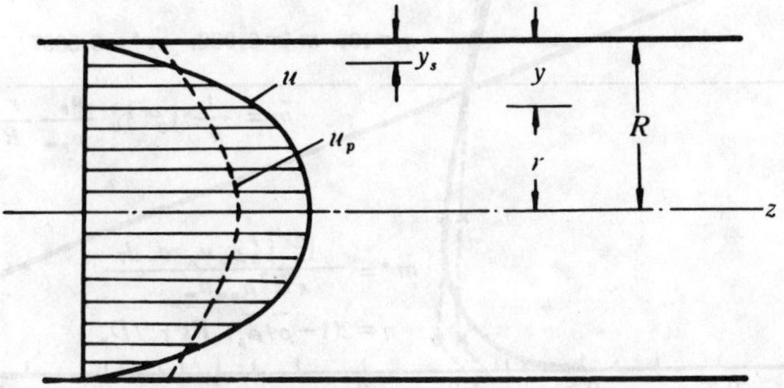

Fig. 6. 5

where u_0 is the maximum velocity at the axis and y_s is the thickness of the viscous sublayer,

$$y_s = 60\,(u_0 R/\nu)^{-7/8} \cdot R$$

As $\bar{\rho}$, $\bar{\rho}_p$, g, τ_r and v_j are given, we can find the particle velocity v_{pj} from Eq. (6. 91). On the other hand, combining Eq. (6. 90) and Eq. (6. 91) gives the particle concentration equation as:

$$- D_p \left(\frac{\partial \rho_p}{\partial x_j} \right) = \rho_p \tau_r g_j (1 - \bar{\rho}/\bar{\rho}_p) \qquad (6.92)$$

For a horizontal pipe, we have

$$- D_p \left(\frac{d \rho_p}{dr} \right) = \rho_p (1 - \bar{\rho}/\bar{\rho}_p)\, g \tau_r \qquad (6.93)$$

Assuming a constant particle diffusivity D_p, integrating Eq. (6. 93) from $-R$ to $+R$ gives the particle concentration distribution as:

$$\ln \frac{\rho_p (R, \pi)}{\rho_p (R, 0)} = \frac{2Rg\tau_r}{D_p} \left(1 - \frac{\bar{\rho}}{\bar{\rho}_p} \right) \qquad (6.94)$$

Where $\rho_p (R, \pi)$ denotes the concentration at the bottom, and $\rho_p (R, 0)$ denotes the concentration at the top. By processing the experimental data of hydraulic conveying of glass beads and magnesia powder in pipes of 51 — 127 mm diameter and using Eq. (6. 94), the particle diffusivity is found as $D_p = 0.5 - 1.2\ D_f$, where D_p and D_f are particle and fluid diffusivity respectively (Fig. 6. 6).

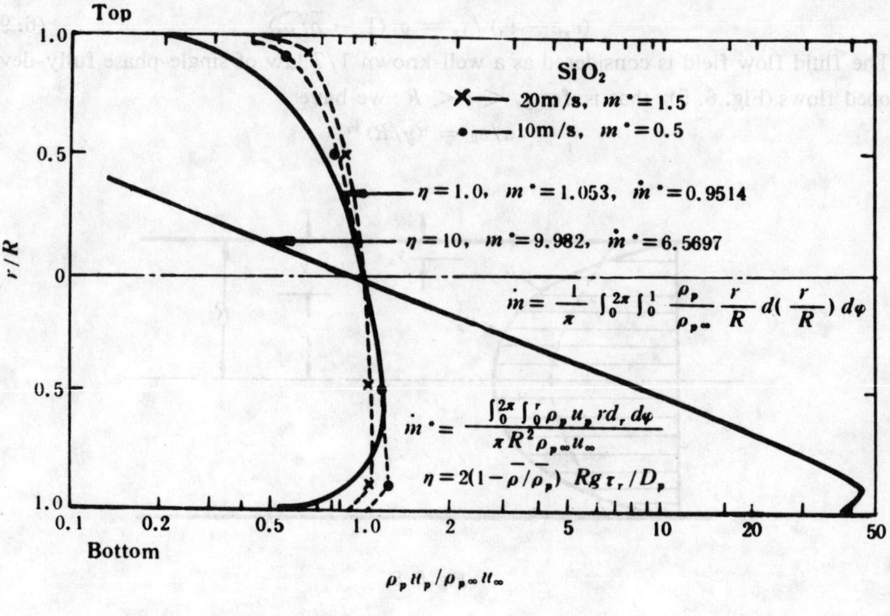
Top
Bottom
1.0
0.5
0
0.5
1.0
r/R
0.1
0.2
0.5
1.0
2
5
10
20
50
SiO_2
× 20m/s, $m^*=1.5$
• 10m/s, $m^*=0.5$
$\eta=1.0$, $m^*=1.053$, $\dot{m}^*=0.9514$
$\eta=10$, $m^*=9.982$, $\dot{m}^*=6.5697$
$m = \frac{1}{\pi}\int_0^{2\pi}\int_0^1 \frac{\rho_p}{\rho_{p\infty}} \frac{r}{R} d(\frac{r}{R}) d\varphi$
$\dot{m}^* = \frac{\int_0^{2\pi}\int_0^r \rho_p u_p r d_r d\varphi}{\pi R^2 \rho_{p\infty} u_\infty}$
$\eta = 2(1-\bar{\rho}/\rho_p) Rg\tau_r/D_p$
$\rho_p u_p/\rho_{p\infty} u_\infty$

Fig. 6. 6

PART Ⅱ

MODELING OF TURBULENT GAS-PARTICLEFLOWS AND COMBUSTION

CHAPTER 7 MODELING OF SINGLE-PHASE TURBULENT FLOWS

7.1 BACKGROUND

Turbulence is one of the most difficult, unsolved theoretical problems in fluid mechanics. Neverthless, we have to use different approaches in dealing with engineering problems. The simplest way is to take the so-called turbulent viscosity or turbulent diffusivity as a constant property, which was used in some earlier hydraulic and combustor flow predictions. For many years the integral method based on empirical velocity and temperature profiles was adopted in predicting jet-type flows, where we need not the knowledge of turbulence. Obviously, more rigorous treatment is needed for predicting complex engineering turbulent flows, such as three-dimensional recirculating or swirling flows, most of which are not simple jet flows, and do not have a constant turbulent viscosity.

The fundamental approach to turbulence studies is the direct numerical simulation (DNS) by solving the transient three-dimensional Navier-Stokes equation in the grid size of Kolmogorov scales without using any turbulence models. However, the DNS needs extremely large computer capacity and CPU time, that in the present state of its development it cannot solve any practical engineering problem. An alternative approach is the large-eddy simulation (LES), which is to solve the N-S equation in the grid size of large eddies, but still needs to model the small-scale turbulence (subgrid modeling). The LES still needs sufficiently large capacity and computational time, therefore it is also far from application in engineering predictions. In the opinion of many fluid-dynamists, the only practical method for engineering predictions in this century is the method of turbulence modeling based on solving the Reynolds time-averaged equations and transport equations of correlation terms.

The basic idea of turbulence modeling is to simulate the unknown higher-order correlation terms by using lower-order correlations or mean-flow properties, in order to close the time-averaged equations or the transport equations of correlation terms. This treatment is acceptable, since in engineering application what we need to know is only the time-averaged velocity, temperature and turbulence characteristics, but not the detail of turbulence production and development. Later it will be shown that these methods of turbulence modeling even are simplest in the theory of turbulence, but are still sufficiently complex in engineering application. In short, the method of turbulence modeling based on solving Reynolds time-averaged equations is the most effective, economical and reasonable method in dealing with engineering problems.

7. 2 REYNOLDS TIME-AVERAGED EQUATIONS

On the basis of the instantaneous conservation equations of single-phase multicomponent flows with heat and mass transfer and chemical reactions described in Chapter 1 and by using the Reynolds expansion, after time averaging, the Reynolds time-averaged equations for turbulent flows can be obtained in the following generalized form:

$$\frac{\partial}{\partial t}(\rho\varphi) + \frac{\partial}{\partial x_j}(\rho v_j \varphi) = \frac{\partial}{\partial x_j}\left(\frac{\mu}{\sigma_\varphi}\frac{\partial \varphi}{\partial x_j}\right) - \frac{\partial}{\partial x_j}(\rho\,\overline{v_j' \varphi'}) + S_\varphi \tag{7.1}$$

where φ is the generalized dependent variable, S_φ is the source term, $\varphi = 1, v_i, Y_s, h$ and $S_\varphi = 0,\ -\frac{\partial p}{\partial x_j} + \rho g_i + \frac{\partial}{\partial x_j}\left(\mu\frac{\partial v_j}{\partial x_i}\right),\ -\dot{w}_s,\ -q_r$ for continuity, momentum, species and energy equations respectively. Equation (7. 1) is not a closed system due to the unknown correlation term $\overline{v_j' \varphi'}$, expressing the transport of momentum, mass and energy caused by turbulent fluctuation, or the Reynolds stress, mass flux and heat flux respectively. In the case of isotropic turbulent flows, the concept of an isotropic scalar turbulent viscosity (or eddy viscosity) can be introduced, and the correlation term in Eq. (7. 1) can be expressed in the so-called Boussinesq form as:

$$\begin{cases} -\rho\,\overline{v_j' v_i'} = \mu_T\left(\frac{\partial v_j}{\partial x_i} + \frac{\partial v_i}{\partial x_j}\right) - \frac{2}{3}\rho k\delta_{ij} \\ -\rho\,\overline{v_j' \varphi'} = \frac{\mu_T}{\sigma_\varphi}\left(\frac{\partial \varphi}{\partial x_j}\right) \qquad (\varphi = Y_s, h) \end{cases} \tag{7.2}$$

and in this case Eq. (7. 1) can be rewritten as

$$\frac{\partial}{\partial t}(\rho\varphi) + \frac{\partial}{\partial x_j}(\rho v_j \varphi) = \frac{\partial}{\partial x_j}\left(\Gamma_\varphi \frac{\partial \varphi}{\partial x_j}\right) + S_\varphi \tag{7.3}$$

where $\Gamma_\varphi = \mu_e/\sigma_\varphi$ denotes the generalized transport coefficient, σ_φ stands for the turbulent Prandtl number, and $\mu_e = \mu + \mu_T = (\nu + \nu_T)\rho$ is called effective viscosity, μ_T or ν_T is called turbulent viscosity. The task of turbulence modeling is either to directly simulate the correlation term $v_j' \varphi'$ in Eq. (7. 1), or to simulate the turbulent viscosity μ_T in Eq. (7. 2), Eq. (7. 3).

7. 3 REYNOLDS STRESS AND FLUX TRANSPORT EQUATIONS

In order to close Eq. (7. 1), one approach is to further derive the transport equations of $\overline{v_j' \varphi'}$, that is the Reynolds stress and flux transport equations. To do this, first write down the N-S equations for velocity components v_i and v_j

$$\frac{\partial}{\partial t}(\rho v_i) + \frac{\partial}{\partial x_k}(\rho v_k v_i) = -\frac{\partial p}{\partial x_i} + \rho g_i \beta \Delta T + \frac{\partial \tau_{ik}}{\partial x_k} \tag{7.4}$$

$$\frac{\partial}{\partial t}(\rho v_j) + \frac{\partial}{\partial x_k}(\rho v_k v_j) = -\frac{\partial p}{\partial x_j} + \rho g_j \beta \Delta T + \frac{\partial \tau_{jk}}{\partial x_k} \tag{7.5}$$

where $\rho g_i \beta \Delta T$ is the buoyancy term and $\beta = -\frac{1}{\rho}\left(\frac{\partial \rho}{\partial T}\right)$, is the volumetric expansion coefficient, and τ_{ik}, τ_{jk} are viscous stress tensors, for example,

$$\tau_{ik} = \mu\left(\frac{\partial v_i}{\partial x_k} + \frac{\partial v_k}{\partial x_i}\right) - \frac{2}{3}\mu\frac{\partial v_l}{\partial x_l}\delta_{ik}$$

The following procedure is:

(1) $v_j \times$Eq. (7. 4) $+ v_i \times$Eq. (7. 5) gives the equation for $v_i v_j$.

(2) $\overline{v}_j \times$Eq. (7. 1) for $\overline{v}_i + \overline{v}_i \times$Eq. (7. 1) for $\overline{v}_j$ gives the equation for $\overline{v}_i\overline{v}_j$.

(3) Time-averaged equation $\overline{v_i v_j}$ -equation for $\overline{v}_i\overline{v}_j$ gives the exact equation for $\overline{v_i' v_j'}$.

Finally, the exact transport equation of Reynolds stresses can be obtained as:

$$\frac{\partial}{\partial t}(\rho\,\overline{v_i' v_j'}) + \frac{\partial}{\partial x_k}(\rho v_k\,\overline{v_i' v_j'}) = -\frac{\partial}{\partial x_k}\Big[\rho\,\overline{v_i' v_j' v_k'} + (\overline{p' v_j'})\,\delta_{ik} + (\overline{p' v_i'})\,\delta_{jk}$$

$$-\mu\left(\frac{\partial}{\partial x_k}\overline{v_i' v_j'}\right)\Big] - \rho\left(\overline{v_i' v_k'}\frac{\partial v_j}{\partial x_k} + \overline{v_j' v_k'}\frac{\partial v_i}{\partial x_k}\right) + \beta\rho\,(g_i\,\overline{v_j' T'} + g_j\,\overline{v_i' T'})$$

$$-2\mu\left(\overline{\frac{\partial v_i'}{\partial x_k}\frac{\partial v_j'}{\partial x_k}}\right) + \overline{p'\left(\frac{\partial v_i'}{\partial x_j} + \frac{\partial v_j'}{\partial x_i}\right)} \tag{7. 6}$$

where the two terms on the left-hand side express the rate of change in time and convection due to mean motion respectively, and the five terms on the right-hand side, denoted by D_{ij}, P_{ij}, G_{ij}, ε_{ij} , and Π_{ij} , express the diffusion, stress production, buoyancy production, dissipation and pressure-strain correlation respectively. The summation of three normal stress components ($i = j$) is related to the turbulent kinetic energy, which is defined as:

$$k \equiv \frac{1}{2}\,\overline{v_i'^{\,2}} = \frac{1}{2}\,(\overline{u'^{\,2}} + \overline{v'^{\,2}} + \overline{w'^{\,2}})$$

The transport equation of k can easily be derived from Eq. (7. 6) by setting $i = j$ and multiplying it by 1/2. Hence the exact k-equation is

$$\frac{\partial}{\partial t}(\rho k) + \frac{\partial}{\partial x_k}(\rho v_k k) = -\frac{\partial}{\partial x_k}\left(\rho\,\overline{v_k' v_i'^{\,2}}/2 + \overline{p' v_k'} - \mu\frac{\partial k}{\partial x_k}\right)$$

$$\rho\,\overline{v_i' v_k'}\frac{\partial v_i}{\partial x_k} + \beta\rho g_i\,\overline{v_i' T'} \quad \mu\,\overline{\left(\frac{\partial v_i'}{\partial x_k}\right)^2} \tag{7. 7}$$

where the six terms from left to right express the change in time, convection, diffusion, stress production, buoyancy production, and dissipation of the turbulent kinetic energy respectively.

In a similar way the transport equation of the heat flux $\overline{v_i' T'}$ can be derived as:

$$\frac{\partial}{\partial t}(\rho\,\overline{v_i' T'}) + \frac{\partial}{\partial x_k}(\rho v_k\,\overline{v_i' T'}) = -\frac{\partial}{\partial x_k}\left(\rho\,\overline{v_k' v_i' T'} + \delta_{ik}\,\overline{p' T'} - \frac{\lambda}{c_p}\,\overline{v_i'\frac{\partial T'}{\partial x_k}} \quad \mu\,\overline{T'\frac{\partial v_i'}{\partial x_k}}\right)$$

$$-\rho\left(\overline{v_i' v_k'}\frac{\partial T}{\partial x_k} + \overline{v_k' T'}\frac{\partial v_i}{\partial x_k}\right) - \beta\rho g_i\,\overline{T'^{\,2}} - \left(\frac{\lambda}{c_p} + \mu\right)\overline{\frac{\partial v_i'}{\partial x_k}\frac{\partial T'}{\partial x_k}} + \overline{p'\frac{\partial T'}{\partial x_i}} \tag{7. 8}$$

and the equation of $\overline{T'^{\,2}}$ can also be derived as:

$$\frac{\partial}{\partial t}(\rho\,\overline{T'^{\,2}}) + \frac{\partial}{\partial x_k}(\rho v_k\,\overline{T'^{\,2}}) = -\frac{\partial}{\partial x_k}\left(\rho\,\overline{v_k' T'^{\,2}} - \frac{\lambda}{c_p}\frac{\partial\,\overline{T'^{\,2}}}{\partial x_k}\right)$$

$$-2\,\overline{v_k' T'}\frac{\partial T}{\partial x_k} - 2\,\frac{\lambda}{c_p}\overline{\left(\frac{\partial T'}{\partial x_k}\right)^2} \tag{7. 9}$$

Equations (7. 6) — (7. 9) do not constitute a closed system. This system needs further closure models.

7.4 ISOTROPIC-VISCOSITY MODEL— k-ε MODEL

7.4.1 Zero-equation Model

The earliest and simplest turbulence model is the mixing-length model or algebraic model (zero-equation model) proposed by Prandtl in 1925, which is to directly simulate $\overline{v_j' \varphi'}$ by an algebraic expression of mean-flow properties. The model is based on the concept of analogy between laminar viscosity and turbulent viscosity, mean motion and turbulent fluctuation. The result can be stated by the following expression

$$\mu_T = \rho l_m^2 \left| \frac{\partial v_j}{\partial x_i} + \frac{\partial v_i}{\partial x_j} \right| \tag{7.10}$$

or

$$\begin{cases} -\rho \overline{v_i' v_j'} = \rho l_m^2 \left| \dfrac{\partial v_j}{\partial x_i} + \dfrac{\partial v_i}{\partial x_j} \right| \left(\dfrac{\partial v_j}{\partial x_i} + \dfrac{\partial v_i}{\partial x_j} \right) \\ -\rho \overline{v_j' \varphi'} = \dfrac{\rho l_m^2}{\sigma_\varphi} \left| \dfrac{\partial v_j}{\partial x_i} + \dfrac{\partial v_i}{\partial x_j} \right| \dfrac{\partial \varphi}{\partial x_j} \qquad (\varphi = Y_s, h) \end{cases} \tag{7.11}$$

where l_m is called mixing length, which is determined by intuition or experiments. For example, in jets we can take $l_m = cx$, where x is the distance along the mean-flow direction, c is a constant.

The merits of the mixing-length model are simple and intuitive, without adding differential equations. It has been successfully used in predicting jets, boundary layers, pipe and nozzle flows. However, it has serious shortcomings. For example, in pipe flows at the axis and in convergent channel flows behind a grid, where the mean velocity gradient is zero, the model gives zero turbulent viscosity. This is not true, because in pipe flows the turbulence generated near the wall diffuses to the near-axis region, while in flows behind the grid the turbulence generated immediately adjacent to the grid is transported downstream by convection. In other words, the turbulence convection and diffusion have been neglected in the mixing-length model. Furthermore, it is difficult to specify the mixing length l_m for complex flows, such as recirculating or swirling flows. Attempts have been made by the present author and other investigators to modify the mixing-length model by taking into account the effect of convection and giving rather complex expressions for l_m for sudden-expansion flows, but these modifications seem to be highly empirical and they have less generality.

7.4.2 Modeled Turbulent Kinetic Energy Equation

Actually, Eq. (7.7) indicates that the turbulent flutuation carries part of the total kinetic energy, obeying the transport or conservation laws, or the turbulent kinetic energy has its convection, diffusion, production and dissipation. The method of predicting turbulent viscosity by solving differential equations was first proposed by Kolmogorov (1942) and Prandtl (1945), defining

$$\nu_T = c_\mu k^{\frac{1}{2}} l, \qquad \mu_T = c_\mu \rho k^{\frac{1}{2}} l$$

Historically, a one-equation or energy-equation model was developed. It is to use a modeled turbulent kinetic energy equation and a presumed algebraic expression of the turbulence scale l .

To close Eq. (7. 7), it is necessary to simulate the unkown second-order and third-order correlations in terms of lower-order correlations or mean-flow properties. The first idea is to use the gradient modeling. Therefore, the diffusion, stress production, buoyancy production terms in Eq. (7. 7) can be modelled as:

$$-\rho\,\overline{v_k'\,(p'/\rho+v_i'^2/2)}+\mu\frac{\partial k}{\partial x_k}=\frac{\mu_T+\mu}{\sigma_k}\frac{\partial k}{\partial x_k}=\frac{\mu_e}{\sigma_k}\frac{\partial k}{\partial x_k}$$

$$-\rho\,\overline{v_i'\,v_k'}\,\frac{\partial v_i}{\partial x_k}=\mu_T\left(\frac{\partial v_k}{\partial x_i}+\frac{\partial v_i}{\partial x_k}\right)\frac{\partial v_i}{\partial x_k}$$

$$\beta\rho g_i\,\overline{v_i'T'}=-\,\beta g_k\,\frac{\mu_T}{\sigma_T}\frac{\partial T}{\partial x_k}$$

Difficulty arises when simulating the dissipation term, since no simple idea, such as gradient modeling, can be used here. Use can be made only with the Kolmogorov's concept or the dimensional analysis by defining

$$\mu\overline{\left(\frac{\partial v_i'}{\partial x_k}\right)^2}=c_D\rho\varepsilon$$

and $\rho\varepsilon$ should have the dimension of $\mu_T k/l^2$, or

$$\rho\varepsilon\sim\mu_T k/l^2\sim\rho k^{\frac{1}{2}}l\cdot k/l^2\sim\rho k^{\frac{3}{2}}l,$$

$$\varepsilon\sim k^{\frac{3}{2}}/l$$

Hence, the dissipation term can be modeled as

$$-\mu\overline{\left(\frac{\partial v_i'}{\partial x_k}\right)^2}=-\,c_D\rho\varepsilon=-\,c_D\rho k^{\frac{3}{2}}/l$$

Finally, the modeled transport equation of turbulent kinetic energy equation is

$$\frac{\partial}{\partial t}(\rho k)+\frac{\partial}{\partial x_k}(\rho v_k k)=\frac{\partial}{\partial x_k}\left(\frac{\mu_e}{\sigma_k}\frac{\partial k}{\partial x_k}\right)+G_k+G_b-c_D\rho k^{\frac{3}{2}}/l \qquad (7.12)$$

where

$$\mu_e=\mu+\mu_T,\qquad \mu_T=c_\mu\rho k^{\frac{1}{2}}l$$

$$G_k=\mu_T\left(\frac{\partial v_i}{\partial x_k}+\frac{\partial v_k}{\partial x_i}\right)\frac{\partial v_i}{\partial x_k},\qquad G_b=-\,\beta g_k\frac{\mu_T}{\sigma_T}\frac{\partial T}{\partial x_k}$$

It can easily be shown that in the case of neglecting the convection and diffusion terms, that is the case of local equilibrium of production with dissipation, the k -equation will lead to the expression of mixing-length model. Although the one-equation model is more reasonable than the mixing-length model, but for simple flows the mixing-length model is sufficient, and we need not to use the one-equation model. For complex flows where it is difficut to give a generalized expression of the turbulence scale l , we cannot use the one-equation model. Therefore, the k -equation model can be considered as an intermediate step which leads to the development of two-equation models.

7. 4. 3 The Second Turbulence parameter

In fact, not only the turbulent kinetic energy, but also other turbulence properties are transported in the flow field. For example, the length scale has its own convection, diffusion, production and destruction. The stretch of large eddies (energy-containing eddies) leads to formation of small eddies, and the dissipation of small eddies (Kolmogorov eddies) leads to the formation of large eddies. Spalding and Launder summarized the second turbulence parameter as $Z = k^m l^n$, proposed by different authors. There have been different forms of two equations, such as k-f (Kolmogorov, $f = k^{1/2}/l$), k-ε (P. Y. Zhou, Harlow, Nukayama, $\varepsilon = k^{3/2}/l$), k-l (Rodi, Spalding), k-kl (Ng, Spalding), k-w (Spalding, $w = k/l^2$) equations. The conservation equation of Z takes the form

$$\frac{\partial}{\partial t}(\rho Z) + \frac{\partial}{\partial x_k}(\rho v_k Z) = \frac{\partial}{\partial x_k}\left(\frac{\mu_e}{\sigma_z}\frac{\partial Z}{\partial x_k}\right) + S_Z \tag{7.13}$$

7.4.4 The Turbulent Kinetic Energy Dissipation Rate Equation

A vast amount of predictions indicates that all of the two-equation models give almost the same results. Among them the well-known k-ε model is most widely used and tested. To derive the ε-equation, using the procedure

$$\overline{2\nu\frac{\partial v_k}{\partial x_k} \times \frac{\partial}{\partial x_k}(\text{N-S eq.})} - 2\nu\frac{\partial v_k}{\partial x_k} \times (\text{Renolds eq.})$$

and assuming that the turbulence dissipation is isotropic, the original form of ε-equation can be obtained as:

$$\frac{\partial}{\partial t}(\rho\varepsilon) + \frac{\partial}{\partial x_k}(\rho v_k \varepsilon) = -\frac{\partial}{\partial x_k}(\rho\,\overline{v_k'\varepsilon'}) + \frac{\partial}{\partial x_k}\left(\mu\frac{\partial \varepsilon}{\partial x_k}\right)$$
$$- 2\mu\overline{\frac{\partial v_i'}{\partial x_k}\frac{\partial v_i'}{\partial x_l}\frac{\partial v_k'}{\partial x_l}} - 2\overline{\left(\nu\frac{\partial^2 v_i'}{\partial x_k \partial x_l}\right)^2} \tag{7.14}$$

where the six terms from left to right express the rate of change in time, convection, turbulent diffusion, molecular diffusion, production and viscous dissipation. The turbulent diffusion term can be simulated by the gradient modeling, that is

$$-\rho\,\overline{v_k'\varepsilon'} = \frac{\mu_T}{\sigma_\varepsilon}\frac{\partial \varepsilon}{\partial x_k}$$

There is some trouble with the simulation of production and dissipation terms, since we have no physical understanding of these terms. A general treatment is to assume the source terms of ε to be proportional to the source term of k by taking

$$S_\varepsilon \sim S_k = G_k - \rho\varepsilon$$

The dimensional analysis gives

$$S_\varepsilon = \frac{\varepsilon}{k}(c_1 G_k - c_2\rho\varepsilon)$$

Therefore, the modeled transport equation of turbulent kinetic energy dissipation rate ε is

$$\frac{\partial}{\partial t}(\rho\varepsilon) + \frac{\partial}{\partial x_k}(\rho v_k \varepsilon) = \frac{\partial}{\partial x_k}\left(\frac{\mu_e}{\sigma_\varepsilon}\frac{\partial \varepsilon}{\partial x_k}\right) + \frac{\varepsilon}{k}(c_1 G_k - c_2\rho\varepsilon) \tag{7.15}$$

7.4.5 Basic Equations of Turbulent Flows Containing the k-ε Turbulence Model

When using the k-ε model, the basic equations of turbulent flows can be expressed in the following generalized form

$$\frac{\partial}{\partial t}(\rho\varphi) + \frac{\partial}{\partial x_k}(\rho v_k \varphi) = \frac{\partial}{\partial x_k}\left(\Gamma_\varphi\frac{\partial \varphi}{\partial x_k}\right) + S_\varphi \tag{7.16}$$

For two-dimensional flows it takes the form

$$\frac{\partial}{\partial t}(\rho\varphi) + \frac{\partial}{\partial x}(\rho u\varphi) + \frac{\partial}{\partial y}(\rho v\varphi) = \frac{\partial}{\partial x}\left(\Gamma_\varphi \frac{\partial \varphi}{\partial x}\right) + \frac{\partial}{\partial y}\left(\Gamma_\varphi \frac{\partial \varphi}{\partial y}\right) + S_\varphi \quad (7.17)$$

where φ is the generalized dependent variable, Γ_φ is the transport coefficient, and S_φ is the source term. The contents of φ, Γ_φ and S_φ are given in Table 7. 1.

Table 7. 1 Generalized Equations

Equation	φ	Γ_φ	S_φ
Continuity	1	0	0
x -momentum	u	μ_e	$-\frac{\partial p}{\partial x} + \frac{\partial}{\partial x}\left(\mu_e \frac{\partial u}{\partial x}\right) + \frac{\partial}{\partial y}\left(\mu_e \frac{\partial v}{\partial x}\right) + \Delta\rho g_x$
y -momentum	v	μ_e	$-\frac{\partial p}{\partial y} + \frac{\partial}{\partial y}\left(\mu_e \frac{\partial v}{\partial y}\right) + \frac{\partial}{\partial x}\left(\mu_e \frac{\partial u}{\partial y}\right) + \Delta\rho g_y$
Turbulent kinetic energy	k	$\frac{\mu_e}{\sigma_k}$	$G_k + G_b - \rho\varepsilon$
TKE dissipation rate	ε	$\frac{\mu_e}{\sigma_\varepsilon}$	$\frac{\varepsilon}{k}(c_1 G_k - c_2\rho\varepsilon)$
Species	Y_s	$\frac{\mu_e}{\sigma_Y}$	$-w_s$
Energy	h	$\frac{\mu_e}{\sigma_h}$	$-q_r$

where

$$G_k = \mu_T\left[2\left(\frac{\partial u}{\partial x}\right)^2 + 2\left(\frac{\partial v}{\partial y}\right)^2 + \left(\frac{\partial u}{\partial y} + \frac{\partial v}{\partial x}\right)^2\right]$$

$$G_b = -\beta\rho\left(g_x \frac{\mu_T}{\sigma_T}\frac{\partial T}{\partial x} + g_y \frac{\mu_T}{\sigma_T}\frac{\partial T}{\partial y}\right)$$

$$\mu_e = \mu + \mu_T, \qquad \mu_T = c_\mu \rho k^2/\varepsilon$$

7. 4. 6 Application and Assessment of the k-ε Model

The k-ε model has been widely used for more than 20 years. A large amount of predictions and their comparison with experiments indicate that the k-ε model is successful or fairly successful in predicting the turbulent flows of the following cases: (1) non-buoyant plain jets; (2) boundary layers near plain walls; (3) pipe, channel and nozzle flows; (4) two-dimensional and three-dimensional non-swirling (or weakly swirling) recirculating flows. However, this model is unsuccessful in the predictions of: (1) strongly swirling flows (with swirl number greater than unity); (2) buoyant flows; (3) gravity-stratified flows; (4) boundary layers near curved walls; (5) low Reynolds-number flows; (6) round jets.

Figures 7. 1—7. 3 are predicted axial velocity field of non-swirling and weakly swirling coaxial jets discharged into a sudden-expansion chamber by Khalil and its comparison with experiments. The predicted velocity profiles are in good agreement with those measured. There is fairly good agreement between the predicted and measured shape, location and size of recirculation zones, but the predicted length of the near-wall

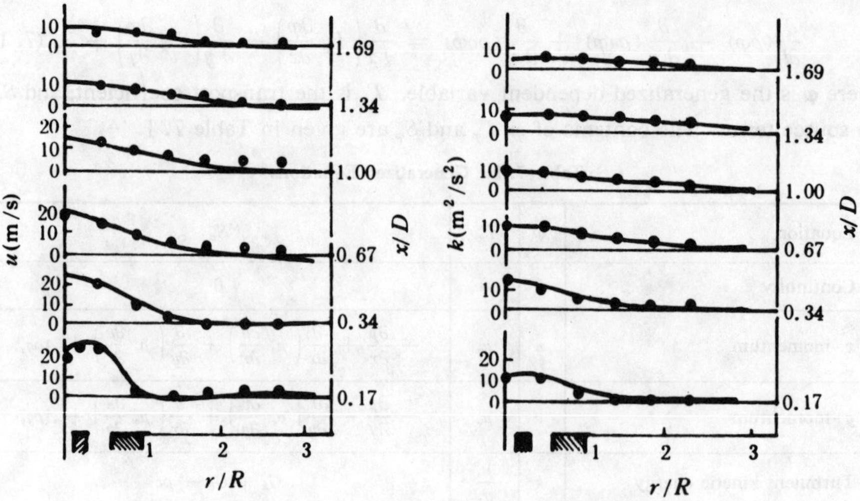

u(m/s)
x/D
k(m²/s²)
x/D
1.69
1.34
1.00
0.67
0.34
0.17
r/R
r/R

Fig. 7. 1

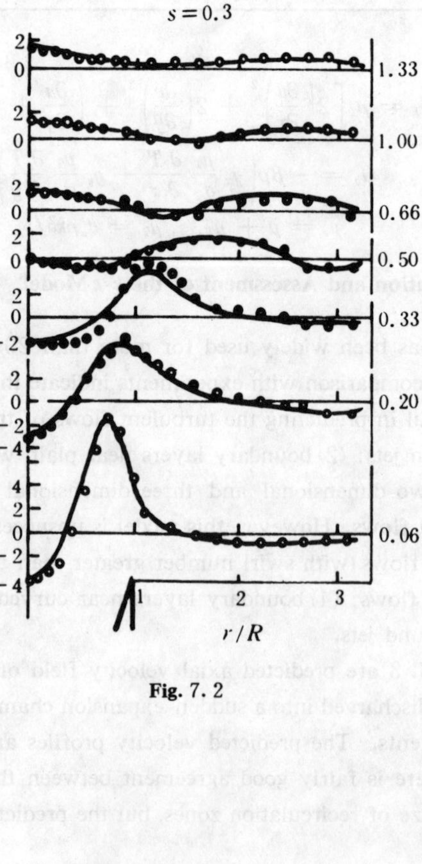

s = 0.3
u(m/s)
x/D
1.33
1.00
0.66
0.50
0.33
0.20
0.067
r/R

Fig. 7. 2

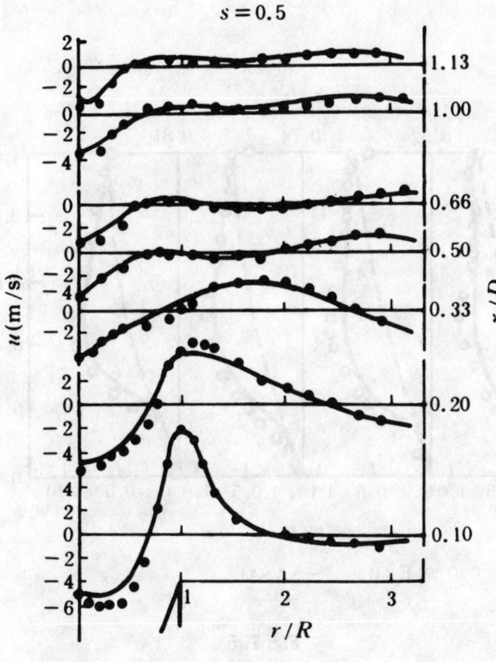
s = 0.5
u(m/s)
x/D
r/R
1.13
1.00
0.66
0.50
0.33
0.20
0.10

Fig. 7. 3

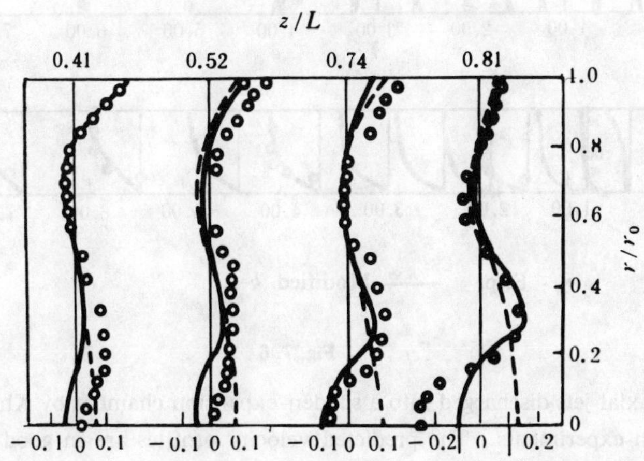
z/L
0.41
0.52
0.74
0.81
r/r0
u/wm
Exp.
ASM
k-ε

Fig. 7. 4

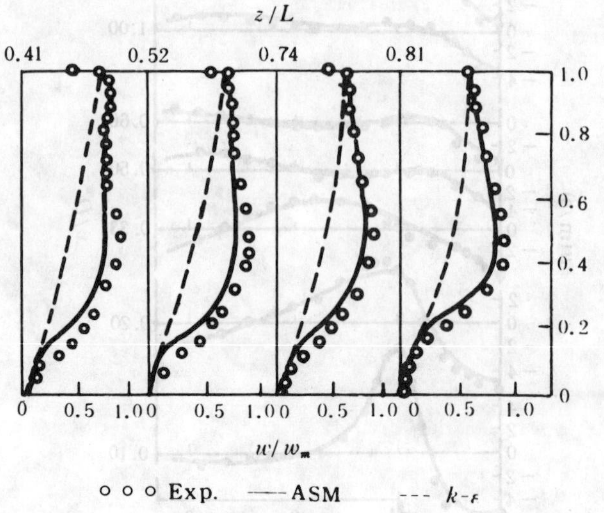

Fig. 7.5

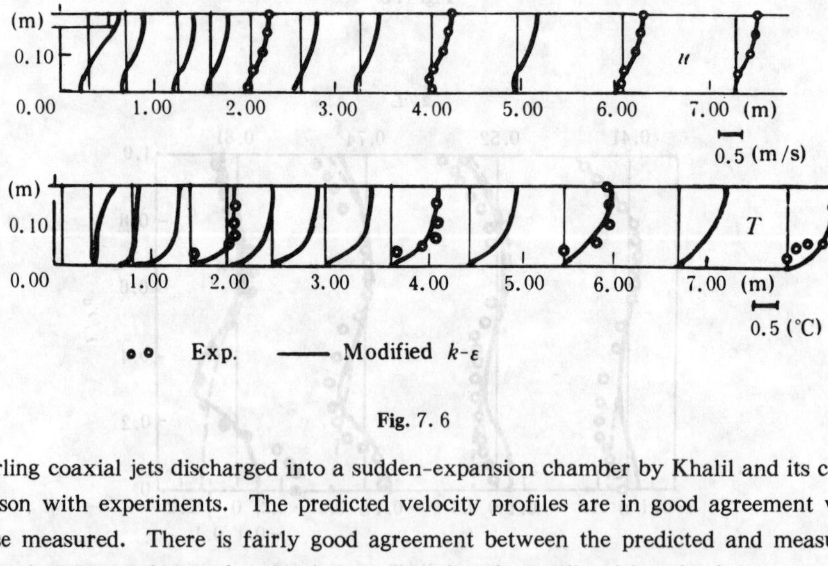

Fig. 7.6

swirling coaxial jets discharged into a sudden-expansion chamber by Khalil and its comparison with experiments. The predicted velocity profiles are in good agreement with those measured. There is fairly good agreement between the predicted and measured shape, location and size of recirculation zones, but the predicted length of the near-wall recirculation zone is smaller than that measured by 5—15%. This discrepancy may be caused by many factors, including: the deficiency of the model, numerical diffusion, inaccuracy in specifying inlet conditions and errors in measurements. Figures 7.4 and 7.5 are predicted axial and tangential velocity profiles of strongly swirling flows in a cyclone chamber by Swithenbank et al. and their comparison with experiments. It can be seen that there is qualitative difference between experiments and predictions by using the k-ε model (dashed line). The k-ε predictions eliminated the measured central recirculation zone and the Rankine vortex structure (solid-body rotation plus free vortex) which is essential to a separator or combustor performance. Figure 7.6 shows that the k-ε model predictions, even with the buoyancy term in the k-equation, cannot predict the thermocline layer in the strongly buoyant or gravity-stratified flows (heated water discharged into a cold water body).

7.4.7 Modified k-ε Model

For buoyant flows, besides the buoyancy production term G_b in the k equation which leads to reduction of k, it is still necessary to modify the ε-equation by using a modified source term

$$S_\varepsilon = \frac{\varepsilon}{k}\left[c_1 (G_k + G_b)(1 + c_3 R_f) - c_2 \rho\varepsilon\right] \tag{7.18}$$

where $R_f = - G_b/G_k$ is called flux Richardson number, and $c_3 = 0.8$. Equation (7.18) implies that the buoyancy force increases ε-production or increases ε, hence further reduces k. For the turbulent heat conduction, the Prandtl number should be modified by

$$\sigma_T/\sigma_{T0} = (1 + 3.33\mathrm{Ri})^{1.5}/(1 + 10\mathrm{Ri})^{0.5} \tag{7.19}$$

where

$$\mathrm{Ri} = - \frac{g}{\rho}\frac{\partial \rho}{\partial y}\Big/\left(\frac{\partial u}{\partial y}\right)^2$$

is the gradient Richardson number. Eq. (7.19) indicates that the reduction of turbulent heat conduction by the buoyancy force is stronger than the reduction of turbulent viscosity. Predictions of the gravity-stratified flows by using the modified k-ε model can better simulate the thermocline layer, but still cannot give satisfactory results.

For strongly swirling flows, Rodi, Morse and Smith proposed three different modifications:

Rodi's modification

$$S_\varepsilon = \frac{\varepsilon}{k}\left[c_1 G_k (1 + c_3 R_f) - c_2 \rho\varepsilon\right] \tag{7.20}$$

where

$$R_f = \frac{\text{Production of } k \text{ due to swirl}}{\text{Total production}}$$

Morse's modification

$$S_\varepsilon = \frac{\varepsilon}{k}(c_1 G_k - c_2 \rho\varepsilon) \tag{7.21}$$

where $c_1 = 1.44 + 0.8 R_f \rho\varepsilon/G_k$

$$R_f = \left[\overline{v'w'}\,\frac{\partial}{\partial r}\left(\frac{w}{r}\right)\right]/\varepsilon$$

P. J. Smith's modification

$$S_\varepsilon = \frac{\varepsilon}{k}\left[c_1 G_k - c_2 (1 - c_3 R_f)\rho\varepsilon\right] \tag{7.22}$$

where

$$R_f = \frac{k^2}{\varepsilon^2}\left(\frac{w}{r^2}\right)\left[\frac{\partial (wr)}{\partial r}\right]$$

According to Eq. (7. 20) and (7. 21), swirl increases ε, and therefore reduces k. However, according to Eq. (7. 22), only in the near-axis region of solid-body rotation where the tangential velocity w increases with the increase of r, swirl increases ε and hence reduces k. An opposite result would be obtained in the near-wall free-vortex region. In the prediction of strongly swirling flows, the modified k-ε model can only sometimes give slightly improved results, and in many cases it cannot give satisfactory results.

7. 5 REYNOLDS STRESS/FLUX TRANSPORT EQUATION MODEL AND ALGEBRAIC MODEL OF STRESSES/FLUXES

The exact form of Reynolds stress/flux transport equations, Eqs. (7. 6), (7. 8) and (7. 9) has already been derived before. Actually, most of turbulent flows are non-isotropic, and the turbulent viscosity μ_T is a tensor, but not a scalar. For non-isotropic turbulent flows, we must abandon the Boussinesq expression and directly solve the Reynolds stress equation by using the so-called second-order moment closure which can automatically account for buoyancy or swirling effect. This model is a simplest one in the theory of turbulence, but is the most complex one in engineering predictions.

7. 5. 1 Second-order Moment Closure (DSM)

The second-order moment closure is to simulate the unknown third-order correlations and other correlations by using the second-order correlations. The basic principles of modeling are: (1) considering the physical significance of each term; (2) using the dimensional analysis; (3) the modeled and original terms having the same character in coordinate transformation; (4) allowing gradient modeling for third-order correlations; and (5) isotropic dissipation.

Based on these principles, the modeling approximations made in the stress transport equation, Eq. (7. 6) are:

Turbulent diffusion term (Daley-Harlow model)

$$-\left(\rho\,\overline{v_i' v_j' v_k'} + \overline{p' v_j'}\,\delta_{ik} + \overline{p' v_i'}\,\delta_{jk}\right) = c_s \rho \frac{k}{\varepsilon}\,\overline{v_k' v_l'}\,\frac{\partial}{\partial x_l}\left(\overline{v_i' v_j'}\right)$$

Dissipation term

$$2\mu\overline{\left(\frac{\partial v_i'}{\partial x_k}\cdot\frac{\partial v_j'}{\partial x_k}\right)} = \frac{2}{3}\delta_{ij}\rho\varepsilon$$

Pressure-strain term (Launder-Rotta model)

$$\Pi_{ij} = \Pi_{ij,1} + \Pi_{ij,2} + \Pi_{ij,3}$$

$$\Pi_{ij,1} = -\,c_1\,(\varepsilon/k)\,\rho\left(\overline{v_i' v_j'} - \frac{2}{3}\delta_{ij}k\right)$$

$$\Pi_{ij,2} = -\,c_2\left(P_{ij} - \frac{2}{3}\delta_{ij}G_k\right)$$

$$\Pi_{ij,3} = -\,c_3\left(G_{ij} - \frac{2}{3}\delta_{ij}G_b\right)$$

Similar closure method can be used for equation of $\overline{v_i' T'}$ and $\overline{T'^2}$, Eqs. (7. 8) and (7. 9).

7.5.2 The Modeled Reynolds Stress/Flux Transport Equations

Neglecting the molecular transport, the whole system of Reynolds stress/flux transport equations in its closed form can be written as

$$\frac{\partial}{\partial t}(\rho\,\overline{v_i' v_j'}) + \frac{\partial}{\partial x_k}(\rho v_k\,\overline{v_i' v_j'}) = \frac{\partial}{\partial x_k}\left[c_s\rho\frac{k}{\varepsilon}\overline{v_k' v_l'}\frac{\partial}{\partial x_l}(\overline{v_i' v_j'})\right]$$
$$-c_1\frac{\varepsilon}{k}\rho\left(\overline{v_i' v_j'} - \frac{2}{3}\delta_{ij}k\right) - c_2\left(P_{ij} - \frac{2}{3}\delta_{ij}G_k\right)$$
$$-c_3\left(G_{ij} - \frac{2}{3}\delta_{ij}G_b\right) - \frac{2}{3}\delta_{ij}\rho\varepsilon + P_{ij} + G_{ij} \tag{7.23}$$

$$\frac{\partial}{\partial t}(\rho\,\overline{v_i' T'}) + \frac{\partial}{\partial x_k}(\rho v_k\,\overline{v_i' T'}) = \frac{\partial}{\partial x_k}\left[c_{sT}\rho\frac{k}{\varepsilon}\overline{v_k' v_l'}\frac{\partial}{\partial x_l}(\overline{v_i' T'})\right]$$
$$-\rho\left(\overline{v_i' v_k'}\frac{\partial T}{\partial x_k} + \overline{v_k' T'}\frac{\partial v_i}{\partial x_k}\right) - \beta\rho g_i\overline{T'^2} - c_{1T}\frac{\varepsilon}{k}\rho\,\overline{v_i' T'}$$
$$+c_{2T}\rho\,\overline{v_k' T'}\frac{\partial v_i}{\partial x_k} + c_{3T}\beta\rho g_i\overline{T'^2} \tag{7.24}$$

$$\frac{\partial}{\partial t}(\rho\overline{T'^2}) + \frac{\partial}{\partial x_k}(\rho v_k\overline{T'^2}) = \frac{\partial}{\partial x_k}\left(c_T\rho\frac{k}{\varepsilon}\overline{v_k' v_l'}\frac{\partial\overline{T'^2}}{\partial x_l}\right) - 2\,\overline{v_k' T'}\frac{\partial T}{\partial x_k} - \frac{1}{R}\overline{T'^2}\frac{\varepsilon}{k} \tag{7.25}$$

$$\frac{\partial}{\partial t}(\rho k) + \frac{\partial}{\partial x_k}(\rho v_k k) = \frac{\partial}{\partial x_k}\left(c_s\rho\frac{k}{\varepsilon}\overline{v_k' v_l'}\frac{\partial k}{\partial x_l}\right) + G_k + G_b - \rho\varepsilon \tag{7.26}$$

$$\frac{\partial}{\partial t}(\rho\varepsilon) + \frac{\partial}{\partial x_k}(\rho v_k\varepsilon) = \frac{\partial}{\partial x_k}\left(c_\varepsilon\rho\frac{k}{\varepsilon}\overline{v_k' v_l'}\frac{\partial\varepsilon}{\partial x_l}\right)$$
$$+c_{\varepsilon1}\frac{\varepsilon}{k}(G_k + G_b)(1 + c_{\varepsilon2}R_f) - c_{\varepsilon3}\rho\frac{\varepsilon^2}{k} \tag{7.27}$$

where

$$P_{ij} = -\rho\left(\overline{v_i' v_k'}\frac{\partial v_j}{\partial x_k} + \overline{v_j' v_k'}\frac{\partial v_i}{\partial x_k}\right)$$
$$G_{ij} = \beta\rho(g_i\overline{v_j' T'} + g_j\overline{v_i' T'})$$

Among Eqs. (7.23) — (7.27) the k-equation is not an independent equation, since k is the summation of three normal stress components divided by a factor of two. The model constants are listed in Table 7.2.

Table 7.2 Empirical Constants for the Reynolds Stress Transport Equation Model

c_s	c_1	c_2	c_3	c_{1T}	c_{2T}	c_{3T}	R	c_{sT}	c_T	c	$c_{\varepsilon1}$	$c_{\varepsilon2}$	$c_{\varepsilon3}$
0.24	2.2	0.55	0.55	3.0	0.5	0.5	0.8	0.11	0.13	0.15	1.44	1.92	0.8

7.5.3 Application and Assessment of DSM

In recent years the Reynolds stress transport equation model has been applied to predict strongly swirling and buoyant flows in combustors and furnaces, and in many cases it gives better results than that given by the k-ε model, showing the merits in its ability to account for the swirling effect, buoyancy effect, curvature effect, near-wall effect, etc. However, the model has its drawback. First, it is still too complex for engineering predictions, because only for the turbulence model itself we need to solve 11 equations instead of two equations in the k-ε model, so the computer storage and CPU time for three-dimensional problems are very large. Second, it is difficult to determine large number of empirical constants, 14 constants instead of three constants in the k-ε model. Next, it is difficult to specify the boundary conditions for each stress and flux. Besides, there is still arguement in modeling of the pressure-strain term. Actually, although this model can account for various non-isotropic effects, but the general accuracy of the model is not always higher than other models, because the k and ε equations in this model are obtained by using the same method as those in the k-ε model, in particular the modeling of source terms in these two equations. All this is the reason why the Reynolds stress model is not widely used nowadays. These problems could be solved in future development of computers and computational practice.

7. 5. 4 Algebraic Model of Stresses/Fluxes (ASM) —Extended k-ε model

It can be seen from the discussions in 7. 4 and 7. 5. 1 to 7. 5. 3 that the k-ε model is rather simple but not sufficiently general. On the other hand, the DSM has better generality, but is rather complex, inconvenient for engineering applications. Launder and Rodi proposed an algebraic model of stresses/fluxes as something compromise between the k-ε model and DSM, trying to combine the generality, simplicity and economy. The ASM consists of algebraic stress and flux expressions and k -equation, ε -equation with non-isotropic diffusion terms, hence it is also called an extended k-ε model. The main idea is to reduce the stress or flux transport equations into algebraic expressions by using some simplifications, while retaining the basic features of non-isotropic turbulence.

Rodi's first approximation is to assume the convection minus diffusion of stresses be proportional to the convection minus diffusion of turbulent kinetic energy, that is:

$$\frac{\partial}{\partial t}(\rho\,\overline{v_i' v_j'}) + \frac{\partial}{\partial x_k}(\rho v_k\,\overline{v_i' v_j'}) - \mathrm{Diff}\,(\overline{v_i' v_j'}) \sim \frac{\partial}{\partial t}(\rho k) + \frac{\partial}{\partial x_k}(\rho v_k k) - \mathrm{Diff}\,(k)$$

or

$$(P_{ij} + G_{ij} + \Pi_{ij} - \rho\varepsilon_{ij}) \sim (G_k + G_b - \rho\varepsilon)$$

The dimensional analysis gives

$$P_{ij} + G_{ij} + \Pi_{ij} - \rho\varepsilon_{ij} = \frac{\overline{v_i' v_j'}}{k}(G_k + G_b - \rho\varepsilon)$$

which leads to

$$\overline{v_i' v_j'} = k\left[\frac{2}{3}\delta_{ij} + \frac{(1-c_2)\left(P_{ij} - \frac{2}{3}\delta_{ij}G_k\right) + (1-c_3)\left(G_{ij} - \frac{2}{3}\delta_{ij}G_b\right)}{G_k + G_b + (c_1 - 1)\,\rho\varepsilon}\right] \tag{7. 28}$$

where

$$P_{ij} = \overline{v_i' v_k'}\frac{\partial v_j}{\partial x_k} + \overline{v_j' v_k'}\frac{\partial v_i}{\partial x_k}, \qquad G_{ij} = \beta\rho\,(g_j\,\overline{v_i' T'} + g_i\,\overline{v_j' T'})$$

The algebraic expression of heat flux can be obtained in a similar way as:

$$\rho\,\overline{v_i'T'} = \frac{2k}{\varepsilon}\left[\frac{\rho\,\overline{v_i'v_k'}\,\frac{\partial T}{\partial x_k} + (1 - c_{2T})\,\rho\,\overline{v_k'T'}\,\frac{\partial v_i}{\partial x_k} + (1 - c_{3T})\,\beta\rho g_i\,\overline{T'^2}}{G_k + G_b + 2\,(c_{1T} - 1)\,\rho\varepsilon}\right] \tag{7.29}$$

Rodi's second approximation is a local equilibrium assumption, that is, the convection minus diffusion is assumed to be zero, or the production is assumed to be equal to the dissipation. In this case we have

$$P_{ij} + G_{ij} + \Pi_{ij} - \rho\varepsilon_{ij} = 0$$

This expression, together with further simplifications of $c_2 = c_3$ and $G_k + G_b = \rho\varepsilon$, leads to

$$\overline{v_i'v_j'} = \frac{2}{3}\lambda k\delta_{ij} - (1 - \lambda)\;\frac{k}{\varepsilon}\left[\overline{v_i'v_k'}\,\frac{\partial v_j}{\partial x_k} + \overline{v_j'v_k'}\,\frac{\partial v_i}{\partial x_k} + \beta g_i\,\overline{v_j'T'} + \beta g_j\,\overline{v_i'T'}\right] \tag{7.30}$$

where

$$\lambda = (c_1 + c_2 - 1)\,/c_1$$

Comparison of Eq. (7.30) with the Boussinesq expression for isotropic turbulence

$$-\rho\,\overline{v_i'v_j'} = \frac{2}{3}\rho k\delta_{ij} + c_\mu\rho\,\frac{k^2}{\varepsilon}\left(\frac{\partial v_j}{\partial x_i} + \frac{\partial v_i}{\partial x_j}\right)$$

shows that for non-isotropic turbulent flows, except the difference in empirical constants and buoyancy term, the original scalar viscosity $c_\mu\rho\,k^2/\varepsilon$ is substituted by a tensor viscosity $\lambda\frac{k}{\varepsilon}\,\overline{v_i'v_k'}$ or $\lambda\frac{k}{\varepsilon}\,\overline{v_j'v_k'}$, or in other words, the coefficient c_μ is not a constant, but a tensor

$$c_\mu \sim \overline{v_i'v_k'}/k \quad \text{or} \quad c_\mu \sim \overline{v_j'v_k'}/k$$

which depends on buoyancy or centrifugal forces. The expressions for $\overline{v_i'T'}$ and $\overline{T'^2}$ can be obtained in a similar way as

$$\overline{v_i'T'} = \frac{k}{\varepsilon c_{1T}}\left[\overline{v_i'v_k'}\,\frac{\partial T}{\partial x_k} + (1 - c_{2T})\;\overline{v_k'T'}\,\frac{\partial v_i}{\partial x_k} + (1 - c_{3T})\,\beta g_i\,\overline{T'^2}\right] \tag{7.31}$$

$$\overline{T'^2} = -\,2R\,\frac{k}{\varepsilon}\,\overline{v_k'T'}\,\frac{\partial T}{\partial x_k} \tag{7.32}$$

It can also be shown that the turbulent Prandtl number σ_T in this case is also a tensor depending on buoyancy or swirling effect.

The final equations and expressions of ASM or the extended k-ε model are:

$$\frac{\partial}{\partial t}(\rho k) + \frac{\partial}{\partial x_k}(\rho v_k k) = \frac{\partial}{\partial x_k}\left(\rho c_s\,\frac{k}{\varepsilon}\,\overline{v_k'v_l'}\,\frac{\partial k}{\partial x_l}\right) + G_k + G_b - \rho\varepsilon \tag{7.26}$$

$$\frac{\partial}{\partial t}(\rho\varepsilon) + \frac{\partial}{\partial x_k}(\rho v_k\varepsilon) = \frac{\partial}{\partial x_k}\left(\rho c_\varepsilon\,\frac{k}{\varepsilon}\,\overline{v_k'v_l'}\,\frac{\partial \varepsilon}{\partial x_l}\right) + c_{\varepsilon 1}\,\frac{\varepsilon}{k}\,(G_k + G_b) \times (1 + c_{\varepsilon 3}R_f) - c_{\varepsilon 2}\rho\,\frac{\varepsilon^2}{k} \tag{7.27}$$

$$\overline{v_i'v_j'} = \frac{2}{3}\lambda k\delta_{ij} - (1 - \lambda)\,\frac{k}{\varepsilon}\left[\overline{v_i'v_k'}\,\frac{\partial v_j}{\partial x_k} + \overline{v_j'v_k'}\,\frac{\partial v_i}{\partial x_k} + \beta g_i\,\overline{v_j'T'} + \beta g_j\,\overline{v_i'T'}\right] \tag{7.30}$$

$$\overline{v_i'T'} = \frac{k}{\varepsilon c_{1T}}\left[\overline{v_i'v_k'}\frac{\partial T}{\partial x_k} + (1 - c_{2T})\,\overline{v_k'T'}\frac{\partial v_i}{\partial x_k} + (1 - c_{3T})\,\beta g_i\,\overline{T'^2}\right] \tag{7.31}$$

$$\overline{T'^2} = -2R\frac{k}{\varepsilon}\,\overline{v_k'T'}\frac{\partial T}{\partial x_k} \tag{7.32}$$

Obviously, the ASM or extended k-ε model reflects the basic features of non-isotropic turbulent flows, relating to buoyancy or swirling effects, while comparing with DSM the number of equations is remarkably reduced and it is not necessary to separately give the inlet and boundary conditions of stresses and fluxes.

7.5.5 Application of ASM and Its Comparison with Other Models

Figure 7.7 is the predicted temperature, velocity, turbulent kinetic energy and its dissipation rate profiles for a turbulent buoyant recirculating flow caused by heated water discharged into cold water by using the modified k-ε model and the ASM and their comparison with experiments. The results indicate that in predicting the temperature stratification (thermocline layer) and velocity stratification the ASM is much better than all modified k-ε models. Figure 7.8 gives the predicted velocity profiles for nonswirling coaxial jets discharged into a sudden-expansion chamber by using k-ε model, ASM and DSM and their comparison with experiments. Obviously, the results given by different models have only slight difference and are in good agreement with experiments. However, for strongly swirling coaxial jets discharged into the same chamber, predicted axial and tangential velocity profiles by using different models are different (Figs. 7.9 and

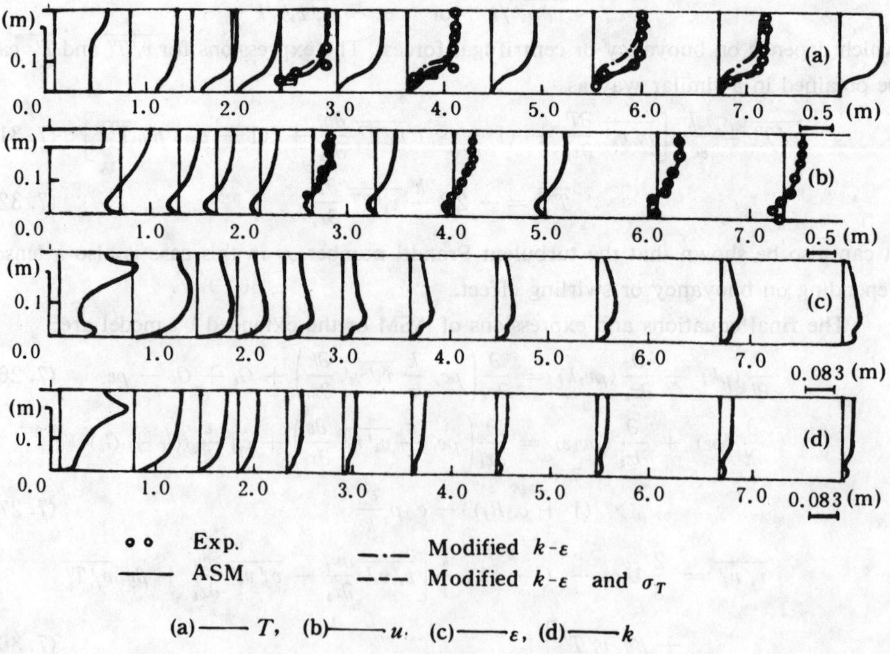

Fig. 7.7

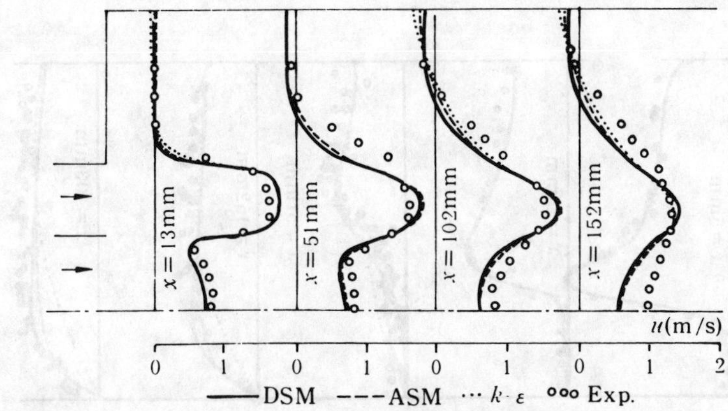

Fig. 7. 8

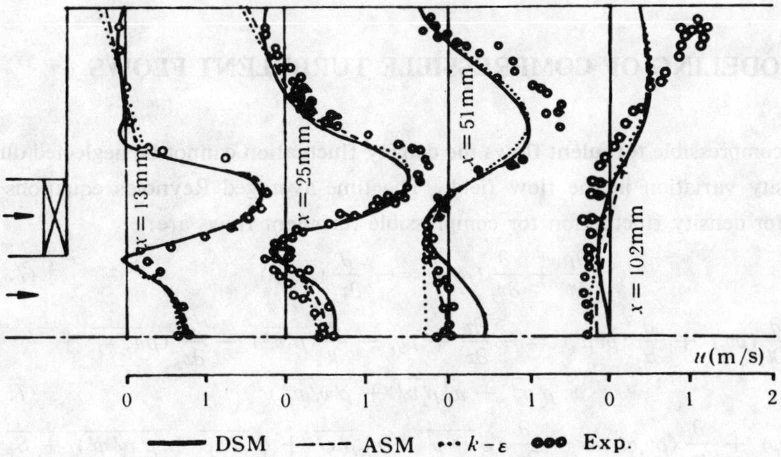

Fig. 7. 9

7. 10), and among them the DSM and ASM give better results. The solid lines in Fig. 7. 4 and 7. 5 are predicted axial and tangential velocity profiles for very strongly swirling flows in a cyclone chamber. These lines show that the ASM can properly predict the central recirculation zone and the Rankine vortex structure, hence the ASM predictions are in much better agreement with experiments than the k-ε model predictions.

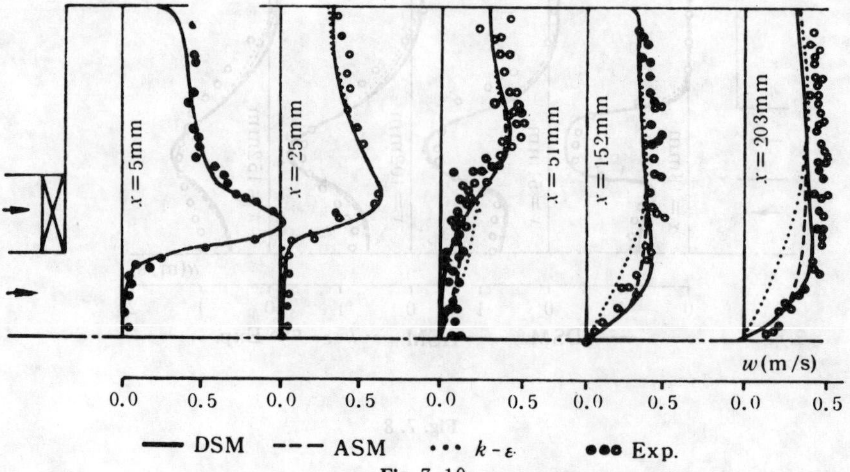

Fig. 7.10

7.6 MODELING OF COMPRESSIBLE TURBULENT FLOWS

For compressible turbulent flows the density fluctuation cannot be neglected due to large density variation in the flow field. The time-averaged Reynolds equations accounting for density fluctuation for compressible turbulent flows are:

$$\frac{\partial \rho}{\partial t} + \frac{\partial}{\partial x_j}(\rho v_j) = -\frac{\partial}{\partial x_j}(\overline{\rho' v_j'}) \tag{7.33}$$

$$\frac{\partial}{\partial t}(\rho v_i) + \frac{\partial}{\partial x_j}(\rho v_j v_i) = -\frac{\partial p}{\partial x_i} + \rho g_i - \frac{\partial}{\partial t}(\overline{\rho' v_i'}) - \frac{\partial}{\partial x_j}(\rho \overline{v_i' v_j'} + v_i \overline{\rho' v_j'} + v_j \overline{\rho' v_i'} + \overline{\rho' v_i' v_j'}) \tag{7.34}$$

$$\frac{\partial}{\partial t}(\rho \varphi) + \frac{\partial}{\partial x_j}(\rho v_j \varphi) = -\frac{\partial}{\partial x_j}(\rho \overline{v_j' \varphi'} + v_j \overline{\rho' \varphi'} + \varphi \overline{\rho' v_j'} + \overline{\rho' v_j' \varphi'}) + \overline{S_\varphi} \tag{7.35}$$

The simplest way is using the gradient modeling to account for the effect of compressibility on turbulence, taking

$$-(\rho \overline{v_i' v_j'} + \overline{\rho' v_i' v_j'}) = \mu_T \left(\frac{\partial v_j}{\partial x_i} + \frac{\partial v_i}{\partial x_j} \right)$$

$$-(\rho \overline{v_j' \varphi'} + \overline{\rho' v_j' \varphi'}) = \frac{\mu_T}{\sigma_\varphi} \frac{\partial \varphi}{\partial x_j}, \quad \mu_T = c_\mu \rho \frac{k^2}{\varepsilon}$$

$$\overline{\rho' v_i'} = -\frac{\mu_e}{\sigma} \frac{\partial \rho}{\partial x_i}, \quad \overline{\rho' v_j'} = -\frac{\mu_e}{\sigma} \frac{\partial \rho}{\partial x_j}, \quad \overline{\rho' \varphi'} = c \frac{k^3}{\varepsilon^2} \frac{\partial \rho}{\partial x_j} \frac{\partial \varphi}{\partial x_j}$$

In this case the conventional k-ε model can still be used. More detailed modeling is to derive the transport equations of $\overline{\rho' v_i'}$, $\overline{\rho' \varphi'}$ and $\overline{\rho'^2}$. Furthermore, the k and ε equations can be derived on the basis of Eq. (7.34) to obtain some new source terms accounting for the effect of density fluctuation. Predictions of temperature and NO_x profiles during gas turbulent combustion accounting for and not accounting for density fluctuation, given by Khalil, indicate that predictions accounting for density fluctuation are in better agreement with experiments. Obviously, combustion can either reduce or enhance turbulence in different regions and for different cases.

Some further developments in engineering turbulence models have been made in recent years, such as the modified ASM partly accounting for the stress convection (retaining the nonderivative term), two-fluid model, nonlinear k-ε model (a constitutive relation between the stress and strain rate based on the principle of rational mechanics), multiscale k-ε model trying to increase the accuracy of ε-equation, third-order moment closure and others. Other models, such as the discrete vortex model (random vortex model), the pdf transport equation model, can avoid gradient modeling assumption, but they are difficult to use in complex engineering turbulent flows, such as three-dimensional recirculating and swirling flows.

In short, there is no general turbulence model both reasonable and economical. For simple flows, such as boundary-layer or jet flows we can use the mixing-length model. The k-ε model or other two-equation models are appropriate for predicting 2-D or 3-D nonbuoyant and nonswirling recirculating flows. For strongly swirling or buoyant flows it is necessary to use ASM or DSM or other higher-order moment closure models. In doing so, many problems still remain to be studied, such as how to improve the modeling of diffusion term and pressure-strain term in Reynolds stress transport equation.

CHAPTER 8 MODELING OF TURBULENT GAS-PARTICLE FLOWS

8.1 TIME-AVERAGED EQUATIONS OF TURBULENT GAS-PARTICLE FLOWS

The basic equations of gas-particle flows obtained in Chapter 6 are applicable only to laminar gas-particle flows, or they can be considered as the instantaneous equations of turbulent gas-particle flows. For turbulent gas-particle flows, it is needed to derive the time-averaged equations. Using the method similar to that used in single-phase turbulent flows, taking

$$\rho = \bar{\rho} + \rho', \quad v_i = \bar{v}_i + v_i', \quad v_j = \bar{v}_j + v_j', \quad T = \bar{T} + T', \quad n_k = \bar{n}_k + n_k'$$

$$v_{ki} = \bar{v}_{ki} + v_{ki}', \quad v_{kj} = \bar{v}_{kj} + v_{kj}', \quad T_k = \bar{T}_k + T_k', \quad \dot{m}_k = \bar{\dot{m}}_k + \dot{m}_k'$$

and substituting them into instantaneous equations, we can obtain the time-averaged equations: (dropping the symbol "——" for mean quantities)

Fluid continuity

$$\frac{\partial \rho}{\partial t} + \frac{\partial}{\partial x_j}(\rho v_j) = -\frac{\partial}{\partial x_j}(\overline{\rho' v_j'}) + S - \sum \overline{n_k' \dot{m}_k'} \tag{8.1}$$

Particle continuity

$$\frac{\partial \rho_k}{\partial t} + \frac{\partial}{\partial x_j}(\rho_k v_{kj}) = n_k \dot{m}_k - \frac{\partial}{\partial x_j}(\overline{\rho_k' v_{kj}'}) + \overline{n_k' \dot{m}_k'} \tag{8.2}$$

$$\frac{\partial n_k}{\partial t} + \frac{\partial}{\partial x_j}(n_k v_{kj}) = -\frac{\partial}{\partial x_j}(\overline{n_k' v_{kj}'}) \tag{8.3}$$

Fluid momentum

$$\begin{aligned}\frac{\partial}{\partial t}(\rho v_i) + \frac{\partial}{\partial x_j}(\rho v_j v_i) = &-\frac{\partial p}{\partial x_i} + \frac{\partial \tau_{ji}}{\partial x_j} + \Delta\rho g_i + \sum \rho_k (v_{ki} - v_i)/\tau_{rk} \\ &+ v_i S + F_{Mi} - \frac{\partial}{\partial x_j}(\rho \overline{v_j' v_i'} + v_i \overline{\rho' v_j'} + v_j \overline{\rho' v_i'} + \overline{\rho' v_j' v_i'}) \\ &+ \sum \frac{m_k}{\tau_{rk}}(\overline{n_k' v_{ki}'} - \overline{n_k' v_i'}) - v_i \sum \overline{n_k' \dot{m}_k'} - \sum n_k \overline{v_i' \dot{m}_k'} - \sum \dot{m}_k \overline{n_k' v_i'} \\ &- \sum \overline{v_i' n_k' \dot{m}_k'}\end{aligned} \tag{8.4}$$

Particle momentum

$$\begin{aligned}\frac{\partial}{\partial t}(n_k v_{ki}) + \frac{\partial}{\partial x_j}(n_k v_{kj} v_{ki}) = &\, n_k g_i + n_k (v_i - v_{ki})\left(\frac{1}{\tau_{rk}} + \frac{\dot{m}_k}{m_k}\right) + F_{k,Mi}/m_k \\ &- \frac{\partial}{\partial x_j}(n_k \overline{v_{kj}' v_{ki}'} + v_{kj} \overline{n_k' v_{ki}'} + v_{ki} \overline{n_k' v_{kj}'} + \overline{n_k' v_{kj}' v_{ki}'}) + (\overline{n_k' v_i'} - \overline{n_k' v_{ki}'})/\tau_{rk} \\ &+ (v_i \overline{n_k' \dot{m}_k'} + n_k \overline{v_i' \dot{m}_k'} + \dot{m}_k \overline{n_k' v_i'} + \overline{n_k' v_i' \dot{m}_k'} - v_{ki} \overline{n_k' \dot{m}_k'} - n_k \overline{v_{ki}' \dot{m}_k'}\end{aligned}$$

$$- \dot{m}_k \overline{n_k' v_{ki}'} - \overline{n_k' v_{ki}' \dot{m}_k'}) / m_k - \frac{\partial}{\partial t} \overline{(n_k' v_{ki}')} \tag{8.5}$$

Fluid energy

$$\frac{\partial}{\partial t}(\rho h) + \frac{\partial}{\partial x_j}(\rho v_j h) = \frac{\partial}{\partial x_j}\left(\lambda \frac{\partial T}{\partial x_j}\right) - q_r + \sum n_k Q_k + hS - h \sum \overline{n_k' \dot{m}_k'}$$

$$- \sum n_k \overline{h' \dot{m}_k'} - \sum \dot{m}_k \overline{n_k' h'} - h' \sum \overline{n_k' \dot{m}_k'} - \frac{\partial}{\partial t} \overline{(\rho' h')}$$

$$- \frac{\partial}{\partial x_j}(\rho \overline{v_j' h'} + v_j \overline{\rho' h'} + h \overline{\rho' v_j'} + \overline{\rho' h' v_j'}) \tag{8.6}$$

Particle energy

$$\frac{\partial}{\partial t}(n_k c_k T_k) + \frac{\partial}{\partial x_j}(n_k v_{kj} c_k T_k) = n_k (Q_h - Q_k - Q_{rk}) / m_k + (c_p T - c_k T_k) \frac{n_k \dot{m}_k}{m_k}$$

$$- \frac{\partial}{\partial t}(c_k \overline{n_k' T_k'}) - \frac{\partial}{\partial x_j}(n_k c_k \overline{v_{kj}' T_k'} + c_k v_{kj} \overline{n_k' T_k'} + c_k T_k \overline{n_k' v_{kj}'} + c_k \overline{v_{kj}' n_k' T_k'})$$

$$+ (c_p T \overline{n_k' \dot{m}_k'} + c_p n_k \overline{T' \dot{m}_k'} + c_p \dot{m}_k \overline{n_k' T'} + c_p \overline{n_k' \dot{m}_k' T'}) / m_k$$

$$- (c_k T_k \overline{n_k' \dot{m}_k'} + c_k n_k \overline{T_k' \dot{m}_k'} + c_k \dot{m}_k \overline{n_k' T_k'} + c_k \overline{n_k' \dot{m}_k' T_k'}) / m_k \tag{8.7}$$

Fluid species

$$\frac{\partial}{\partial t}(\rho Y_s) + \frac{\partial}{\partial x_j}(\rho v_j Y_s) = \frac{\partial}{\partial x_j}\left(D\rho \frac{\partial Y_s}{\partial x_j}\right) - w_s - \alpha_s n_k \dot{m}_k$$

$$- \frac{\partial}{\partial x_j}(\rho \overline{v_j' Y_s'} + Y_s \overline{\rho' v_j'} + v_j \overline{\rho' Y_s'} + \overline{\rho' v_j' Y_s'})$$

$$- \alpha_s \overline{n_k' \dot{m}_k'} - \frac{\partial}{\partial t} \overline{(\rho' Y_s')} \tag{8.8}$$

Equations (8. 1) — (8. 8) are general description of turbulent gas-particle flows, they do not constitute a closed system because of unknown correlation terms in these equations. In different particle-phase models the closure problem is solved by using either modeling approximations, or different simplifications.

8.2 SINGLE-FLUID MODEL (NO-SLIP MODEL)

In the early 1970's numerical modeling of gas-particle two-phase or multiphase flows was developed, on the basis of modeling of single-phase turbulent flows. The simplest model for simulating gas-particle flows is the single-fluid model which is also called no-slip model. The basic assumptions are: (1) the time-averaged particle velocity of each size group is equal to the local time-averaged gas velocity, i. e. $v_{ki} = v_i$ (momentum equilibrium, or no-slip); (2) the particle temperature either keeps constant (T_k = const, energy frozen), or is equal to the local gas temperature (T_k = T, energy equilibrium); (3) particle phase, as a gas species, diffuses with the same rate as that of other gas species ($v_k = v_T$, diffusion equilibrium); (4) particle groups may be identified either by their initial size distribution, or by their local size distribution.

Using the time-averaged gas-phase and particle-phase equations derived in 8. 1, neglecting the gas density fluctuation, the phase change fluctuation and the drag force (no slip), the gas-phase conservation equations can be expressed in the following generalized form:

$$\frac{\partial}{\partial t}(\rho\varphi) + \frac{\partial}{\partial x_j}(\rho v_j \varphi) = \frac{\partial}{\partial x_j}\left(\Gamma_\varphi \frac{\partial \varphi}{\partial x_j}\right) + S_\varphi + S_{p\varphi} \tag{8. 9}$$

Where φ is the generalized dependent variable, Γ_φ is the transport coefficient, S_φ is the source term of gas-phase itself, $S_{p\varphi}$ is the source term due to gas-particle interaction. The meaning of φ, Γ_φ, S_φ and $S_{p\varphi}$ for each equation is given in Table 8. 1, and the meaning of μ_e and G_k is the same as that given in Chapter 7. For particle phase no momentum and energy equations are needed due to no-slip assumptions. What we need is only particle continuity equation or particle diffusion equation

$$\frac{\partial \rho_k}{\partial t} + \frac{\partial}{\partial x_j}(\rho_k v_j) = \frac{\partial}{\partial x_j}\left(\frac{\gamma_e}{\sigma_{kp}} m_k \frac{\partial n_k}{\partial x_j}\right) + n_k \dot{m}_k \tag{8. 10}$$

$$\frac{\partial n_k}{\partial t} + \frac{\partial}{\partial x_j}(n_k v_j) = \frac{\partial}{\partial x_j}\left(\frac{\gamma_e}{\sigma_{kp}} \frac{\partial n_k}{\partial x_j}\right) \tag{8. 11}$$

Table 8. 1 Single-fluid Model Equations

Equation	φ	Γ_φ	S_φ	$S_{p\varphi}$
Gas continuity	1	0	0	$S = -\sum n_k \dot{m}_k$
Gas momentum	v_i	μ_e	$\frac{\partial p}{\partial x_i} + \Delta\rho g_i + \frac{\partial}{\partial x_j}\left[\mu_e \frac{\partial v_j}{\partial x_i}\right]$	$v_i S$
Gas turbulent kinetic energy	k	$\frac{\mu_e}{\sigma_k}$	$G_k - \rho\varepsilon$	0
TKE dissipation rate	ε	$\frac{\mu_e}{\sigma_\varepsilon}$	$\frac{\varepsilon}{k}(c_1 G_k - c_2\rho\varepsilon)$	0
Gas species	Y_s	$\frac{\mu_e}{\sigma_Y}$	$-w_s$	$\alpha_s S$
Gas enthalpy	h	$\frac{\mu_e}{\sigma_h}$	$-q_r$	$\sum n_k Q_k + hS$

The particle-gas interaction term in Eqs. (8. 9) and (8. 10) is written on the basis of the particle groups identified by their initial size distribution. This term can also be written on the basis of the particle groups identified by their local size distribution. For the latter case, the source term can be obtained from the statistical conservation equation of a particle group described in Chapter 6. Assume that the local size distribution can be approximated by a stepwise distribution (Fig. 8. 1). If we define

$$s = d^2 \qquad (d \text{ is particle diameter})$$

$$\dot{s} = \frac{ds}{dt} = \frac{d(d^2)}{dt} \qquad (\dot{s} \text{ is independent of } s)$$

f_p' = particle mass fraction per unit s

ρ_p' = particle apparent density per unit s

n' = particle number density per unit s

then we should have

$$f_k = \int_{s_{k+1}}^{s_k} f_p' ds = \rho_k/\rho_m, \quad \rho_m = \rho + \sum \rho_k$$

$$\rho_k = \int_{s_{k+1}}^{s_k} \rho_p' ds = n_k m_k, \quad n_k = \int_{s_{k+1}}^{s_k} n' ds$$

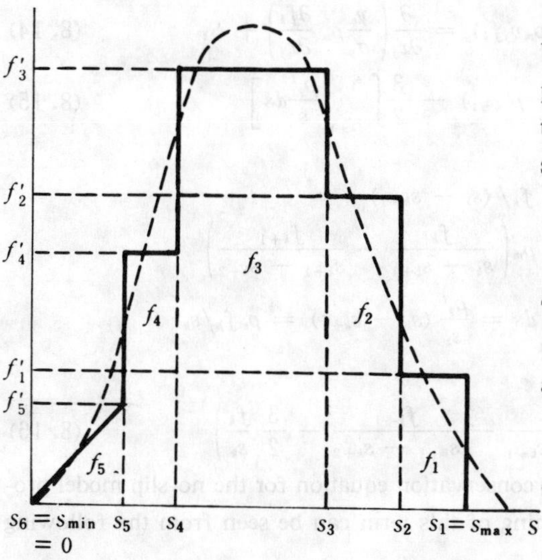

Fig. 8. 1

$$f_p'\rho_m = \rho_p' = n'm$$

Using Eq. (6. 32), considering that the particle number density is only a function of time, location and local particle size, we can write

$$\frac{\partial n'}{\partial t} + \frac{\partial}{\partial s}(n'\dot{s}) + \frac{\partial}{\partial x_j}(n'v_{pj}) = 0 \tag{8. 12}$$

Taking

$$v_{pj} = v_j + V_{pj}$$

and

$$-\rho_p' V_{pj} = \frac{\gamma_e}{\sigma_\rho}\rho_m \frac{\partial f_p'}{\partial x_j}$$

we have

$$\frac{\partial n'}{\partial t} + \frac{\partial}{\partial s}(n'\dot{s}) + \frac{\partial}{\partial x_j}(n'v_j) = \frac{\partial}{\partial x_j}\left(\frac{\gamma_e}{m\sigma_p}\rho_m\frac{\partial f_p'}{\partial x_j}\right) \tag{8. 13}$$

Notice here that v_{pj} is different from v_j by a drift velocity V_{pj} . Actually this is contradictory to the basic concept of no-slip, and it is equivalent to a small-slip concept. However, no slip is considered in particle momentum treatment, so it is not identical to the small-slip model. Furthermore, the particle diffusivity used here is assumed to be $D_p = \gamma_e/\sigma_p$, which is also different from that used in the small-slip model. Since in phase space $m = \overline{\rho}_p\pi d^3/6$ depends only on the independent variable $s = d^2$ and is independent of other independent variables x_i and t , after dividing each term by m , Eq. (8. 13) can be transformed into

$$\frac{\partial \rho_p'}{\partial t} + \frac{\partial}{\partial x_j}\left(\rho_p' v_j - \frac{\gamma_e}{\sigma_p}\rho_m\frac{\partial f_p'}{\partial x_j}\right) = -m\frac{\partial}{\partial s}\left(\dot{s}\frac{\rho_p'}{m}\right)$$

Considering that $\dot{s}$ is independent of s , we have

$$m\frac{\partial}{\partial s}\left(\dot{s}\frac{\rho_p'}{m}\right) = -\dot{s}\frac{\partial}{\partial s}\left(\frac{\rho_p'}{m}\right) = -\frac{\dot{s}}{m}\left(\frac{\partial \rho_p'}{\partial s} - \frac{\rho_p'}{m}\frac{\partial m}{\partial s}\right)$$

and

$$\frac{\rho_p'}{m}\frac{\partial m}{\partial s} = \left[6\rho_p'/(\overline{\rho}_p\pi d^3)\right]\frac{\partial(\overline{\rho}_p\pi d^3/6)}{\partial(d^2)} = \frac{3}{2}\frac{\rho_p'}{s}$$

and hence

$$\frac{\partial \rho_p'}{\partial t} + \frac{\partial}{\partial x_j}\left(\rho_p' v_j - \frac{\gamma_e}{\sigma_p}\rho_m\frac{\partial f_p'}{\partial x_j}\right) = -\dot{s}\left(\frac{\partial \rho_p'}{\partial s} - \frac{3}{2}\frac{\rho_p'}{s}\right)$$

Integration from s_{k+1} to s_k gives

$$\frac{\partial \rho_k}{\partial t} + \frac{\partial}{\partial x_j}\left(\rho_k v_j - \frac{\gamma_e}{\sigma_p}\rho_m\frac{\partial f_k}{\partial x_j}\right) = -\dot{s}\left[(\rho_k' - \rho'_{k+1}) - \frac{3}{2}\int_{s_{k+1}}^{s_k}\frac{\rho_p'}{s}ds\right]$$

or

$$\begin{cases} \frac{\partial}{\partial t}(\rho_m f_k) + \frac{\partial}{\partial x_j}(\rho_m v_j f_k) = \frac{\partial}{\partial x_j}\left(\frac{\gamma_e}{\sigma_p}\rho_m \frac{\partial f_k}{\partial x_j}\right) + S_k & (8.14) \\ S_k = -\dot{s}\left[(\rho_k' - \rho'_{k+1}) - \frac{3}{2}\int_{s_{k+1}}^{s_k} \frac{\rho_p'}{s} ds\right] & (8.15) \end{cases}$$

By using Fig. 8. 1, we have

$$f_k' = f_k / (s_k - s_{k+1})$$

$$\dot{s}\,(\rho_k' - \rho'_{k+1}) = \dot{s}\,\rho_m\left(\frac{f_k}{s_k - s_{k+1}} - \frac{f_{k+1}}{s_{k+1} - s_{k+2}}\right)$$

$$\int_{s_{k+1}}^{s_k} \frac{\rho_p'}{s} ds \approx \frac{\rho_p'}{s_k}\int_{s_{k+1}}^{s_k} ds = \frac{\rho_p'}{s_k}(s_k - s_{k+1}) = \rho_m f_k / s_k$$

and therefore

$$S_k = -\dot{s}\,\rho_m\left(\frac{f_k}{s_k - s_{k+1}} - \frac{f_{k+1}}{s_{k+1} - s_{k+2}} - \frac{3}{2}\frac{f_k}{s_k}\right) \quad (8.16)$$

This is the source term in particle mass conservation equation for the no-slip model proposed by Spalding. The physical meaning of this term can be seen from the following expression

$$S_k = \rho_k' - \left[\frac{d(d^2)}{dt}\right] - \rho'_{k+1}\left[-\frac{d(d^2)}{dt}\right] + \int_{s_{k+1}}^{s_k} n' \frac{dm}{dt} ds$$

This expression implies that when particle material density keeps constant, the particle mass loss due to evaporation, devolatilization or heterogeneous reaction will be caused by their size reduction. For particle groups identified by their local size distribution, the mass loss of k-th particle phase consists of three parts: mass addition of group k due to the size reduction of group $k+1$ to the size of group k; mass loss of group k due to the size reduction of group k to the size of group $k-1$; mass loss of group k itself.

The solution procedure of the single-fluid model for gas-particle flows is almost the same as that for single-phase flows except that several particle-phase continuity equations (like gas species diffusion equations) and particle-source terms in gas-phase equations are added. Therefore, the computer code for solving single-phase turbulent flows can still be used, with only some minor modifications.

The single-fluid model has been used to simulate turbulent gas-particle jet with carbon reaction and gas-phase diffusion combustion, two-dimensional recirculating gas-droplet flows with combustion in after-burners, three-dimensional recirculating and swirling gas-droplet flows with combustion in gas-turbine combustors, two-dimensional recirculating and swirling gas-particle flows with coal combustion in sudden-expansion combustors. The prediction results will be given in Chapter 10. The merits of the single-fluid model are its simplicity and convenience for using the computer code for single-phase flows to predict two-phase flows. However, the discrepancy between the predictions based on this model and experiments is considerably large, because actually there is velocity and temperature slip between gas and particles. Therefore, up to now, the single-fluid model is seldom used in solving practical engineering problems, but it can be used for preliminary predictions and for educational purpose.

8.3 PARTICLE TRAJECTORY MODEL

The single particle dynamics model is a treatment of particle phase in Lagrangian coordinate, and in the small-slip model and no-slip model the particle phase is treated in Eulerian coordinate. In the particle trajectory model the particle phase is once again treated in Lagrangian coordinate. Unlike the single particle dynamics model, the particle trajectory model fully accounts for the interaction between gas and particle phases. On the other hand, unlike the small-slip and no-slip models, this model accounts for large velocity and temperature slip between two phases, which is not related to particle diffusion.

8.3.1 Basic Assumptions

In the particle trajectory model it is assumed that: (1) the particle phase is a discrete system, and there is velocity and temperature slip between gas and particle phases (the slip may be small or large), that is, $v_{ki} \neq v_i$ (dynamic nonequilibrium) and $T_k \neq T$ (energy nonequilibrium); (2) no particle diffusion in deterministic trajectory model, that is, $\nu_k = D_k = \lambda_k = 0$ (diffusion frozen); (3) particle groups are identified by their initial size distribution, each having the same size, velocity and temperature at any time instant; (4) each particle group moves along its own trajectory from a certain initial position, and particle mass, velocity and temperature change are tracked along the trajectory; and (5) equivalent particle mass, momentum and energy sources (sinks) distributed in gas-phase cells are considered.

8.3.2 Basic Conservation Equations

Taking all of the correlation terms of particle phase to be zero and the fluctuation of gas density and mass source to be zero in Eqs. (8.1) — (8.8), the time-averaged equations of gas phase and particle phase in Eulerian coordinate for a deterministic particle trajectory model can be obtained as:

Gas continuity

$$\frac{\partial \rho}{\partial t} + \frac{\partial}{\partial x_j}(\rho v_j) = S, \quad S = -\sum n_k \dot{m}_k = -\sum S_k \tag{8.17}$$

Particle continuity

$$\frac{\partial \rho_k}{\partial t} + \frac{\partial}{\partial x_j}(\rho_k v_{kj}) = S_k, \quad S_k = n_k \dot{m}_k \tag{8.18}$$

Gas momentum

$$\frac{\partial}{\partial t}(\rho v_i) + \frac{\partial}{\partial x_j}(\rho v_j v_i) = -\frac{\partial p}{\partial x_i} + \frac{\partial}{\partial x_j}\left[\mu_e\left(\frac{\partial v_j}{\partial x_i} + \frac{\partial v_i}{\partial x_j}\right)\right] + \Delta \rho g_i + \sum \rho_k (v_{ki} - v_i)/\tau_{rk} + v_i S + F_{Mi} \tag{8.19}$$

Particle momentum

$$\frac{\partial}{\partial t}(\rho_k v_{ki}) + \frac{\partial}{\partial x_j}(\rho_k v_{kj} v_{ki}) = \frac{\rho_k}{\tau_{rk}}(v_i - v_{ki}) + \rho_k g_i + F_{k,Mi} + v_i S_k \tag{8.20}$$

Gas energy

$$\frac{\partial}{\partial t}(\rho c_p T) + \frac{\partial}{\partial x_j}(\rho v_j c_p T) = \frac{\partial}{\partial x_j}\left(\frac{\mu_e}{\sigma_T}\frac{\partial T}{\partial x_j}\right) + w_s Q_s - q_r + \sum n_k Q_k + c_p T S \tag{8.21}$$

Particle energy

$$\frac{\partial}{\partial t}(\rho_k c_k T_k) + \frac{\partial}{\partial x_j}(\rho_k v_{kj} c_k T_k) = n_k (Q_h - Q_k - Q_{rk}) + c_p T S_k \tag{8.22}$$

Gas species

$$\frac{\partial}{\partial t}(\rho Y_s) + \frac{\partial}{\partial x_j}(\rho v_j Y_s) = \frac{\partial}{\partial x_j}\left(\frac{\mu_e}{\sigma_Y}\frac{\partial Y_s}{\partial x_j}\right) - w_s + \alpha_s S \tag{8.23}$$

Obviously, Eqs. (8. 17) — (8. 23) have the same form as that of the instantaneous equations (6. 43), (6. 44) (6. 45) (6. 46) (6. 49), (6. 50), (6. 52). Compared with the equations of the single-fluid model, there is particle drag term in the gas-phase momentum equation and there are particle momentum and energy equations in the particle trajectory model. Besides, the particle mass source terms may be different based on different ways to identify the particle groups. Compared with the equations of the small-slip model, there are no particle diffusion, particle viscosity and particle bulk thermo-conductivity in the deterministic particle trajectory model.

Different approaches exist in treating the energy distribution between two phases. One approach is to assume all of the heating effects added to the particle phase and the gas phase heated by the particle phase through convective heat transfer between two phases. This approach will result in unbelievably high particle temperature. The other is to assume all of the heating effects added to the gas phase and the particle phase heated by the gas phase through convective heat transfer between two phases. This second approach will result in unbelievably high gas temperature. Experiments have shown that in the case of stable combustion the temperature of several combusting particles or of a single combusting particle is either slightly higher, or slightly lower than that of the gas temperature. Therefore, the approach taken in Eqs. (8. 21) and (8. 22) is that the heat of gas-phase reaction is added to the gas phase, while the heat of evaporation, condensation, devolatilization or surface reaction is added to the particle phase.

The gas-phase equations can still be expressed in the generalized form Eq. (8. 9), and the meaning of φ, Γ_φ, and S_φ is the same as that given in Table 8. 1. However, here $S_{p\varphi}$ is different, and $S_{p\varphi}$ will be

$$S_{pm} = S = -\sum n_k \dot{m}_k$$

$$S_{pvi} = \sum \rho_k (v_{ki} - v_i)/\tau_{rk} + v_i S$$

$$S_Y = \alpha_s S$$

$$S_h = \sum n_k Q_k + c_p T S$$

for gas continuity, gas momentum, gas species and gas energy equations, and $S_{p\varphi}$ will be zero for other gas-phase equations.

Owing to no particle diffusion terms in the deterministic particle trajectory model, the time-averaged particle-phase equations, as their instantaneous equations, can be transformed into the equations in Lagrangian coordinate

Particle continuity

$$N_k = \int_A n_k v_{kn} dA = \text{const} \tag{8.24}$$

Particle momentum

$$\frac{dv_{ki}}{dt} = (v_i - v_{ki})/\tau_{rk} + g_i + \frac{F_{M,ki}}{n_k m_k} + (v_i - v_{ki})\dot{m}_k/m_k \tag{8.25}$$

Particle energy

$$\frac{dT_k}{dt} = [Q_h - Q_k - Q_{rk} + \dot{m}_k(c_p T - c_k T_k)]/(m_k c_k) \tag{8.26}$$

Eqs. (8. 24), (8. 25) and (8. 26) have the same form as that of Eqs. (6. 55) — (6. 57), but now Eqs. (8. 24) — (8. 26) are time-averaged particle-phase equations. The advantage of the particle trajectory model is that the treatment of particle phase may be directly based on ordinary differential equations (8. 24) — (8. 26), but not necessarily on partial differential equations (8. 18), (8. 20) and (8. 22).

8. 3. 3 Modifications for Particle Turbulent Diffusion

In the original deterministic particle trajectory model it is assumed that the total particle number flux keeps constant along the trajectories with no particle turbulent diffusion. In fact, experiments have shown that in many cases the particle turbulent diffusion cannot be neglected. For example, in numerical modeling of sudden-expansion gas-particle flows based on the deterministic trajectory model, the prediction results give the particle trajectories concentrated near the axis, but actually particles disperse almost throughout the whole flow field. Some modifications are made in the deterministic particle trajectory model to account for the particle turbulent diffusion. The simplest method is to consider the change in trajectory by particle diffusion, introducing the concept of "particle drift velocity" or "particle drift force", proposed by Smoot et al. and Lockwood et al.. It can be considered that the particle velocity consists of two parts, i. e.

$$v_{kj} = v_{kc,j} + v_{kd,j} \tag{8.27}$$

where $v_{kc,j}$ is the particle convective velocity determined by the time-averaged momentum equation (8. 25), and $v_{kd,j}$ is the particle diffusive drift velocity determined by the Fick's type diffusion law

$$-\rho_k v_{kd,j} = -n_k m_k v_{kd,j} = D_k m_k \frac{\partial n_k}{\partial x_j}$$

or

$$v_{kd,j} = -\frac{D_k}{n_k}\frac{\partial n_k}{\partial x_j} \tag{8.28}$$

Actually, it is worthwhile to note that this modification is based on the concept of pseudo-fluid model of particle phase in Eulerian coordinate, which will be discussed later. It has been mentioned earlier in Sec. 6. 5. 1 that in the small-slip model the velocity slip is considered as a result of diffusion drift. In the full pseudo-fluid model of particle phase, diffusion is not related to slip.

Consider the time-averaged particle continuity equation (8. 3)

$$\frac{\partial n_k}{\partial t} + \frac{\partial}{\partial x_j}(n_k v_{kj}) = -\frac{\partial}{\partial x_j}\overline{(n_k' v_{kj}')}$$

If we take a gradient modeling as

$$-\overline{n_k' v_{kj}'} = -n_k v_{kd,j} = D_k \frac{\partial n_k}{\partial x_j}$$

or

$$v_{kd,j} = -\frac{D_k}{n_k} \frac{\partial n_k}{\partial x_j}$$

and putting

$$v_{kj,0} = v_{kj} + v_{kd,j}$$

as a "total" particle velocity, then we have

$$\frac{\partial n_k}{\partial t} + \frac{\partial}{\partial x_j}(n_k v_{kj}) = \frac{\partial}{\partial x_j}\left(D_k \frac{\partial n_k}{\partial x_j} \right) \tag{8.29}$$

and

$$\frac{\partial n_k}{\partial t} + \frac{\partial}{\partial x_j}(n_k v_{kj,0}) = 0 \tag{8.30}$$

or

$$N_{k0} = \int_A n_k v_{kn,0} \, dA = \text{const} \tag{8.31}$$

Equations (8. 29) and (8. 30) imply that the concept of "drift velocity" is based just on the diffusion term of particle continuity equation in the pseudo-fluid model of particle phase. However, this concept has not taken the diffusion term of particle momentum equation into account.

To determine $v_{kd,j}$, we need to know the particle diffusivity D_k, which is taken by using the Hinze-Tchen formula

$$\nu_k / \nu_T = (1 + \tau_{rk}' / \tau_T)^{-1} \tag{8.32}$$

and

$$D_k = \nu_k / \sigma_{kp}, \quad \sigma_{kp} = 0.35 \sim 0.7$$

where ν_T is the gas turbulent viscosity, $\nu_T = c_\mu k^2 / \varepsilon$, τ_T is the gas fluctuation time

$$\tau_r = l/u' = \sqrt{3/2}\, c_\mu^{3/4}\, k/\varepsilon \tag{8.33}$$

τ_{rk}' is the particle relaxation time, determined by the Stokes drag

$$\tau_{rk}' = \overline{\rho_p} d_k^2 / (18\mu) \tag{8.34}$$

Alternatively, another modification is to introduce the concept of "drift force". The time-averaged particle momentum equation is modified by adding to it a "drift force" term:

$$\frac{dv_{ki}}{dt} = (v_i - v_{ki}) / \tau_{rk} + g_i + F_{TDi} \tag{8.35}$$

where the drift force is determined by

$$F_{TDi} = v_{kd,i} / \tau_{rk} \tag{8.36}$$

and $v_{kd,i}$ is determined by

$$v_{kd,i} = -\frac{\nu_k}{\sigma_k n_k} \frac{\partial n_k}{\partial x_i} \tag{8.37}$$

The remaining problem is to find the particle number density gradient $\partial n_k / \partial x_i$, However, the particle number density cannot be predicted by the deterministic particle trajectory model. An approximate treatment is to use the no-slip model, solving the equation

$$\frac{\partial n_k}{\partial t} + \frac{\partial}{\partial x_j}(n_k v_j) = \frac{\partial}{\partial x_j}\left(\frac{\gamma_e}{\sigma_{kp}} \frac{\partial n_k}{\partial x_j} \right) \tag{8.38}$$

to find the particle number density and its gradient, where σ_{kp} is taken as 0.35.

8.3.4 Stochastic Particle Trajectory Model

The above-described modification is only a simple method to account for the particle diffusion. It can give only the modified trajectories, but cannot give the particle velocity and concentration field. Furthermore, in this modification it is still needed to introduce the concept of concentration gradient and particle diffusivity from the pseudo-fluid model. In recent years many investigators used the stochastic trajectory model to account for particle diffusion. Computation of the stochastic particle trajectories is based on the instantaneous particle momentum equations

$$\frac{du_k}{dt} = (\bar{u} + u' - u_k)/\tau_{rk} \tag{8.39}$$

$$\frac{dv_k}{dt} = (\bar{v} + v' - v_k)/\tau_{rk} + w_k^2/r_k + g \tag{8.40}$$

$$\frac{dw_k}{dt} = (\bar{w} + w' - w_k)/\tau_{rk} - v_k w_k/r_k \tag{8.41}$$

where u_k, v_k, w_k are instantaneous particle axial, radial and tangential velocity respectively, $\bar{u}$, $\bar{v}$, $\bar{w}$ and u', v', w' are time-averaged and fluctuation gas velocity in these three directions respectively. Assume isotropic and local homogeneous turbulence and Gaussian PDF of stochastic velocity distribution, and take the stochastic examples of gas velocity

$$u' = \zeta(\overline{u'^2})^{\frac{1}{2}},\ v' = \zeta(\overline{v'^2})^{\frac{1}{2}},\ w' = \zeta(\overline{w'^2})^{\frac{1}{2}} \quad (\zeta = 0, 1, 2, 3, \cdots)$$

where

$$(\overline{u'^2})^{\frac{1}{2}} = (\overline{v'^2})^{\frac{1}{2}} = (\overline{w'^2})^{\frac{1}{2}} = \left(\frac{2}{3}k\right)^{\frac{1}{2}}$$

and ζ is a random number. Substitute the random u, v and w into Eqs. (8.39) — (8.41) to calculate u_k, v_k and w_k and

$$x_k = \int u_k dt, \quad r_k = \int v_k dt, \quad \theta_k = \int (w_k/r_k)\, dt$$

The interaction time of particles with random eddies may be taken as:

$$\tau_{\text{int}} = \min\left[\tau_{rk}, \tau_T\right] \tag{8.42}$$

The stochastic trajectories are calculated by using the Monte-Carlo method. The computation is time consuming, because it is needed to calculate several thousands, even tens of thousands of trajectories.

8.3.5 Solution Procedure

Taking an example of axisymmetric recirculating and swirling turbulent gas-particle flows with reacting particles, the generalized gas-phase conservation equations in cylindrical coordinates can be taken in the form

$$\frac{\partial}{\partial x}(\rho u \varphi) + \frac{1}{r}\frac{\partial}{\partial r}(r \rho v \varphi) = \frac{\partial}{\partial x}\left(\Gamma_\varphi \frac{\partial \varphi}{\partial x}\right) + \frac{1}{r}\frac{\partial}{\partial r}\left(r\Gamma_\varphi \frac{\partial \varphi}{\partial r}\right) + S_\varphi + S_{p\varphi} \tag{8.43}$$

where φ, Γ_φ, S_φ and $S_{p\varphi}$ are given in Table 8.2.

Table 8. 2 Gas-Phase Equations for Particle Trajectory Model

Equation	φ	Γ_φ	S_φ	$S_{p\varphi}$
Continuity	1	0		$S=-\sum n_k \dot{m}_k$
Axial momentum	u	μ_e	$-\frac{\partial p}{\partial x}+\frac{\partial}{\partial x}\left(\mu_e \frac{\partial u}{\partial x}\right)+\frac{1}{r}\frac{\partial}{\partial r}\left(r\mu_e \frac{\partial v}{\partial x}\right)$	$\sum_k \frac{\rho_k}{\tau_{rk}}(u_k - u)+uS$
Radial momentum	v	μ_e	$-\frac{\partial p}{\partial r}+\frac{\partial}{\partial x}\left(\mu_e \frac{\partial u}{\partial r}\right)+\frac{1}{r}\frac{\partial}{\partial r}\left(r\mu_e \frac{\partial v}{\partial r}\right)-2\mu_e \frac{v}{r^2}+\rho\frac{w^2}{r}$	$\sum_k \frac{\rho_k}{\tau_{rk}}(v_k - v)+vS$
Tangential momentum	w	μ_e	$-\frac{\rho v w}{r}-\mu_e\frac{w}{r^2}-\frac{w}{r}\frac{\partial \mu_e}{\partial r}$	$\sum_k \frac{\rho_k}{\tau_{rk}}(w_k - w)+wS$
Turbulent kinetic energy	k	$\frac{\mu_e}{\sigma_k}$	$G_k-\rho\varepsilon$	0
TKE dissipation rate	ε	$\frac{\mu_e}{\sigma_\varepsilon}$	$\frac{\varepsilon}{k}(c_1 G_k - c_2\rho\varepsilon)$	0
Mixture fraction	f	$\frac{\mu_e}{\sigma_f}$	0	0
f-fluctuation	g_f	$\frac{\mu_e}{\sigma_g}$	$c_{g1}\mu_e\left[\left(\frac{\partial f}{\partial x}\right)^2+\left(\frac{\partial f}{\partial r}\right)^2\right]-\left(c_{g2}\rho\varepsilon\frac{g_f}{k}\right)$	0
Mixture fraction of particle-off gases	η	$\frac{\mu_e}{\sigma_\eta}$	0	S
η-fluctuation	g_η	$\frac{\mu_e}{\sigma_g}$	$c_{g1}\mu_e\left[\left(\frac{\partial \eta}{\partial x}\right)^2+\left(\frac{\partial \eta}{\partial r}\right)^2\right]-c_{g2}\rho\varepsilon\frac{g_\eta}{k}$	0
Gas species	Y_s	$\frac{\mu_e}{\sigma_Y}$	$-w_s$	$\alpha_s S$
Gas energy	h	$\frac{\mu_e}{\sigma_h}$	$-q_r$	$\sum_k n_k Q_k + c_p T S$

In Table 8. 2,

$$G_k=\mu_e\left[2\left(\frac{\partial u}{\partial x}\right)^2+2\left(\frac{\partial v}{\partial r}\right)^2+\left(\frac{\partial u}{\partial r}+\frac{\partial v}{\partial x}\right)^2\right]$$

The universal constants are: $c_\mu=0.09$, $c_1=1.44$, $c_2=1.92$, $c_{g1}=2.8$, $c_{g2}=2.0$, $\sigma_k=\sigma_f=\sigma_\eta=\sigma_g=\sigma_Y=\sigma_h=0.9$, $\sigma_\varepsilon=1.3$.

The gas flow field can be solved by using the SIMPLE algorithm, that is, the differential equations are integrated in the control volume to obtain the finite difference equations and then these FDE's are solved by using p-v corrections , TDMA line-by-line iterations and under-relaxations.

The particle-source term, for example, the mass source term in the gas-phase cells for FDE's can be written as:

$$- \Delta V \cdot S = - \Delta V (\sum_k n_k \dot{m}_k) = - \Delta V \left(\sum_k n_k \frac{\Delta m_k}{\Delta t} \right)$$

$$= - \sum_k \frac{\Delta V \cdot n_k}{\Delta t} \Delta m_k = \sum_{k,c} \sum_{j,c} N_{k,j} (m_{k,i} - m_{k,e}) \tag{8.44}$$

where $\sum_{k,c}$ expresses the summation of all size groups whose trajectories are penetrating through the considered cell, $\sum_{j,c}$ denotes the summation of different initial positions or initial directions for the same size group, $N_{k,j}$ is the total particle number flux of k-th group along the j-th trajectory. $N_{k,j}$ keeps constant along its trajectory. Similarly, the momentum and energy source terms can be written as:

$$\Delta V \cdot S_{pu} = \sum_{k,c} \sum_{j,c} N_{kj} (m_{ke} u_{ke} - m_{ki} u_{ki}) \tag{8.45}$$

$$\Delta V \cdot S_{pv} = \sum_{k,c} \sum_{j,c} N_{kj} (m_{ke} v_{ke} - m_{ki} v_{ki}) \tag{8.46}$$

$$\Delta V \cdot S_{pw} = \sum_{k,c} \sum_{j,c} N_{kj} (m_{ke} w_{ke} - m_{ki} w_{ki}) \tag{8.47}$$

$$\Delta V \cdot S_{ph} = \sum_{k,c} \sum_{j,c} N_{kj} [- Q_k + c_p T (m_{ke} - m_{ki})] \tag{8.48}$$

where subscripts i and e express the values corresponding to the entrance and exit of trajectories in the cell of gas phase, and

$$Q_k = \frac{\pi}{2} (d_{ke} + d_{ki}) \mathrm{Nu} \lambda (T - T_k) \ln(1 + B) / B \tag{8.49}$$

For the boundary conditions, besides the gas-phase boundary conditions at the entrance, the exit, the wall and symmetric axis, it is necessary to specify the inlet particle concentration distribution, the initial size distribution, the initial particle velocity distribution, initial location of particles, and the condition of particles impacting on the wall. The gas-phase pressure-correction equation must be based on the gas continuity equation with particle source term.

To calculate particle trajectories and particle history along these trajectories, it is necessary to select appropriate integration method and to treat the problem of determining the location of intersection of trajectories with cell walls of gas phase. We should determine, with which wall the trajectory will intersect, when leaving this cell, in order to promptly stop the trajectory calculation in this cell. Some investigators solved the particle momentum equation, using the Rung-Kutta method, to find the trajectories according to

$$x_k = \int_0^t u_k dt, \quad r_k = \int_0^t v_k dt, \quad \theta_k = \int_0^t (w_k / r_k) dt$$

For selecting Δt, first take

$$\Delta t^{(1)} = \min \left[\frac{\Delta x_i}{u_{ki}}, \frac{r_e - r_i}{v_{ki}} \right]$$

and calculate x_k and r_k in this $\Delta t^{(1)}$, using a fourth-order Rung-Kutta method. If the point (x_{ke}, r_{ke}) is not located in the expected area, as shown in Fig. 8. 2, then we should select

$$\Delta t^{(2)} = \min \left[\frac{\Delta x_c}{(u_{ki} - u_{ke})/2}, \frac{r_e - r_i}{(v_{ki} - v_{ke})/2} \right]$$

and repeat the trajectory calculation until (x_{ke}, r_{ke}) falls into the shadowed area. The method proposed by C. T. Crowe is to assume v_i = const and τ_{rk} = const in the cell, which gives an analytical solution

$$u_{ke} = u - (u - u_{ki}) \exp(-\Delta t/\tau_{rk})$$
$$v_{ke} = v - (v - v_{ki}) \exp(-\Delta t/\tau_{rk})$$

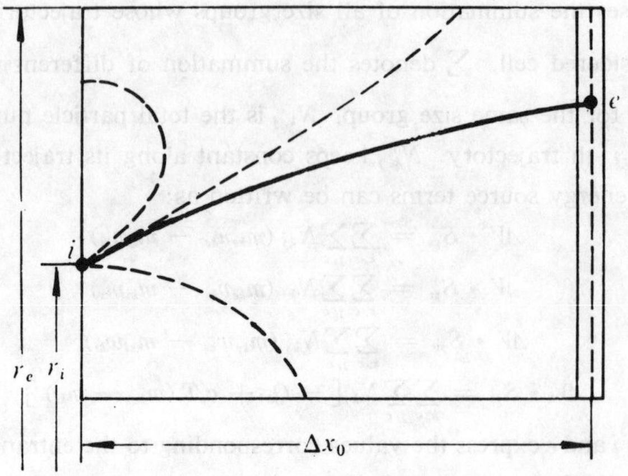

Fig. 8. 2

Then, by using a linear approximation, the trajectory is calculated as

$$x_{ke} = x_{ki} + 0.5(u_{ki} + u_{ke})\Delta t$$
$$r_{ke} = r_{ki} + 0.5(v_{ki} + v_{ke})\Delta t$$

Considering that actually the particle mass, diameter, drag coefficient and relative velocity are changing along its traectory, the following analytical solution can be used

$$\Delta x = x - x_i = t_1[uf_1(\bar{d}) - (u - u_i)f_2(\bar{d})]$$
$$\Delta r = r - r_i = t_1[vf_1(\bar{d}) - (v - v_i)f_2(\bar{d})]$$

where x_i and r_i express the initial position of the trajectory in a cell (Fig. 8. 3), and

$$f_1(\bar{d}) = \int_{\bar{d}}^{1} \frac{\bar{d}}{1 + \Pi \bar{d}^{0.5(m+1)}} d(\bar{d}), \quad f_2(\bar{d}) = \int_{\bar{d}}^{1} \frac{\bar{d}^{(m+1)}}{1 + \Pi \bar{d}^{0.5(m+1)}} d(\bar{d})$$

$$t_1 = \rho_p d_{ki}^2 / [4D\rho \ln(1 + B)]$$

or $$t_1 = \rho_p d_{ki}^2 c_p / [4\lambda \ln(1 + B)]$$

$$\bar{d} = d_k/d_{ki}, \quad m = 3.24/B, \quad \Pi = 0.25(\mathrm{Re}_{ki})^{1/2}$$

For given x_i, r_i, u_{ki}, v_{ki} and d_{ki} in a selected cell, first take a $\bar{d}^{(1)}$ near to unity and calculate the integrals $f_1(\bar{d})$ and $f_2(\bar{d})$, then compare the calculated $\Delta x^{(1)}$ and $\Delta r^{(1)}$ with the size of the cell Δx_g and Δr_g . If $\Delta x^{(1)} < \Delta x_g$ and $\Delta r^{(1)} < \Delta r_g$, we can take $\bar{d}^{(2)}$ smaller than $\bar{d}^{(1)}$ and calculate $\Delta x^{(2)}$ and $\Delta r^{(2)}$, till $\Delta x^{(n)}$ first approaches Δx_g or $\Delta r^{(n)}$ first approaches Δr_g . Thus, the final position of the trajectory in this cell (x_e, r_e) is determined.

The gas-particle coupling method proposed by C. T. Crowe is called PSIC method (Particle Source in Cell Method), whose flow chart is shown in Fig. 8. 4.

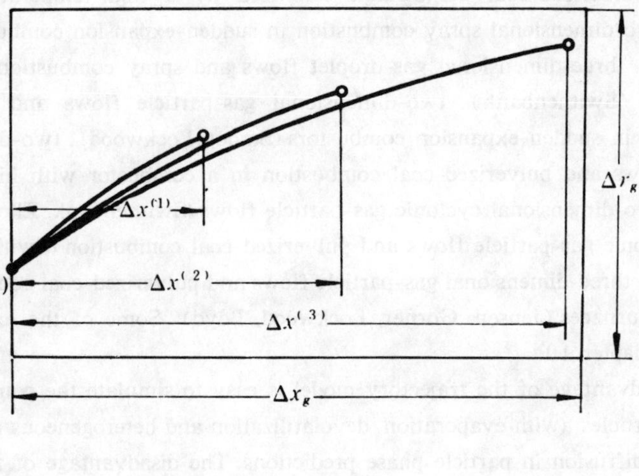

Fig. 8. 3

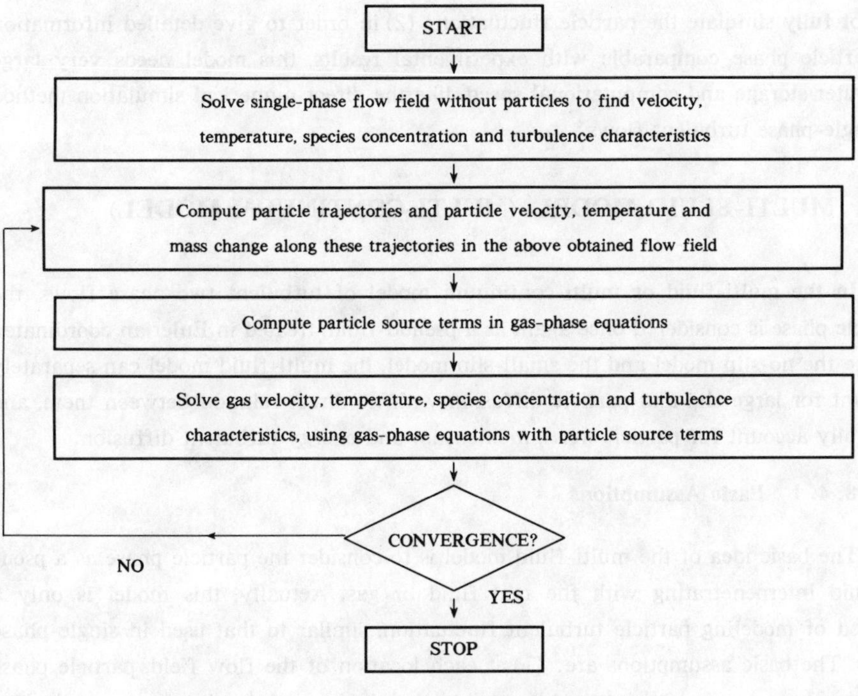

Fig. 8. 4 PSIC Flow Chart

8. 3. 6 Model Application and Assessment

Up to now, the particle trajectory model is most widely used in simulating turbulent gas-particle flows and combustion. It has been used to predict gas-droplet flows in the recirculation zone behind a baffle (Crowe), spray cooling (Crowe), gas-droplet flows with evaporating spray in a pipe and in an afterburner diffusor of turbo-fan engines

(Zhou), pulverized-coal combustion stabilized by a high-temperature plasma jet (Zhou), two-dimensional spray combustion in sudden-expansion combustors (Gosman, Whitelaw), three-dimensional gas-droplet flows and spray combustion in gas turbine combustors (Swithenbank), two-dimensional gas-particle flows and pulverized-coal combustion in sudden-expansion combustors (Smoot, Lockwood), two-dimensional gas-particle flows and pulverized-coal combustion in a combustor with high-velocity jets (Zhou), two-dimensional cyclonic gas-particle flows (Swithenbank, Zhou), two-dimensional cyclonic gas-particle flows and pulverized-coal combustion (Swithenbank, Zhang and Zhou), three-dimensional gas-particle flows and pulverized-coal combustion in utility boiler furnaces (Jensen, Gorner, Lockwood, Boyd). Some of the examples will be given in Chapter 10.

The advantage of the trajectory model is easy to simulate the complex history of reacting particles (with evaporation, devolatilization and heterogeneous reaction) and no numerical diffusion in particle-phase predictions. The disadvantage of the particle trajectory model is that: (1) even the stochastic trajectory model, based on the Monte-Carlo method, does not satisfy the continuity equation for the fluctuational component, so, it cannot fully simulate the particle fluctuation; (2) in order to give detailed information of particle phase comparable with experimental results, this model needs very large computer storage and computational speed, like the direct numerical simulation method in single-phase turbulent flows.

8.4 MULTI-FLUID MODEL (MULTI-CONTINUUM MODEL)

In the multi-fluid or multi-continuum model of turbulent two-phase flows, the particle phase is considered once again as a pseudo-fluid, treated in Eulerian coordinate. Unlike the no-slip model and the small-slip model, the multi-fluid model can separately account for large slip and particle diffusion with no direct relation between them, and can fully account for particle mass, momentum and energy turbulent diffusion.

8.4.1 Basic Assumptions

The basic idea of the multi-fluid model is to consider the particle phase as a pseudo-fluid interpenetrating with the real fluid or gas. Actually, this model is only a method of modeling particle turbulent fluctuation, similar to that used in single-phase flows. The basic assumptions are: (1) at each location of the flow field, particle phase and gas phase coexist and interpenetrate into each other, each having its own velocity, temperature and volume fraction, however, each size group has the same velocity and temperature; (2) each particle phase (size group) has continuous velocity, temperature and volume fraction distribution in space, and the condition of this approximation has been discussed in Sec. 6.3.1.; (3) each particle phase, besides its mass, momentum and energy interaction with gas phase, has its own turbulent fluctuation resulting in particle turbulent transport of mass, momentum and energy, and this particle fluctuation is determined by convection, diffusion, production and the interaction with gas-phase turbulence; (4) it is appropriate to identify the particle group (phase) by its initial size distribution; and (5) for dense particulate suspensions the particle-particle collision will cause additional particle viscosity, diffusion and heat conduction.

8. 4. 2 Time-averaged Equations

For the multi-fluid model, the starting point is the time-averaged equations (8. 1) — (8. 8). For simplicity we can neglect the gas density fluctuation, the fluctuation of particle mass changing rate, the fluctuation of drag-force term and the transient correlation term. In this case, the time-averaged equations can be written as: Gas continuity

$$\frac{\partial \rho}{\partial t} + \frac{\partial}{\partial x_j}(\rho v_j) = S \tag{8. 50}$$

Particle continuity

$$\frac{\partial \rho_k}{\partial t} + \frac{\partial}{\partial x_j}(\rho_k v_{kj}) = -\frac{\partial}{\partial x_j}(\overline{\rho'_k v'_{kj}}) + n_k \dot{m}_k \tag{8. 51}$$

$$\frac{\partial n_k}{\partial t} + \frac{\partial}{\partial x_j}(n_k v_{kj}) = -\frac{\partial}{\partial x_j}(\overline{n'_k v'_{kj}}) \tag{8. 52}$$

Gas momentum

$$\frac{\partial}{\partial t}(\rho v_i) + \frac{\partial}{\partial x_j}(\rho v_j v_i) = -\frac{\partial p}{\partial x_i} + \frac{\partial}{\partial x_j}\left[\mu\left(\frac{\partial v_j}{\partial x_i} + \frac{\partial v_i}{\partial x_j}\right)\right] + \rho g_i + v_i S$$

$$+ F_{Mi} + \sum_k \rho_k (v_{ki} - v_i)/\tau_{rk} - \frac{\partial}{\partial x_j}(\rho \overline{v'_j v'_i}) \tag{8. 53}$$

Particle momentum

$$\frac{\partial}{\partial t}(n_k v_{ki}) + \frac{\partial}{\partial x_j}(n_k v_{kj} v_{ki}) = n_k g_i + n_k (v_i - v_{ki})\left(\frac{1}{\tau_{rk}} + \frac{\dot{m}_k}{m_k}\right)$$

$$+ F_{k,Mi}/m_k - \frac{\partial}{\partial x_j}(n_k \overline{v'_{kj} v'_{ki}} + v_{kj}\overline{n'_k v'_{ki}} + v_{ki}\overline{n'_k v'_{kj}}) \tag{8. 54}$$

Gas energy

$$\frac{\partial}{\partial t}(\rho h) + \frac{\partial}{\partial x_j}(\rho v_j h) = \frac{\partial}{\partial x_j}\left(\lambda \frac{\partial T}{\partial x_j}\right) - q_r + \sum n_k Q_k + hS - \frac{\partial}{\partial x_j}(\rho \overline{v'_j h'}) \tag{8. 55}$$

Particle energy

$$\frac{\partial}{\partial t}(n_k c_k T_k) + \frac{\partial}{\partial x_j}(n_k v_{kj} c_k T_k) = n_k (Q_h - Q_k - Q_{rk})/m_k$$

$$+ (c_p T - c_k T_k)\frac{n_k \dot{m}_k}{m_k} - \frac{\partial}{\partial x_j}(n_k c_k \overline{v'_{kj} T'_k} + c_k v_{kj}\overline{n'_k T'_k} + c_k T_k \overline{n'_k v'_{kj}}) \tag{8. 56}$$

Gas species

$$\frac{\partial}{\partial t}(\rho Y_s) + \frac{\partial}{\partial x_j}(\rho v_j Y_s) = \frac{\partial}{\partial x_j}\left(D\rho \frac{\partial Y_s}{\partial x_j}\right) - w_s + \alpha_s S - \frac{\partial}{\partial x_j}(\rho \overline{Y'_s v'_j}) \tag{8. 57}$$

8. 4. 3 Unified Second-order Moment Closure Model

Equations (8. 50) to (8. 57) do not constitute a closed system because of unkown correlations $\overline{v'_j v'_i}$, $\overline{v'_j T'}$, $\overline{v'_j Y'_s}$, $\overline{\rho'_k v'_{kj}}$, $\overline{n'_k v'_{kj}}$, $\overline{n'_k v'_{ki}}$, $\overline{v'_{kj} v'_{ki}}$, $\overline{v'_{kj} T'_k}$, $\overline{n'_k T'_k}$. Besides, it is still necessary to know the time-averaged reaction rate w_s, which will be determined by the interaction of chemical kinetics with turbulence, and will be discussed in Chapter 9.

The unified Second-order moment closure model is an extension of the second-order moment closure model for single-phase flows to that for gas-particle flows. It can be shown that the Reynolds stress transport equation for the gas phase in gas-particle flows is

$$\frac{\partial}{\partial t}(\rho\,\overline{v'_jv'_i}) + \frac{\partial}{\partial x_k}(\rho v_k\,\overline{v'_jv'_i}) = D_{ij} + P_{ij} + G_{ij} + \Pi_{ij} - \rho\varepsilon_{ij} + G_{p,ij} + G_{R,ij} \tag{8.58}$$

where D_{ij}, P_{ij}, G_{ij}, Π_{ij} and ε_{ij} are the same as those for single-phase flows, and

$$G_{p,ij} = \sum_p \frac{m_p}{\tau_{rp}}\left[n_p\left(c_p^k\sqrt{kk_p}\,\delta_{ij} - 2\,\overline{v'_iv'_j}\right) + (v_{pi} - v_i)\,\overline{n'_pv'_{pj}} + (v_{pj} - v_j)\,\overline{n'_pv'_{pi}}\right]$$

$$G_{R,ij} = \overline{v'_jv'_iS}$$

are the stress production due to particle-gas interaction. Consider the instantaneous particle momentum equations for v_{pi} and v_{pj} in the form:

$$\frac{\partial v_{pi}}{\partial t} + v_{pk}\frac{\partial v_{pi}}{\partial x_k} = (v_i - v_{pi})\left(\frac{1}{\tau_r} + \frac{\dot{m}_p}{m_p}\right) + g_i$$

$$\frac{\partial v_{pj}}{\partial t} + v_{pk}\frac{\partial v_{pj}}{\partial x_k} = (v_j - v_{pj})\left(\frac{1}{\tau_r} + \frac{\dot{m}_p}{m_p}\right) + g_j$$

and take the similar procedure as that used in single-phase flows, the Reynolds stress transport equation for particle phase can be obtained as:

$$\begin{aligned}\frac{\partial}{\partial t}(\overline{v'_{pj}v'_{pi}}) + v_{pk}\frac{\partial}{\partial x_k}(\overline{v'_{pj}v'_{pi}}) &= -\frac{\partial}{\partial x_k}(\overline{v'_{pk}v'_{pj}v'_{pi}}) \\ &- \left(\overline{v'_{pj}v'_{pk}}\frac{\partial v_{pi}}{\partial x_k} + \overline{v'_{pi}v'_{pk}}\frac{\partial v_{pj}}{\partial x_k}\right) + \left(\frac{1}{\tau_r} + \frac{\dot{m}_p}{m_p}\right) \\ &\cdot\left(\overline{v'_iv'_{pj}} + \overline{v'_jv'_{pi}} - 2\,\overline{v'_{pj}v'_{pi}}\right)\end{aligned} \tag{8.59}$$

To avoid too large computer capacity and CPU time, Eqs. (8.58) and (8.59) can be simplified to algebraic expressions by dropping the convection and diffusion terms, while retaining the turbulent kinetic energy equations and gas-phase TKE dissipation rate equation with tensor transport coefficients, similar to that we have done for single-phase flows. The obtained algebraic expressions and transport equations are:

$$\begin{aligned}\overline{v'_jv'_i} &= (1-\lambda)\frac{2}{3}k\delta_{ij} - \lambda\frac{k}{\varepsilon}\left[\overline{v'_iv'_k}\frac{\partial v_j}{\partial x_k} + \overline{v'_jv'_k}\frac{\partial v_i}{\partial x_k} + \beta\left(g_j\,\overline{v'_iT'} + g_i\,\overline{v'_jT'}\right)\right] \\ &+ \frac{k}{c_1\rho\varepsilon}\left\{\sum_p\frac{m_p}{\tau_{rp}}\left[n_p\left(c_p^k\sqrt{kk_p}\,\delta_{ij} - 2\,\overline{v'_iv'_j}\right) + (v_{pi} - v_i)\,\overline{n'_pv'_{pj}}\right.\right. \\ &\left.\left. + (v_{pi} - v_j)\,\overline{n'_pv'_{pi}} + \overline{v'_jv'_i}\cdot 2S\right\}\right.\end{aligned} \tag{8.60}$$

$$\begin{aligned}\overline{v'_jT'} &= \frac{k}{c_{1T}\varepsilon}\left[\overline{v'_jv'_k}\frac{\partial T}{\partial x_k} + (1 - c_{2T})\,\overline{v'_kT'}\frac{\partial v_j}{\partial x_k} + (1 - c_{3T})\,\beta g_j\,\overline{T'^2}\right. \\ &+ v_j\sum_p A\,\overline{n'_pT'_p} + h\sum_p\frac{m_p}{\tau_{rp}}\,\overline{n'_pv'_{pj}} \\ &\left.- \overline{v'_jT'}\left(c_p\sum\frac{m_pn_p}{\tau_{rp}} + \sum_p n_pA + 2c_p\sum n_p\dot{m}_p\right)\right]\end{aligned} \tag{8.61}$$

$$\overline{T'^2} = -c_T\frac{k}{\varepsilon}\overline{v'_kT'}\frac{\partial T}{\partial x_k} \tag{8.62}$$

$$\overline{Y'_sv'_j} = \frac{k}{c_{1Y}\varepsilon}\left[\overline{v'_jv'_k}\frac{\partial Y_s}{\partial x_k} + (1 - c_{2Y})\,\overline{v'_kY'_s}\frac{\partial v_j}{\partial x_k} + (1 - c_{3Y})\,\beta g_j\,\overline{Y'^2_s}\right.$$

$$+ Y_s \sum_p \frac{m_p}{\tau_{rp}} \overline{n'_p v'_{pj}} - \overline{Y'_s v'_j} \left(\sum_p \frac{m_p n_p}{\tau_{rp}} + \sum_p n_p \dot{m}_p \right)] \tag{8.63}$$

$$\overline{Y'^2_s} = -2c_Y \frac{k}{\varepsilon} \overline{Y'_s v'_k} \frac{\partial Y_s}{\partial x_k} \tag{8.64}$$

$$\frac{\partial}{\partial t}(\rho k) + \frac{\partial}{\partial x_k}(\rho v_k k) = \frac{\partial}{\partial x_k}\left(\rho c_s \frac{k}{\varepsilon} \overline{v'_k v'_l} \frac{\partial k}{\partial x_l} \right) + G_k + G_b + G_p + G_R - \rho\varepsilon \tag{8.65}$$

$$\frac{\partial}{\partial t}(\rho \varepsilon) + \frac{\partial}{\partial x_k}(\rho v_k \varepsilon) = \frac{\partial}{\partial x_k}\left(\rho c_\varepsilon \frac{k}{\varepsilon} \overline{v'_k v'_l} \frac{\partial \varepsilon}{\partial x_l} \right) + \frac{\varepsilon}{k}[c_{\varepsilon 1}(G_k + G_b + G_p + G_R) - c_{\varepsilon 2}\rho\varepsilon] \tag{8.66}$$

where $G_k = -\rho \overline{v'_i v'_k} \frac{\partial v_i}{\partial x_k}, \quad G_b = \beta \rho g_i \overline{v'_i T'}$

$$G_p = \sum_p \sum_i \frac{\rho_p}{\tau_{rp}} \left[(c_p^k \sqrt{k k_p} - 2k) + (v_{pi} - v_i) \overline{n'_p v'_{pi}} / n_p \right]$$

$$G_R = -2k\left(\sum_p n_p \dot{m}_p \right) = 2kS$$

$$A = \frac{Q_p}{(T_p - T)} = \pi d_p \mathrm{Nu} \lambda \ln(1 + B)/B = c_p \dot{m}_p / B$$

$$\overline{v_{pj}' v_{pi}'} = c_p^k \sqrt{k k_p} \delta_{ij} - \frac{\tau_{rp} m_p}{2(m_p + \dot{m}_p \tau_{rp})} \left(\overline{v_{pj}' v_{pk}'} \frac{\partial v_{pi}}{\partial x_k} + \overline{v_{pj}' v_{pk}'} \frac{\partial v_{pj}}{\partial x_k} \right) \tag{8.67}$$

$$\overline{v_{pj}' T_p'} = c_{p1} \frac{\tau_{rp} m_p}{m_p + \dot{m}_p \tau_{rp}} \left[\overline{v_{pj}' v_{pk}'} \frac{\partial T_p}{\partial x_k} + (1 - c_{p2}) \overline{v_{pk}' T_p'} \frac{\partial v_{pj}}{\partial x_k} \right] \tag{8.68}$$

$$\overline{n_p' v_{pj}'} = c_{n1} \frac{\tau_{rp} m_p}{m_p + \dot{m}_p \tau_{rp}} \left(\overline{v_{pj}' v_{pk}'} \frac{\partial n_p}{\partial x_k} + c_{n2} \overline{n_p' v_{pk}'} \frac{\partial v_{pj}}{\partial x_k} \right) \tag{8.69}$$

$$\overline{n_p' v_{pi}'} = c_{n1} \frac{\tau_{rp} m_p}{m_p + \dot{m}_p \tau_{rp}} \left(\overline{v_{pi}' v_{pk}'} \frac{\partial n_p}{\partial x_k} + c_{n2} \overline{n_p' v_{pk}'} \frac{\partial v_{pi}}{\partial x_k} \right) \tag{8.70}$$

$$\overline{\rho_p' v_{pj}'} = m_p \overline{n_p' v_{pj}'} \tag{8.71}$$

$$\overline{n_p' T_p'} = c_{n3} \left(\frac{\tau_p m_p}{m_p + \dot{m}_p \tau_{rp}} \right)^2 k_p \left(\frac{\partial n_p}{\partial x_i} \right) \left(\frac{\partial T_p}{\partial x_i} \right) \tag{8.72}$$

$$\frac{\partial}{\partial t}(\rho_p k_p) + \frac{\partial}{\partial x_k}(\rho_p v_{pk} k_p) = \frac{\partial}{\partial x_k}\left(\rho_p c_{kp} \frac{k_p}{\varepsilon_p} \overline{v_{pk}' v_{pl}'} \frac{\partial k_p}{\partial x_l} \right) + G_{kp} - \rho_p \varepsilon_p \tag{8.73}$$

where $G_{kp} = -\rho_p \overline{v_{pi}' v_{pk}'} \frac{\partial v_{pi}}{\partial x_k}$

$$\varepsilon_p = -2\left(\frac{m_p + \dot{m}_p \tau_{rp}}{m_p \tau_{rp}} \right) \left[(c_p^k \sqrt{k k_p} - k_p) + \frac{1}{n_p}(v_i - v_{pi}) \overline{n_p' v_{pi}'} \right]$$

8.4.4 k-ε-k_p Model

For non-swirling and non-buoyant isotropic turbulent gas-particle flows, the unified second-order moment closure model can be reduced to a k-ε-k_p model. In this case, the algebraic expressions and transport equations for closure are:

$$-\rho \overline{v_j' v_i'} = \mu_T \left(\frac{\partial v_j}{\partial x_i} + \frac{\partial v_i}{\partial x_j} \right) + \frac{2}{3}\rho k \delta_{ij} \tag{8.74}$$

$$-\rho \overline{v_j' T'} = \frac{\mu_T}{\sigma_T} \frac{\partial T}{\partial x_j} \tag{8.75}$$

$$-\rho \overline{Y_s' v_j'} = \frac{\mu_T}{\sigma_Y} \frac{\partial Y_s}{\partial x_j} \tag{8.76}$$

$$\frac{\partial}{\partial t}(\rho k) + \frac{\partial}{\partial x_k}(\rho v_k k) = \frac{\partial}{\partial x_k}\left(\frac{\mu_e}{\sigma_k}\frac{\partial k}{\partial x_k}\right) + G_k + G_b + G_p + G_R - \rho\varepsilon \tag{8.77}$$

$$\frac{\partial}{\partial t}(\rho \varepsilon) + \frac{\partial}{\partial x_k}(\rho v_k \varepsilon) = \frac{\partial}{\partial x_k}\left(\frac{\mu_c}{\sigma_\varepsilon}\frac{\partial \varepsilon}{\partial x_k}\right) + \frac{\varepsilon}{k}\left[c_{\varepsilon 1}(G_k + G_b + G_p + G_R) - c_{\varepsilon 2}\rho\varepsilon\right] \tag{8.78}$$

$$-\overline{v_{pj}'v_{pi}'} = \nu_p\left(\frac{\partial v_{pj}}{\partial x_i}\frac{\partial v_{pi}}{\partial x_j}\right) \tag{8.79}$$

$$-\overline{v_{pj}'T_p'} = \frac{\nu_p}{\sigma_p}\frac{\partial T_p}{\partial x_j} \tag{8.80}$$

$$-\overline{n_p'v_{pj}'} = \frac{\nu_p}{\sigma_p}\frac{\partial n_p}{\partial x_j} \tag{8.81}$$

$$-\overline{n_p'v_{pi}'} = \frac{\nu_p}{\sigma_p}\frac{\partial n_p}{\partial x_i} \tag{8.82}$$

$$-\overline{\rho_p'v_{pj}'} = m_p\frac{\nu_p}{\sigma_p}\frac{\partial n_p}{\partial x_j} \tag{8.83}$$

$$\overline{n_p'T_p'} = c_{n3}\left(\frac{m_p\tau_{rp}}{m_p + \dot{m}_p\tau_{rp}}\right)^2 k_p\frac{\partial n_p}{\partial x_i}\frac{\partial T_p}{\partial x_i} \tag{8.84}$$

$$\frac{\partial}{\partial t}(\rho_p k_p) + \frac{\partial}{\partial x_k}(\rho_p v_{pk} k_p) = \frac{\partial}{\partial x_k}\left(\frac{\mu_p}{\sigma_p}\frac{\partial k_p}{\partial x_k}\right) + G_{kp} - \rho_p\varepsilon_p \tag{8.85}$$

where

$$\mu_e = \mu + \mu_T, \quad \mu_T = c_\mu\rho k^2/\varepsilon$$

$$\mu_p = \rho_p\nu_p = c_{\mu p}\rho_p k_p^2/|\varepsilon_p|$$

$$G_k = \mu_T\left(\frac{\partial v_i}{\partial x_k} + \frac{\partial v_k}{\partial x_i}\right)\frac{\partial v_i}{\partial x_k}, \quad G_b = -\beta g_k\frac{\mu_T}{\sigma_T}\frac{\partial T}{\partial x_k}$$

$$G_p = \sum_p\sum_i\frac{\rho_p}{\tau_{rp}}\left[(c_p^k\sqrt{kk_p} - 2k) - \frac{(v_{pi} - v_i)}{n_p}\frac{\nu_p}{\sigma_p}\frac{\partial n_p}{\partial x_i}\right]$$

$$G_R = 2kS$$

$$G_{kp} = \mu_p\left(\frac{\partial v_{pk}}{\partial x_i} + \frac{\partial v_{pi}}{\partial x_k}\right)\frac{\partial v_{pi}}{\partial x_k}$$

$$\varepsilon_p = -2\left(\frac{m_p + \dot{m}_p\tau_{rp}}{m_p\tau_{rp}}\right)\left[(c_p^k\sqrt{kk_p} - k_p) - \frac{(v_i - v_{pi})}{n_p}\frac{\nu_p}{\sigma_p}\frac{\partial n_p}{\partial x_i}\right]$$

For the k-ε-k_p model, the time-averaged equations of gas and particle phases become

$$\frac{\partial\rho}{\partial t} + \frac{\partial}{\partial x_j}(\rho v_j) = S \tag{8.50}$$

$$\frac{\partial n_p}{\partial t} + \frac{\partial}{\partial x_j}(n_p v_{pj}) = \frac{\partial}{\partial x_j}\left(\frac{\nu_p}{\sigma_p}\frac{\partial n_p}{\partial x_j}\right) \tag{8.86}$$

$$\frac{\partial\rho_p}{\partial t} + \frac{\partial}{\partial x_j}(\rho_p v_{pj}) = \frac{\partial}{\partial x_j}\left(m_p\frac{\nu_p}{\sigma_p}\frac{\partial n_p}{\partial x_j}\right) + n_p\dot{m}_p \tag{8.87}$$

$$\frac{\partial}{\partial t}(\rho v_i) + \frac{\partial}{\partial x_j}(\rho v_j v_i) = -\frac{\partial p}{\partial x_i} + \frac{\partial}{\partial x_j}\left[\mu_e\left(\frac{\partial v_j}{\partial x_i} + \frac{\partial v_i}{\partial x_j}\right)\right] + \rho g_i + v_i S + F_{Mi} + \sum\rho_p(v_{pi} - v_i)/\tau_{rp} \tag{8.88}$$

$$\frac{\partial}{\partial t}(n_p v_{pi}) + \frac{\partial}{\partial x_j}(n_p v_{pj} v_{pi}) = n_p g_i + (v_i - v_{pi})\left(\frac{1}{\tau_{rp}} + \frac{\dot{m}_p}{m_p}\right)n_p + \frac{\partial}{\partial x_j}\left[n_p\nu_p\left(\frac{\partial v_{pj}}{\partial x_i} + \frac{\partial v_{pi}}{\partial x_j}\right)\right] + \frac{\partial}{\partial x_j}\left[\frac{\nu_p}{\sigma_p}\left(v_{pj}\frac{\partial n_p}{\partial x_i} + v_{pi}\frac{\partial n_p}{\partial x_j}\right)\right] \tag{8.89}$$

$$\frac{\partial(\rho h)}{\partial t} + \frac{\partial}{\partial x_j}(\rho v_j h) = \frac{\partial}{\partial x_j}\left[\frac{\mu_e}{\sigma_h}\frac{\partial h}{\partial x_j}\right] - q_r + \sum_p n_p Q_p + hS \tag{8.90}$$

$$\frac{\partial}{\partial t}(n_p c T_p) + \frac{\partial}{\partial x_j}(n_p v_{pj} c T_p) = n_p (Q_h - Q_p - Q_{rp}) / m_p$$
$$+ (c_p T - c T_p) \frac{n_p \dot{m}_p}{m_p} + \frac{\partial}{\partial x_j}\left[n_p c \frac{\nu_p}{\sigma_p} \frac{\partial T_p}{\partial x_j} + c v_{pj} c_{n3} \left(\frac{m_p \tau_{rp}}{m_p + \dot{m}_p \tau_{rp}} \right)^2 \right.$$
$$\left. \times k_p \frac{\partial n_p}{\partial x_j} \frac{\partial T_p}{\partial x_j} + c T_p \frac{\nu_p}{\sigma_p} \frac{\partial n_p}{\partial x_j} \right] \tag{8.91}$$
$$\frac{\partial}{\partial t}(\rho Y_s) + \frac{\partial}{\partial x_j}(\rho v_j Y_s) = \frac{\partial}{\partial x_j}\left(\frac{\mu_e}{\sigma_Y} \frac{\partial Y_s}{\partial x_j} \right) - w_s + \alpha_s S \tag{8.92}$$

Notice that the term on the right-hand side of Eq. (8. 86) expresses the particle mass turbulent diffusion, the last two terms on the right-hand side of Eq. (8. 89) denote the particle momentum transport due to turbulent viscosity and particle mass diffusion, and the last three terms on the right-hand side of Eq. (8. 91) stand for the particle energy transport due to turbulent fluctuation.

Comparing the particle continuity equation (8. 87) with that of the small-slip model Eq. (6. 75) and that of the particle trajectory model Eq. (8. 18), it can be seen that Eq. (8. 87) has a particle diffusion term which is not related to the velocity slip. The particle momentum and energy transport terms due to turbulent fluctuation in Eqs. (8. 89) and (8. 91) also show the specific features of the multi-fluid model, while no such terms appear in the particle momentum and energy equations (8. 20) and (8. 22) of the particle trajectory model. Some investigators account for only the particle turbulent viscosity and heat conduction terms and neglect the particle momentum and energy transport due to mass diffusion. Others account for only the diffusion term in the particle continuity equation, but neglect the diffusion terms in particle momentum and energy equations. In some so-called "multi-fluid" models even there are entirely no diffusion terms in particle continuity, momentum and energy equations. Actually, these are not multi-fluid models, but the particle trajectory model expressed in Eulerian coordinate.

There is a simpler way to close Eqs. (8. 86), (8. 87), (8. 89), and (8. 91) by using the so-called Hinze-Tchen's algebraic model of particle turbulent viscosity

$$\frac{\nu_p}{\nu_T} = \left(\frac{k_p}{k} \right)^{\frac{1}{2}} = \left(1 + \frac{\tau_{rp}}{\tau_T} \right)^{-1} \tag{8.93}$$

which is derived from the concept of particle tracking fluid. If we neglect the diffusion and production terms in Eq. (8. 85) and drop the second part of ε_p, taking $c_p^k = 1$, then we have

$$\frac{dk_p}{dt} = \frac{2}{\tau_{rp}} (\sqrt{k k_p} - k_p)$$

Integrating this equation within the time period $\tau_T = c_T (k/\varepsilon)$ with constant k and k_p gives

$$\frac{k_p}{k} = \left(1 + \frac{\tau_{rp}}{2\tau_T} \right)^{-2} \tag{8.94}$$

which is almost the same as the Hinze-Tchen's expression. Clearly, the classical particle-tracking-fluid theory is the limiting case of k_p model.

Equation (8. 85) implies that the particle turbulent kinetic energy depends on not only the effect of local gas turbulence (the term ε_p), but also its own convection, diffusion and production. According to the Hinze-Tchen's algebraic model of particle turbulence, or the particle-tracking-fluid theory, i. e. Eq (8. 94), the particle turbulence should be always smaller than gas turbulence, and the larger the particle size, the smaller its turbulent intensity (or turbulent kinetic energy). However, experiments have shown that larger particles diffuse faster than smaller particles and particle turbulent fluctuation is stronger than gas turbulent fluctuation in enclosed gas-particle jets. Furthermore, in recirculating gas-particle flows the particle turbulent fluctuation is found to be stronger than gas turbulent fluctuation in reverse flow zones. Predictions based on the k-ε-k_p model can well explain this phenomena.

It can also be seen from Eq. (8.77) that particles of the same size can reduce or enhance gas turbulence in different regions of the flow field, depending on whether the source term G_p is greater or smaller than zero. This conclusion has been verified by the LDV measurements in turbulent recirculating gas-particle flows.

8.4.5 Solution Procedure

Spalding has proposed a solution procedure for general two-phase flows including dense suspensions, which is an extension of the SIMPLE procedure to two-phase flows, called IPSA procedure. The flow chart of IPSA procedure is shown in Fig. 8.5.

For dilute suspensions the present anthor and his colleagues have developed a LEAGAP Procedure. The LEAGAP flow chart is shown in Fig. 8.6.

8.4.6 Model Application and Assessment

In recent years the multi-fluid model has been used to simulate 1-D unsteady water-vapor two-phase flows, unsteady 2-D turbulent gas-particle flows in gun barrels, gas-particle jets, evaporating spray-gas jets, gas-spray flow and combustion in enclosed coaxial jets, 2-D and 3-D turbulent recirculating gas-particle flows in a sudden-expansion combustor with and without high-velocity jets, 3-D turbulent swirling and recirculating gas-particle flows in a tangentially fired furnace, 2-D gasification in fluidized bed. Some examples will be discussed in Chapter 10.

The advantage of the multi-fluid model is that this model can fully account for various turbulent transport processes of particle phase, the gas and particle phases can be treated by using a unified numerical method, and the detailed information of particle phase can be obtained for comparison with experiments without excessive storage and computational time. The disadvantage of this model is: (1) the feasibility of extension of closure model for single-phase flows to both gas and particle phases in two-phase flows needs to be further studied; (2) there is numerical diffusion of particle phase; and (3) the problem of how to deal with the particle history effect for particles with evaporation, devolatilization and heterogeneous combustion still remains to be solved.

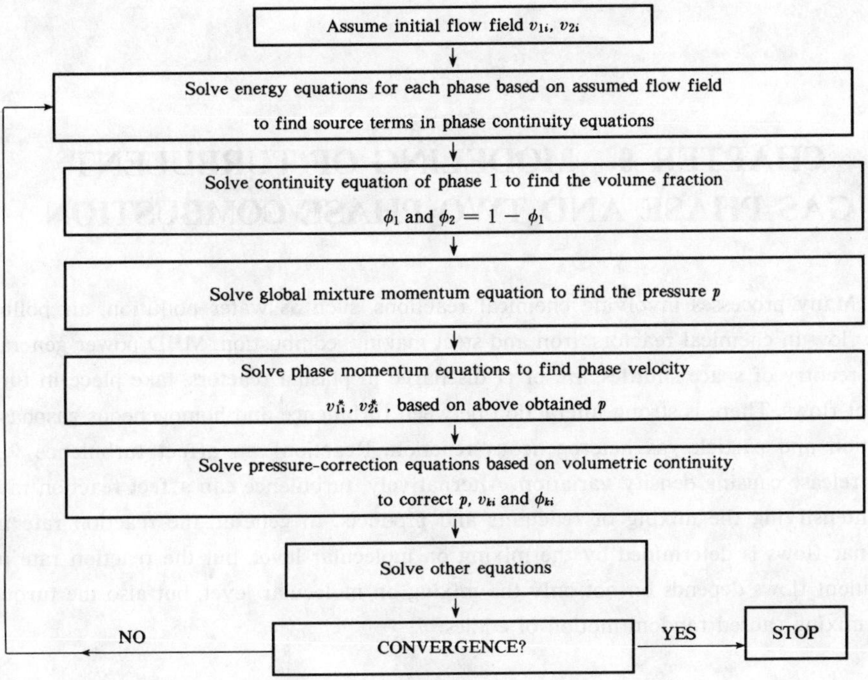

Fig. 8. 5 IPSA Flow Chart

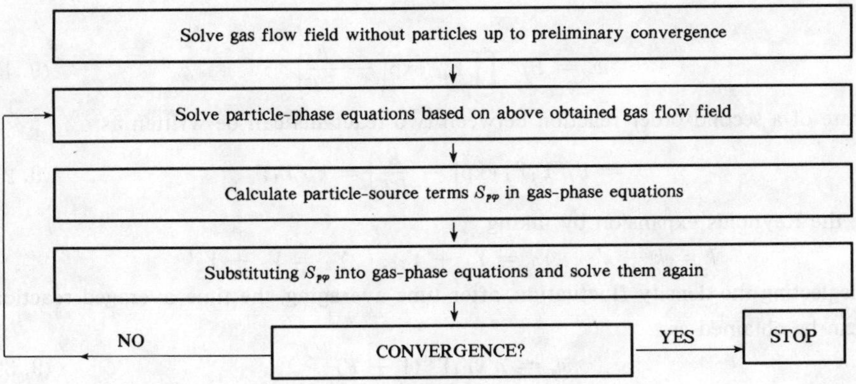

Fig. 8. 6 LEAGAP Flow Chart

CHAPTER 9 MODELING OF TURBULENT GAS-PHASE AND TWO-PHASE COMBUSTION

Many processes involving chemical reactions, such as water pollution, air pollution, flow in chemical reactors, iron and steel making, combustion, MHD power generation, reentry of space shuttles, arc or rf discharge in plasma reactors, take place in turbulent flows. There is strong interaction between turbulence and homogeneous gasphase reaction and particle-gas heterogeneous reaction. Reaction can affect turbulence via heat release causing density variation. Alternatively, turbulence can affect reaction rate by intensifying the mixing of reactants and products. In general, the reaction rate in laminar flows is determined by the mixing on molecular level, but the reaction rate in turbulent flows depends on not only the mixing on molecular level, but also the turbulent mixing caused random motion of eddies.

9.1 TIME-AVERAGED REACTION RATE

The global reaction rate in laminar flows or the instantaneous reaction rate in turbulent flows, as we have discussed in Chapter 2, can be expressed by the Arrhenius formula

$$w_s = B\rho^m \prod_{s=1}^{z} Y_s^{m_s} \exp\left(-\frac{E}{RT}\right) \tag{9.1}$$

The rate of a second-order reaction between two reactants can be written as

$$w_s = B\rho^2 Y_1 Y_2 \exp\left(-\frac{E}{RT}\right) = k\rho^2 Y_1 Y_2 \tag{9.2}$$

Using the Reynolds expansion by taking

$$k = \bar{k} + k', \quad Y_1 = \bar{Y}_1 + Y_1', \quad Y_2 = \bar{Y}_2 + Y_2'$$

and neglecting the density fluctuation, after time averaging, the time-averaged reaction rate can be obtained as

$$\bar{w}_s = \bar{\rho}^2 \bar{k}\bar{Y}_1\bar{Y}_2(1 + F) \tag{9.3}$$

where

$$\bar{k} = \overline{B\exp\left(-\frac{E}{RT}\right)}, \quad F = \frac{\overline{Y'_1Y'_2}}{\bar{Y}_1\bar{Y}_2} + \frac{\overline{k'Y'_1}}{\bar{k}\bar{Y}_1} + \frac{\overline{k'Y'_2}}{\bar{k}\bar{Y}_2} + \frac{\overline{k'Y'_1Y'_2}}{\bar{k}\bar{Y}_1\bar{Y}_2} > 0$$

Obviously, the central problem in modeling of reaction in turbulent flows is that the time-averaged reaction rate is not equal to the reaction rate expressed in the time-averaged variables, or

$$\bar{w}_s \neq \bar{\rho}^2 \bar{k}\bar{Y}_1\bar{Y}_2$$

The difficulties arising in modeling are how to model the correlation terms $\overline{Y_1'Y_2'}$, $\overline{k'Y_1'}$, $\overline{k'Y_2'}$, and $\overline{k'Y_1'Y_2'}$, while one must account for the fact that there is a highly nonlinear exponential function of temperature in the expression of k . One way is to use the closure methods, similar to those used in turbulence modeling, that is, to derive transport equations of these correlations and to solve them. This approach will be discussed later in this chapter, and it needs some further simplifications. However, first we will discuss some of the most widely used models in engineering problems.

9. 2 EBU-ARRHENIUS MODEL WITH FINITE-RATE REACTIONS

The simplest reaction model in turbulent flows is the EBU-Arrhenius (Eddy-Break-Up-Arrhenius) model proposed by Spalding. The simple idea is that two mechanisms—Arrhenius mechanism (laminar reaction) and turbulent fluctuation—control the reaction rate in turbulent flows. If we define

$$\overline{w}_{sA} = B\overline{\rho}^2\overline{Y}_1\overline{Y}_2 \exp\left(-\frac{E}{R\overline{T}}\right) \tag{9. 4}$$

and assume that the correlations in Eq. (9. 3) are proportional to $g = \overline{Y'^2}$, then the time-averaged reaction rate will take the following general form

$$\overline{w}_s = f\,(\overline{w}_{sA},\, g) \tag{9. 5}$$

Actually, as discussed in Chapter 1, there are two characteristic times in turbulent reacting flows

Reaction time $\quad \tau_c = \overline{Y}_1/\overline{w}_{1A}$

Diffusion time (fluctuation time) $\quad \tau_T = \frac{l}{k^{\frac{1}{2}}} = c_T\frac{k}{\varepsilon}$

The limiting case of $\tau_T/\tau_c \gg 1$ is diffusion-controlled reacting flows (flows with fast reaction), and the limiting case of $\tau_T/\tau_c \ll 1$ is the kinetics-controlled reacting flows. Another typical case is the premixed reacting flows or the diffusion-kinetics controlled reacting flows, where the Damkohler number τ_T/τ_c is the order of unity, or both the reaction kinetics and turbulent fluctuation (diffusion) control the reaction rate. Therefore the time-averaged reaction rate in the EBU-Arrhenius model is assumed to be

$$\overline{w}_s = f_1\,(\tau_T,\, \tau_c)$$

or

$$\overline{w}_s = f_2\,(\overline{w}_{sA},\, \overline{w}_{sT})$$

where $\overline{w}_{sA}$ has already been defined, and $\overline{w}_{sT}$ is assumed to be proportional to the mean square value of concentration fluctuation $g = \overline{Y'^2}$ and the fluctuation frequency, that is

$$\overline{w}_{sT} \sim g/\tau_T \qquad or \qquad \overline{w}_{sT} \sim g\varepsilon/k$$

By dimensional analysis we can take

$$\overline{w}_{sT} = c_{EBU}\,\rho g\varepsilon/k \tag{9. 6}$$

where c_{EBU} is an empirical constant, taken to be 0. 35—0. 4, and g may be determined by solving its transport equation or algebraic expression, or by using the simple assumption

$$g \sim \overline{Y_1} \qquad \text{or} \qquad g \sim \overline{Y_2}$$

or

$$g \sim \min\,[\overline{Y_1},\, \overline{Y_2},\, \overline{Y_3}] \tag{9. 7}$$

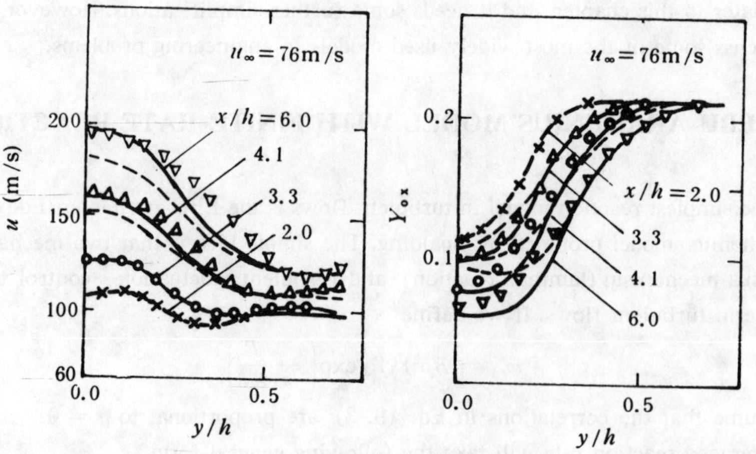

Fig. 9. 1

where $\overline{Y_1}$, $\overline{Y_2}$, and $\overline{Y_3}$ denote the mass fractions of fuel, oxygen and combustion products respectively. The last expression is called Magnusen's modification. Finally, in numerical modeling the time-averaged reaction rate is determined by

$$\overline{w}_s = \min\left[\overline{w}_{sT}, \overline{w}_{sA}\right] \tag{9.8}$$

which means that in the region where $\tau_c < \tau_T$ or $\overline{w}_{sT} < \overline{w}_{sA}$, we take $\overline{w}_s \approx \overline{w}_{sT}$ and in the region where $\tau_T < \tau_c$ or $\overline{w}_{sA} < \overline{w}_{sT}$, we take $\overline{w}_s \approx \overline{w}_{sA}$.

The EBU-Arrhenius model has been used to predict the premixed turbulent gas combustion behind a baffle in a channel (by Spalding, Fig. 9. 1), in sudden-expansion combustors (by the author, Fig. 9. 2). It can be seen that either quantitative (Fig. 9. 1) or qualitative agreement were obtained, but still, there is discrepancy between predictions and experiments. The EBU-Arrhenius model is simple and intuitive, however, later it can be shown that this model underestimates the role of chemical kinetics in premixed turbulent reacting flows.

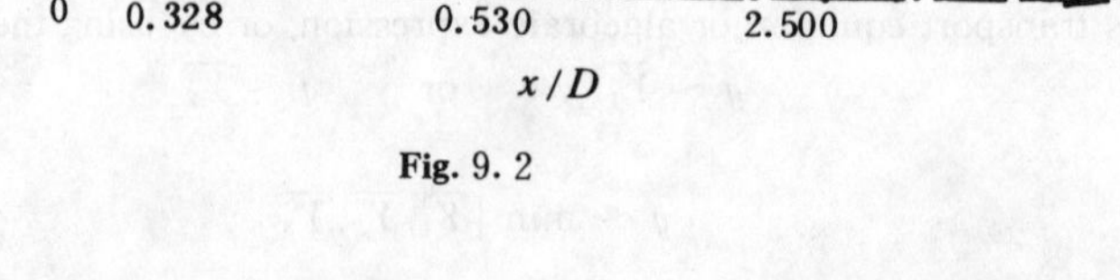

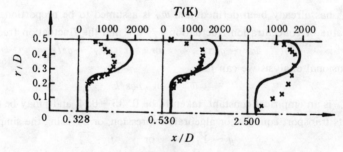

Fig. 9. 2

9.3 SIMPLIFIED-PDF-FAST-REACTION MODEL

For diffusion-controlled turbulent reacting flows the simplest model is the so-called simplified-PDF-fast-reaction model, based on the concept of mixture fraction, fast reaction and the probability density distribution function (PDF).

9.3.1 Conservative Scalar And Mixture Fraction

For a single-step reaction of only two reactants—fuel and oxygen

$$\text{Fuel} + \text{Oxygen} \longrightarrow \text{Products}$$
$$w_F \qquad w_{ox}/\beta \qquad -w_{pr}/(1+\beta)$$

The combined mass fraction by using the Zeldovich transformation

$$X = Y_F - Y_{ox}/\beta$$

is a conservative scalar, as there are no source terms in the conservation equation of X. For diffusion-controlled reacting flows (two reactants previously unmixed), the so-called mixture fraction can be introduced as:

$$f \equiv \frac{X - X_2}{X_1 - X_2}$$

where X_1, X_2 express values of X for fuel side and oxygen side respectively and

$$X_1 = Y_{F1} = 1, \qquad X_2 = -Y_{ox.2}/\beta = -1/\beta$$
$$f_1 = 1, \qquad f_2 = 0$$

Generally, the mixture fraction "f" expresses the "mixedness" or "unmixedness" of two species (spatial or temporal or both).

Accordingly, f is also a conservative scalar and its instantaneous conservation equation is

$$\frac{\partial}{\partial t}(\rho f) + \frac{\partial}{\partial x_j}(\rho v_j f) = \frac{\partial}{\partial x_j}\left(D\rho \frac{\partial f}{\partial x_j}\right) \qquad (9.9)$$

9.3.2 Fast-Reaction Assumption—Local Instantaneous Unmixedness

Assume that: (1) the reaction is sufficiently fast; (2) two components—fuel and oxygen do not coexist instantaneously at the same time (local instantaneous unmixedness), but their time-averaged value can coexist at the same location in a finite reaction zone; (3) the spatial distribution of mass fractions of fuel, oxygen and combustion products is now taken as the instantaneous possible relationship between Y_F and Y_{pr}, or Y_{ox} and Y_{pr} (Fig. 9.3), which can be expressed as:

when $1 > f > f_F$, no oxygen,

$$Y_F = (f - f_F)/(1 - f_F)$$

when $0 < f < f_F$, no fuel,

$$Y_{ox} = 1 - f/f_F$$

Notice that

$$\overline{Y_F} \neq (\bar{f} - f_F)/(1 - f_F) \ !$$
$$\overline{Y}_{ox} \neq (1 - \bar{f}/f_F) \ !$$

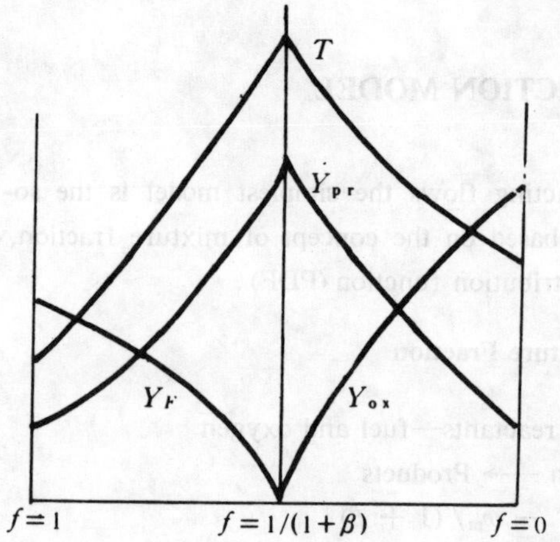

Fig. 9.3

since this instantaneous relationship does not hold all the time, and holds only for certain values of f which will occur with different probability. So, by using the assumption of instantaneous unmixedness, all of the instantaneous variables, including mass fraction and temperature, can be expressed by the same conservative scalar—mixture fraction f. So far we do not know the relationship between the time-averaged mass fraction (or temperature) and the time-averaged mixture fraction.

9.3.3 k-ε-$\bar{f}$-g Equations

The relations discussed in 9.3.2 just express the state inside turbulent eddies. For the whole flow field we should solve the time-averaged equations as

$$\left\{\begin{array}{ll} \dfrac{\partial \rho}{\partial t}+\dfrac{\partial}{\partial x_j}(\rho v_j)=0 & (9.10)\\ \dfrac{\partial}{\partial t}(\rho v_i)+\dfrac{\partial}{\partial x_j}(\rho v_j v_i)=\dfrac{\partial}{\partial x_j}\left(\mu_e \dfrac{\partial v_i}{\partial x_j}\right)+S_{vi} & (9.11)\\ \dfrac{\partial}{\partial t}(\rho k)+\dfrac{\partial}{\partial x_j}(\rho v_j k)=\dfrac{\partial}{\partial x_j}\left(\dfrac{\mu_e}{\sigma_k}\dfrac{\partial k}{\partial x_j}\right)+G_k-\rho\varepsilon & (9.12)\\ \dfrac{\partial}{\partial t}(\rho \varepsilon)+\dfrac{\partial}{\partial x_j}(\rho v_j \varepsilon)=\dfrac{\partial}{\partial x_j}\left(\dfrac{\mu_e}{\sigma_\varepsilon}\dfrac{\partial \varepsilon}{\partial x_j}\right)+\dfrac{\varepsilon}{k}(c_1 G_k-c_2\rho\varepsilon) & (9.13)\\ \dfrac{\partial}{\partial t}(\rho \bar{f})+\dfrac{\partial}{\partial x_j}(\rho v_j \bar{f})=\dfrac{\partial}{\partial x_j}\left(\dfrac{\mu_e}{\sigma_f}\dfrac{\partial \bar{f}}{\partial x_j}\right) & (9.14)\\ \dfrac{\partial}{\partial t}(\rho g)+\dfrac{\partial}{\partial x_j}(\rho v_j g)=\dfrac{\partial}{\partial x_j}\left(\dfrac{\mu_e}{\sigma_g}\dfrac{\partial g}{\partial x_j}\right)+c_{g1}\mu_T\left(\dfrac{\partial \bar{f}}{\partial x_j}\right)^2-c_{g2}\rho g\varepsilon/k & (9.15) \end{array}\right.$$

where the last equation— g equation has been derived by using the method similar to that used in deriving the ε equation. By solving these equations, we can find v_i, k, ε, $\bar{f}$ and g.

9.3.4 Probability Density Distribution Function

In order to find the time-averaged mass fractions and temperature, we need to use the concept of probability density distribution funtion. For the stochastic mixture fraction "f" fluctuating between "0" and "1" with time (Fig. 9.4), its probability in the range of "f" to "$f+df$" can be defined as $p(f)\,df$, where $p(f)$ is called the probability density distribution function (Fig. 9.4). Obviously, we have

$$\int_0^1 p(f)\,df = 1 \tag{9.16}$$

and the time-averaged and root-mean-square values of f will be determined by

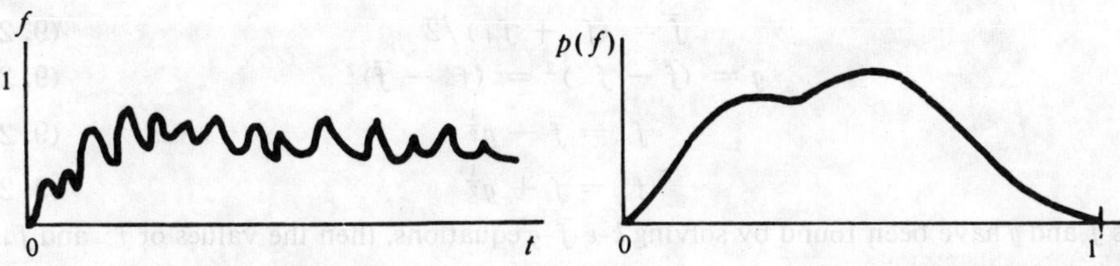

Fig. 9.4

$$\bar{f} = \int_0^1 f p(f)\, df \tag{9.17}$$

$$\overline{f'^2} = \overline{f^2} - (\bar{f})^2 = \int_0^1 f^2 p(f)\, df - (\bar{f})^2 \tag{9.18}$$

For any function $\varphi(f)$, the time-averaged and root-mean-square values of φ are:

$$\bar{\varphi}(x) = \int_0^1 \varphi(f)\, p(f, x)\, df \tag{9.19}$$

$$\overline{\varphi'^2}(x) = \int_0^1 \varphi^2(f)\, p(f, x)\, df - (\bar{\varphi})^2 \tag{9.20}$$

Therefore, the time-averaged mass fractions of fuel and oxygen are:

$$\bar{Y}_F = \int_0^1 Y_F(f)\, p(f)\, df = \int_0^1 \left(\frac{f - f_F}{1 - f_F}\right) p(f)\, df \tag{9.21}$$

$$\bar{Y}_{ox} = \int_0^1 Y_{ox}(f)\, p(f)\, df = \int_0^1 (1 - f/f_F)\, p(f)\, df \tag{9.22}$$

9.3.5 Simplified PDF

The remaining problem is to find PDF. One approach is the full PDF model, which is to solve the PDF transport equation by using the Monte-Carlo method. This needs for much computational efforts and up to now has not yet been used to solve practical engineering problems. The other is to assume a PDF, called simplified PDF, for example, a PDF for a top-hat fluctuation (Fig. 9.5). In this case, $p(f)$ has two peaks at $f = f_-$ and $f = f_+$, and $p(f)$ is equal to zero when $f \neq f_-$ and $f \neq f_+$, or (Fig. 9.5)

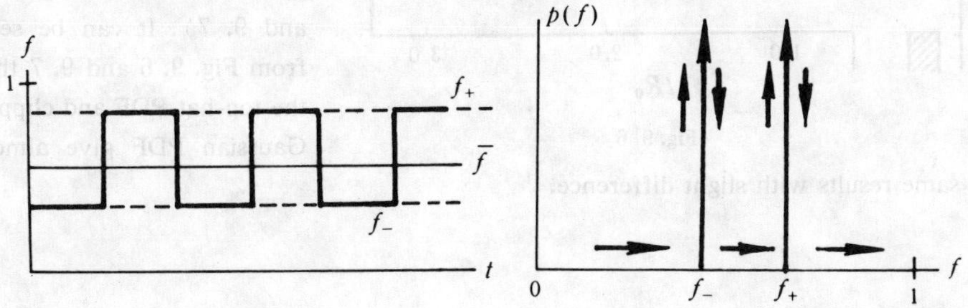

Fig. 9.5

$$p(f) = \alpha\delta(f_-) + (1-\alpha)\delta(f_+) \quad (9.23)$$

When $\alpha = 0.5$, we have

$$\bar{f} = (f_- + f_+)/2 \quad (9.24)$$

$$g = (\bar{f} - f_-)^2 = (f_+ - \bar{f})^2 \quad (9.25)$$

$$f_- = \bar{f} - g^{\frac{1}{2}} \quad (9.26)$$

$$f_+ = \bar{f} + g^{\frac{1}{2}} \quad (9.27)$$

As f and g have been found by solving k-ε-$\bar{f}$-g equations, then the values of f_- and f_+ at each location of the flow field can be predicted by using Eq. (9.26) and (9.27), and finally the time-averaged and root-mean-square values of any scalar φ can be found by using the following expressions

$$\bar{\varphi} = \varphi(f_-)/2 + \varphi(f_+)/2 \quad (9.28)$$

$$\overline{\varphi'^2} = [\varphi(f_-) - \bar{\varphi}]^2 + [\varphi(f_+) - \bar{\varphi}]^2 \quad (9.29)$$

Other forms of PDF, such as clipped Gaussian PDF can also be used, but the computational time is much longer than that needed for the top-hat PDF.

Predictions for a simple diffusion jet-flame are in general agreement with experiments for the length of CO, H_2, C_2H_2 and C_3H_8 flame (by Spalding et al), but there are discrepancies showing the effect of molecular diffusivity and finite kinetics. Predicted velocity profiles and the change of velocity at the axis along axial direction in turbulent diffusion combustion for coaxial jets discharged into sudden-expansion combustors are in good agreement with experiments (Fig. 9. 6 and 9. 7). It can be seen from Fig. 9. 6 and 9. 7 that the top-hat PDF and clipped Gaussian PDF give almost the same results with slight difference.

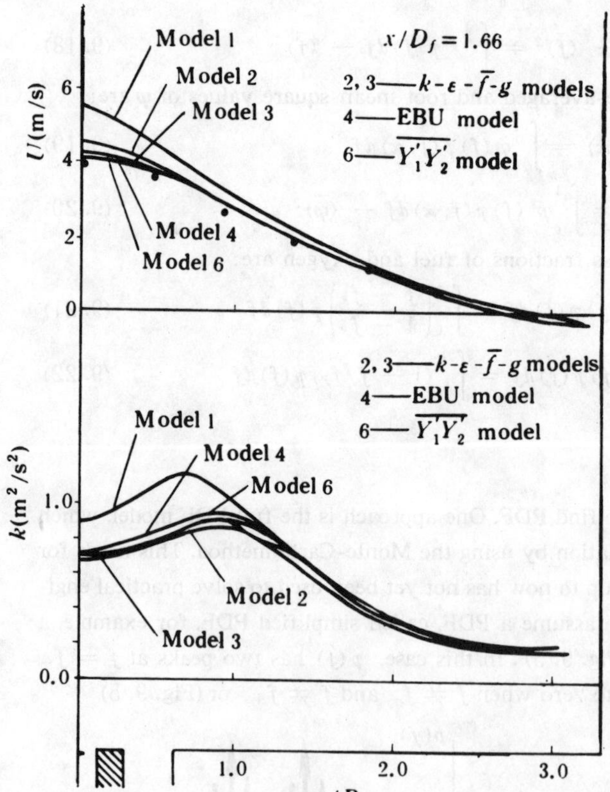

Fig. 9. 6

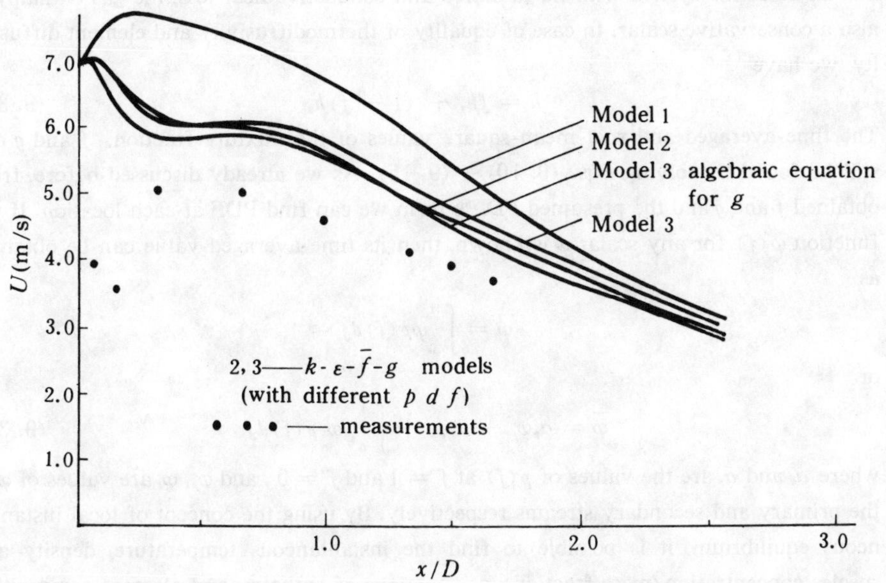

Fig. 9. 7

9. 4 SIMPLIFIED-PDF-LOCAL-INSTANTANEOUS-EQUILIBRIUM MODEL

The above-discussed model is appropriate only for the simple case of two-species reactions. For more complicated diffusion combustion, such as coal-volatile combustion with many species and more than one-step reaction, Smoot et al. proposed a local instantaneous equilibrium model combined with a simplified PDF. For the case of coaxial jets discharged into a sudden-expansion combustor with a central primary jet of fuel and air and an annular jet of pure secondary air, the mixture fraction is defined as

$$f = \frac{m_p}{m_p + m_s} \tag{9.30}$$

where m_p is the mass of the fluid atom coming from the primary stream and m_s is the mass of the fluid atom coming from the secondary stream. The mixture fraction f expresses the mixedness at any instant and at any location, or the local instantaneous stoichiometric ratio. Any other conservative scalar (not including the mass fraction Y_s which is not a conservative scalar due to the reaction source terms in its conservation equation) can be expressed by

$$\varphi = f\varphi_p + (1 - f)\varphi_s \tag{9.31}$$

where φ_p and φ_s are values of φ in primary and secondary streams. Since any chemical element cannot be produced or destructed in ordinary reactions, then the local instantaneous mass fraction of element k, "b_k" is also a conservative scalar. In the case of equal diffusivities of elements (this assumption is questionable) we have

$$b_k = f b_{kp} + (1 - f) b_{ks} \tag{9.32}$$

For an adiabatic system with no radiative and conductive heat loss, the gas enthalpy is also a conservative scalar. In case of equality of thermodiffusivity and element diffusivity, we have

$$h = f h_p + (1 - f) h_s \tag{9.33}$$

The time-averaged and root-mean-square values of the mixture fraction, $\bar{f}$ and g can still be found by solving Eqs. (9.10) — (9.15). As we already discussed before, from obtained $\bar{f}$ and g and the presumed PDF's form we can find PDF at each location. If the function $\varphi(f)$ for any scalar φ is known, then its time-averaged value can be obtained as

$$\bar{\varphi} = \int_0^1 \varphi p(f)\, df$$

or

$$\bar{\varphi} = \alpha_p \varphi_p + \alpha_s \varphi_s + \int_{0^+}^{1^-} \varphi\, p(f)\, df \tag{9.34}$$

where α_p and α_s are the values of $p(f)$ at $f = 1$ and $f = 0$, and φ_p, φ_s are values of φ in the primary and secondary streams respectively. By using the concept of local instantaneous equilibrium, it is possible to find the instantaneous temperature, density and species concentration (mass fraction) as functions of enthalpy and element composition (with minor pressure change)

$$T = T(b_k, h)$$

$$\rho = \rho(b_k, h)$$

$$Y_s = Y_s(b_k, h)$$

These functional relationships can be found by using chemical equilibrium calculations. For an adiabatic system, the enthalpy and element mass fraction are functions only of mixture fraction, that is

$$b_k = b_k(f)$$

$$h = h(f)$$

so, we have

$$T = T(f),\ \rho = \rho(f),\ Y_s = Y_s(f)$$

Therefore, for this case, we can directly obtain the time-averaged temperature, gas density and species mass fraction $\bar{T}(f)$, $\bar{\rho}(f)$ and $\bar{Y}_s(f)$ by using $p(f)$. For non-adiabatic systems we need to solve the energy equation to find the enthalpy h, and we have

$$T = T(f, h),\ \rho = \rho(f, h),\ Y_s = Y_s(f, h)$$

The time-averaged values $\bar{T}$, $\bar{\rho}$, and $\bar{Y}_s$ should be determined by

$$\bar{\varphi} = \alpha_p \varphi_p + \alpha_s \varphi_s + \int_{0^+}^{1^-} \varphi p(f, h)\, df dh \tag{9.34a}$$

However, the joint PDF $p(f, h)$ is difficult to be presumed. Hence an alternative approach is to assume

$$h = h_f + h_r,\quad h_f = h_f(f)$$

$$h_r = \bar{h}_r + h'_r,\quad \bar{h'_r} = 0$$

and to set

$$p(f, h) = p(f)$$

which leads to

$$\overline{\varphi} = \alpha_p \varphi_p + \alpha_s \varphi_s + \int_{0^+}^{1-} \varphi p(f) \, df \tag{9.34}$$

Prediction results of a H_2-air diffusion flame in coaxial jets discharged into a sudden-expansion combustor are shown in Fig. (9. 8). Except the RMS fluctuation velocity, the predicted time-averaged velocity, temperature, H_2, O_2 and H_2O mass fraction arein good agreement with experiments, which indicates that the mechanism of hydrogen-air diffusion flame is near to the local instantaneous equilibrium model. However, predicted species mass fractions for a methane-air diffusion flame in similar flow configuration (Fig. 9. 9) are quite different, even qualitatively different from the experimental results, which shows that the methane-air diffusion flame, including CO combustion, is not in local instantaneous equilibrium.

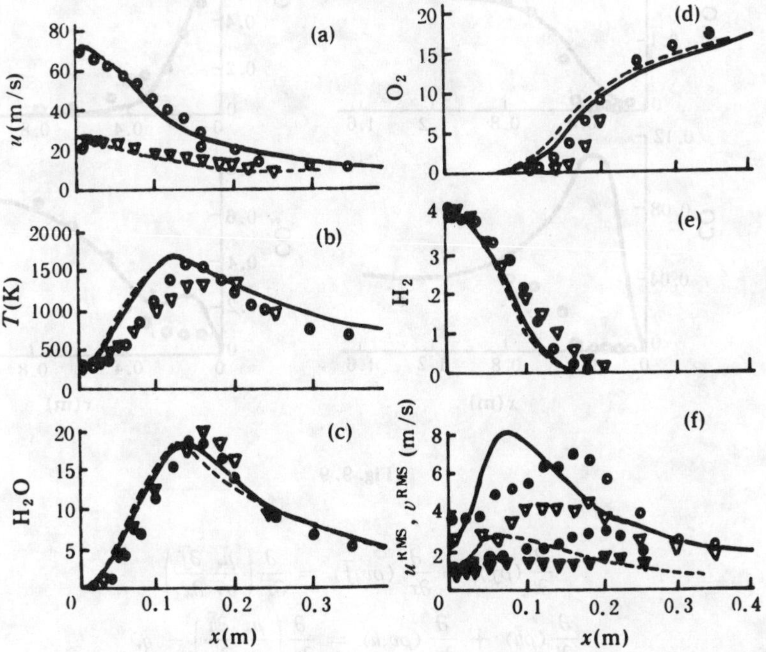

Fig. 9. 8

9. 5 SIMPLIFIED-PDF-FINITE-REACTION-RATE MODEL

It has been found that for some reactions, for example, CH_4-O_2 reaction, CO-O_2 reaction, and NO_x formation in coal-air flames, the reaction is not in equilibrium, or the reaction rate is not infinitively fast. Therefore, models combining the presumed PDF with finite reaction rate have been developed. One of them is the modeling of CO combustion by using a three-dimensional clipped Gaussian joint PDF. For reacting turbulent flows, besides the continuity equation, momentum equation, k-equation and ε-equation, the time-averaged equations to be considered are the mixture fraction equation, enthalpy equation and the CO-species equation:

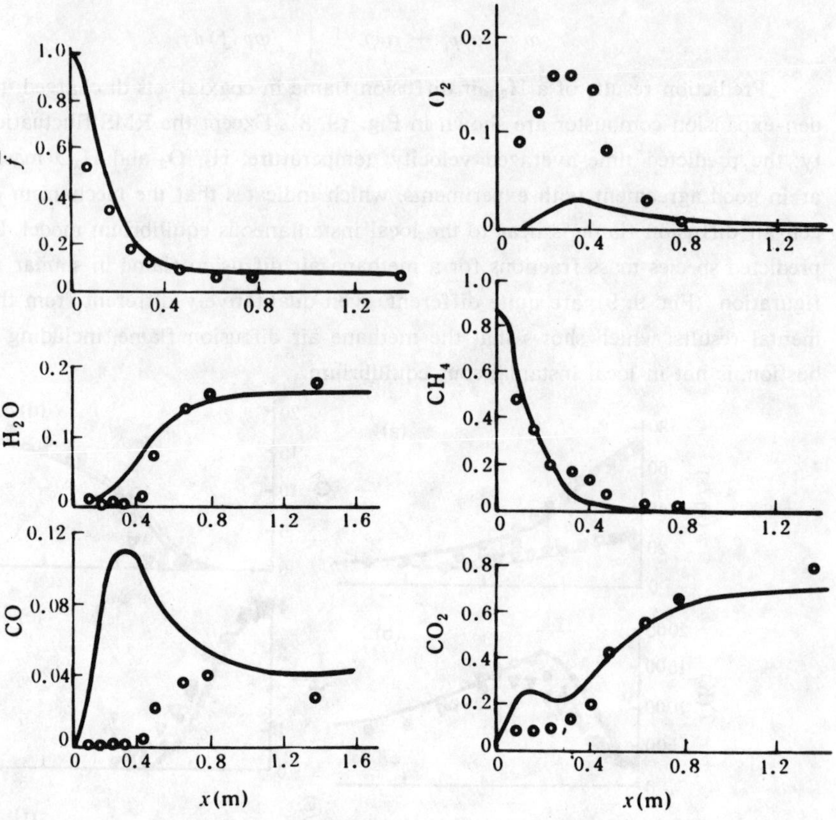

Fig. 9. 9

$$\frac{\partial}{\partial t}(\rho \bar{f}) + \frac{\partial}{\partial x_j}(\rho v_j \bar{f}) = \frac{\partial}{\partial x_j}\left(\frac{\mu_e}{\sigma_f}\frac{\partial \bar{f}}{\partial x_j}\right) \tag{9.35}$$

$$\frac{\partial}{\partial t}(\rho \bar{h}) + \frac{\partial}{\partial x_j}(\rho v_j \bar{h}) = \frac{\partial}{\partial x_j}\left(\frac{\mu_e}{\sigma_h}\frac{\partial \bar{h}}{\partial x_j}\right) - q_r \tag{9.36}$$

$$\frac{\partial}{\partial t}(\rho \bar{Y}_{CO}) + \frac{\partial}{\partial x_j}(\rho v_j \bar{Y}_{CO}) = \frac{\partial}{\partial x_j}\left(\frac{\mu_e}{\sigma_Y}\frac{\partial \bar{Y}_{CO}}{\partial x_j}\right) - \bar{w}_{CO} \tag{9.37}$$

Assuming a reaction of

$$CO + \tfrac{1}{2}O_2 \longrightarrow CO_2$$

the instantaneous mass fractions Y_{H_2O}, Y_{O_2} and temperature can be expressed by the following relations

$$Y_{H_2O} = (1 - f)\, Y_{H_2O,\,in} \tag{9.38}$$

$$Y_{O_2} = (1 - f)\, Y_{O_2,\,in} - \beta\,(fY_{CO,\,in} - Y_{CO}) \tag{9.39}$$

$$T = (h - h_F Y_{CO})\,/c_p \tag{9.40}$$

where β is the stoichiometric ratio of oxygen to CO. The instantaneous reaction rate is

$$w_{CO} = 3.98 \times 10^{17} \rho^{1.75} Y_{CO}\,(Y_{H_2O}/M_{H_2O})^{0.5}\,(Y_{O_2}/M_{O_2})^{0.25} \exp(-E/RT) \tag{9.41}$$

Its general form is

$$w_{CO} = w_{CO}\,(Y_{CO}, f, h)$$

The time-averaged reaction rate can be calculated by using a joint PDF

$$\overline{w}_{CO} = \int_0^1 \int_0^1 \int_{h_{1,in}}^{h_{2,in}} w_{CO}(Y_{CO}, f, h)\, p(Y_{CO}, f, h)\, dh df dY_{CO} \tag{9. 42}$$

where a three-dimensional clipped Gaussian PDF is assumed as

$$p(Y_{CO}, f, h) = C \exp(-Q/2) \tag{9. 43}$$

where C and Q are functions of the variances, covariances and time-averaged values of f, h and Y_{CO}. The variances and covariances are determined by their transport equations

$$\frac{\partial}{\partial t}(\rho \overline{\varphi'^2}) + \frac{\partial}{\partial x_j}(\rho v_j \overline{\varphi'^2}) = \frac{\partial}{\partial x_j}\left(\frac{\mu_e}{\sigma_\varphi} \frac{\partial \overline{\varphi'^2}}{\partial x_j}\right) + c_1 \mu_T \left(\frac{\partial \overline{\varphi}}{\partial x_j}\right)^2 - c_2 \frac{\varepsilon}{k} \rho \overline{\varphi'^2} \tag{9. 44}$$

$$\frac{\partial}{\partial t}(\rho \overline{\varphi' \psi'}) + \frac{\partial}{\partial x_j}(\rho v_j \overline{\varphi' \psi'}) = \frac{\partial}{\partial x_j}\left(\frac{\mu_e}{\sigma_\varphi} \frac{\partial \overline{\varphi' \psi'}}{\partial x_j}\right) + c_1 \mu_T \left(\frac{\partial \overline{\varphi}}{\partial x_j}\right)\left(\frac{\partial \overline{\psi}}{\partial x_j}\right) - c_2 \frac{\varepsilon}{k} \rho \overline{\varphi' \psi'}$$

(9. 45)

where φ, ψ denote f, h and Y_{CO}.

This model has been used to simulate the axisymmetric combustion of a CO jet with oxidizing flow in comparison with other three models as

$$\overline{w}_{CO} = w_{CO}(\overline{Y}_{CO}, \overline{f}, \overline{h}) \tag{9. 46}$$

$$\overline{w}_{CO} = \int_0^1 \int_0^1 \int_{h_{1,in}}^{h_{2,in}} w_{CO}\, p(Y_{CO})\, p(f)\, p(h)\, dh df dY_{CO} \tag{9. 47}$$

$$\overline{w}_{CO} = w_{CO}(\overline{Y}_{CO}, \overline{f}, \overline{h})\left[1 + f(\overline{\varphi'^2}, \overline{\varphi' \psi'}, \cdots)\right] \tag{9. 48}$$

However the comparison of predicted and measured CO concentration radial profiles does not give good agreement (Fig. 9. 10).

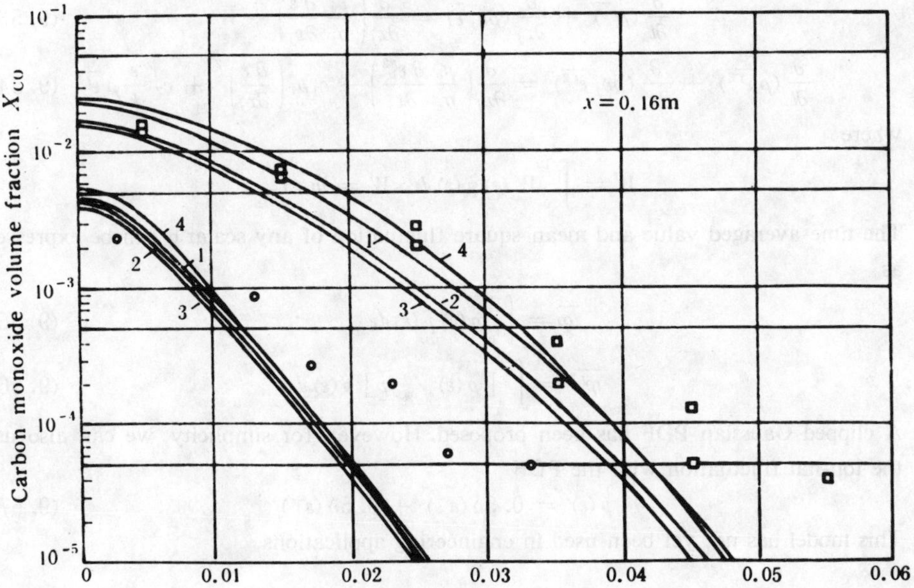

Fig. 9. 10

An alternative approach is to use a one-dimensional PDF. Assume a one-step two-species second-order instantaneous reaction rate

$$w_s = B\rho^2 Y_1 Y_2 \exp(-E/RT)$$

and instantaneous laminar premixed adiabatic combustion in turbulent eddies, which gives a single parameter—reactiveness ε relating the instantaneous mass fractions and temperature by the following expression

$$\varepsilon = 1 - \frac{Y_1}{Y_{1\infty}} = 1 - \frac{Y_2}{Y_{2\infty}} = \frac{T - T_\infty}{T_m - T_\infty} = \frac{c_p(T - T_\infty)}{Y_{1\infty}Q_1} \tag{9.49}$$

hence we have

$$\begin{cases} Y_1 = (1 - \varepsilon) Y_{1\infty} \\ Y_2 = (1 - \varepsilon) Y_{2\infty} \\ T = T_\infty[1 + \varepsilon Y_{1\infty}Q_1/(c_p T_\infty)] = T_\infty(1 + \varepsilon\Lambda) \\ \rho = \rho_\infty(1 + \varepsilon\Lambda)^{-1} \end{cases} \tag{9.50}$$

where $\Lambda = T_m/T_\infty - 1 = Y_{1\infty}Q_1/(c_p T_\infty)$, and the subscripts ∞ and m indicate values corresponding to the oncoming flow and the adiabatic flame temperature. The time-averaged reaction rate can be expressed as

$$\overline{w}_s = \int_0^1 w_s(\varepsilon)\, p(\varepsilon)\, d(\varepsilon) \tag{9.51}$$

The instantaneous transport equations of ε can be obtained from Y_1 or Y_2 species equation or energy equation as

$$\frac{\partial}{\partial t}(\rho\varepsilon) + \frac{\partial}{\partial x_j}(\rho v_j \varepsilon) = \frac{\partial}{\partial x_j}\left(D\rho\frac{\partial \varepsilon}{\partial x_j}\right) + W \tag{9.52}$$

and the time-averaged value and RMS fluctuation of ε are determined by the equations

$$\frac{\partial}{\partial t}(\rho\bar{\varepsilon}) + \frac{\partial}{\partial x_j}(\rho v_j \bar{\varepsilon}) = \frac{\partial}{\partial x_j}\left(\frac{\mu_e}{\sigma_\varepsilon}\frac{\partial \bar{\varepsilon}}{\partial x_j}\right) + \overline{W} \tag{9.53}$$

$$\frac{\partial}{\partial t}(\rho\overline{\varepsilon'^2}) + \frac{\partial}{\partial x_j}(\rho v_j \overline{\varepsilon'^2}) = \frac{\partial}{\partial x_j}\left(\frac{\mu_e}{\sigma_\varepsilon}\frac{\partial \overline{\varepsilon'^2}}{\partial x_j}\right) - c_1\mu_T\left(\frac{\partial \bar{\varepsilon}}{\partial x_j}\right)^2 + c_2\frac{\varepsilon}{k}\rho\overline{\varepsilon'^2} \tag{9.54}$$

where

$$\overline{W} = \int_0^1 W(\varepsilon)\, p(\varepsilon)\, d\varepsilon, \quad W = w_s/Y_{s\infty}$$

The time-averaged value and mean-square fluctuation of any scalar φ can be expressed as

$$\bar{\varphi} = \int_0^1 \varphi(\varepsilon)\, p(\varepsilon)\, d\varepsilon \tag{9.55}$$

$$\overline{\varphi'^2} = \int_0^1 [\varphi(\varepsilon) - \bar{\varphi}]^2 p(\varepsilon)\, d\varepsilon \tag{9.56}$$

A clipped Gaussian PDF has been proposed. However, for simplicity, we can also use the top-hat fluctuation with the PDF

$$p(\varepsilon) = 0.5\delta(\varepsilon^-) + 0.5\delta(\varepsilon^+) \tag{9.57}$$

This model has not yet been used in engineering applications.

9.6 CORRELATION-MOMENT CLOSURE MODEL

It has been pointed out in Sec. 9.1 that there is an approach in modeling turbulent combustion which is to use the correlation momentum closure method similar to that used in turbulence modeling.

9.6.1 Time-averaged Reaction Rate

In order to simulate the correlation terms, it is necessary to make an approximate expansion of the nonlinear exponential term in the expression of the time-averaged reaction rate, for which we may write

$$\exp\left(-\frac{E}{RT}\right) = \exp\left[-\frac{E}{R(\overline{T} + T')}\right] = \exp\left[-\frac{E}{R\overline{T}}\left(1 + \frac{T'}{\overline{T}}\right)^{-1}\right]$$

For $T'/\overline{T} \ll 1$, we have

$$\exp\left(-\frac{E}{RT}\right) \approx \exp\left(-\frac{E}{R\overline{T}}\right) \cdot \exp\left(\frac{E}{R\overline{T}^2}T'\right)$$

In the case where $\frac{E}{R\overline{T}}\frac{T'}{\overline{T}}$ is still a small quantity and only in this case, the following approximation can be made

$$\exp\left(\frac{E}{R\overline{T}^2} \cdot T'\right) \approx 1 + \frac{E}{R\overline{T}}\frac{T'}{\overline{T}} + \frac{1}{2}\left(\frac{E}{R\overline{T}}\frac{T'}{\overline{T}}\right)^2$$

$$\exp\left(-\frac{E}{RT}\right) \approx \exp\left(-\frac{E}{R\overline{T}}\right)\left[1 + \frac{E}{R\overline{T}}\frac{T'}{\overline{T}} + \frac{1}{2}\left(\frac{E}{R\overline{T}} \cdot \frac{T'}{\overline{T}}\right)^2\right]$$

and the time-averaged reaction rate can be expressed as

$$\overline{w}_s = B\rho^2 Y_1 Y_2 \exp\left(-\frac{E}{RT}\right) = B\rho^2 (\overline{Y}_1 + Y'_1)(\overline{Y}_2 + Y'_2)\exp\left(-\frac{E}{R\overline{T}}\right)$$
$$\left[1 + \frac{E}{R\overline{T}}\frac{T'}{\overline{T}} + \frac{1}{2}\left(\frac{E}{R\overline{T}}\frac{T'}{\overline{T}}\right)^2\right]$$

or

$$\overline{w}_s = B\rho^2 \overline{Y}_1 \overline{Y}_2 \exp\left(-\frac{E}{R\overline{T}}\right)\left[1 + \frac{\overline{Y'_1 Y'_2}}{\overline{Y}_1\overline{Y}_2} + \frac{E}{R\overline{T}}\left(\frac{\overline{T'Y'_1}}{\overline{T}\,\overline{Y}_1} + \frac{\overline{T'Y'_2}}{\overline{T}\,\overline{Y}_2}\right)\right.$$
$$\left. + \frac{1}{2}\left(\frac{E}{R\overline{T}}\right)^2 \overline{\left(\frac{T'}{\overline{T}}\right)^2}\right] \tag{9.58}$$

9.6.2 Correlation-Moment Transport Equation Closure

To close Eq. (9.58), the correlations $\overline{Y'_1 Y'_2}$, $\overline{T'Y'_1}$, $\overline{T'Y'_2}$, *and* $\overline{T'^2}$ should be found by solving their transport equations in the form of Eqs. (9.44) and (9.45) as:

$$\frac{\partial}{\partial t}(\rho \overline{Y'_1 Y'_2}) + \frac{\partial}{\partial x_j}(\rho v_j \overline{Y'_1 Y'_2}) = \frac{\partial}{\partial x_j}\left(\frac{\mu_e}{\sigma_Y}\frac{\partial \overline{Y'_1 Y'_2}}{\partial x_j}\right) + c_1\mu_T\left(\frac{\partial \overline{Y}_1}{\partial x_j}\right)\left(\frac{\partial \overline{Y}_2}{\partial x_j}\right)$$
$$- c_2 \frac{\varepsilon}{k}\rho \overline{Y'_1 Y'_2} \tag{9.59}$$

$$\frac{\partial}{\partial t}(\rho \overline{T'Y'_1}) + \frac{\partial}{\partial x_j}(\rho v_j \overline{T'Y'_1}) = \frac{\partial}{\partial x_j}\left(\frac{\mu_e}{\sigma_Y}\frac{\partial \overline{T'Y'_1}}{\partial x_j}\right) + c_1\mu_T\left(\frac{\partial \overline{T}}{\partial x_j}\right)\left(\frac{\partial \overline{Y}_1}{\partial x_j}\right)$$
$$- c_2 \frac{\varepsilon}{k}\rho \overline{T'Y'_1} \tag{9.60}$$

$$\frac{\partial}{\partial t}(\rho \overline{T'Y'_2}) + \frac{\partial}{\partial x_j}(\rho v_j \overline{T'Y'_2}) = \frac{\partial}{\partial x_j}\left(\frac{\mu_e}{\sigma_Y}\frac{\partial \overline{T'Y'_2}}{\partial x_j}\right) + c_1\mu_T\left(\frac{\partial \overline{T}}{\partial x_j}\right)\left(\frac{\partial \overline{Y}_2}{\partial x_j}\right)$$
$$- c_2 \frac{\varepsilon}{k}\rho \overline{T'Y'_2} \tag{9.61}$$

$$\frac{\partial}{\partial t}(\rho \overline{T'^2}) + \frac{\partial}{\partial x_j}(\rho v_j \overline{T'^2}) = \frac{\partial}{\partial x_j}\left(\frac{\mu_e}{\sigma_T}\frac{\partial \overline{T'^2}}{\partial x_j}\right) + c_1\mu_T\left(\frac{\partial \overline{T}}{\partial x_j}\right)^2 - c_2\frac{\varepsilon}{k}\rho \overline{T'^2} \tag{9.62}$$

9. 6. 3 Algebraic Expressions of Correlation Moment

In order to reduce the storage and computational time, while keeping the main feature of the model, the transport equation model can be simplified into an algebraic correlation model, like what we have done in the algebraic stress/flux model of gas turbulence. Dropping the convection and diffusion terms in Eqs. (9. 59) — (9. 62), the algebraic expressions can be obtained as:

$$\overline{Y'_1Y'_2} = c_Y \frac{k^3}{\varepsilon^2} \frac{\partial \overline{Y}_1}{\partial x_j} \frac{\partial \overline{Y}_2}{\partial x_j} \tag{9. 63}$$

$$\overline{T'Y'_1} = c_{Y1} \frac{k^3}{\varepsilon^2} \frac{\partial \overline{T}}{\partial x_j} \frac{\partial \overline{Y}_1}{\partial x_j} \tag{9. 64}$$

$$\overline{T'Y'_2} = c_{Y2} \frac{k^3}{\varepsilon^2} \frac{\partial \overline{T}}{\partial x_j} \frac{\partial \overline{Y}_2}{\partial x_j} \tag{9. 65}$$

$$\overline{T'^2} = c_T \frac{k^3}{\varepsilon^2} \left(\frac{\partial \overline{T}}{\partial x_j} \right)^2 \tag{9. 66}$$

The physical meaning of Eqs. (9. 63) — (9. 66) is obvious. It implies that the RMS fluctuation of concentration, concentration-temperature and temperature is proportional to the turbulence length scale multiplied by the gradient of time-averaged value, like the mixing-length model, in which the RMS fluctuation velocity is proportional to length scale and the mean velocity gradient, since we know $k^3/\varepsilon^2 = l^2$. It should be pointed out that this algebraic correlation closure model is suitable only for shear flows, and it is clear that this model cannot be used to predict, for example, uniform flows, in which the concentration and temperature gradients are zero but the fluctuation may not be zero.

Predictions of turbulent combustion in coaxial jets discharged into a sudden expansion combustor by using different models, including the correlation-moment closure model, and their comparison with experiments, made by Khalil are shown in Fig. 9. 6. No obvious difference is observed. It is argued by some investigators that the correlation-moment closure model may lead to remarkable inaccuracy due to series expansion, since in many cases the ratio $E/(RT)$ is much larger than unity.

9. 7 MODELING OF TURBULENT TWO-PHASE COMBUSTION

What we have discussed above is limited to modeling of single-phase reacting flows. In turbulent reacting two-phase flows, such as liquid-spray combustion, pulverized-coal combustion or gasification, the problem is more complex, and at least we need to further account for: (1) the effect of particle reaction (devolatilization, evaporation, heterogeneous combustion) on gas turbulence; (2) the effect of gas turbulence on particle reaction; (3) the effect of particle reaction on gas turbulent combustion; and (4) modeling of reacting particle phase in turbulent two-phase flows.

9. 7. 1 Effect of Particle Reaction on Gas Turbulence

It has been shown in Sec. 8. 4. 3 that the source term due to particle mass change (reaction) in the k -equation is approximately $2kS$. This term implies that the gas turbulence modification due to particle reaction is proportional to the local turbulent kinetic energy, the particle number density and the particle mass changing rate, while evaporation, devolatilization and char combustion will enhance gas turbulence and condensation will reduce gas turbulence. This conclusion needs experimental verification.

9. 7. 2 Effect of Gas Turbulence on Particle Reaction

The time-averaged source term due to particle reaction in gas-phase continuity equation is

$$\overline{S} = -\overline{\sum n_k \dot{m}_k} = -\left(\sum \overline{n_k}\ \overline{\dot{m}_k} + \sum \overline{n_k' \dot{m}'_k}\right) \qquad (9.67)$$

where the correlation $\overline{n'_k \dot{m}'_k}$ expresses the effect of gas turbulence on particle reaction. It can be shown that for droplet evaporation and coal-char reaction this correlation term can be modeled by

$$\overline{n_k' \dot{m}_k'} = c_{m1} \frac{k^3}{\varepsilon^2} \frac{\partial n_k}{\partial x_j} \frac{\partial T}{\partial x_j} \pi d_k \, \mathrm{Nu} \, \lambda / c_p \qquad (9.68)$$

or

$$\overline{n_k' \dot{m}_k'} = c_{m2} \frac{k^3}{\varepsilon^2} \frac{\partial n_k}{\partial x_j} \frac{\partial Y_{ox}}{\partial x_j} \pi d_k D \rho \qquad (9.69)$$

Clearly, the stronger the gas temperature and oxygen concentration fluctuation, the larger the time-averaged particle-reaction rate.

9. 7. 3 Effect of Particle Reaction on Gas-phase Turbulent Combustion

In most of the presently developed comprehensive models of two-phase combustion the single-phase turbulent combustion model is used to simulate the gas-phase combustion in two-phase flows. However, P. J. Smith and L. D. Smoot have pointed out that it is necessary to account for the effect of particle mass change on the mixture fraction. To do this, a second mixture fraction η is defined as

$$\eta = \frac{\text{Mass of local gases originating from the solid or liquid fuel}}{\text{Mass of total local gases}}$$

Unlike the mixture fraction of the inlet gas fuel f, η is not a conservative scalar, because there is a particle-source term in the time-averaged transport equation of η

$$\frac{\partial}{\partial t}(\rho\, \overline{\eta}) + \frac{\partial}{\partial x_j}(\rho v_j \overline{\eta}) = \frac{\partial}{\partial x_j}\left(\frac{\mu_e}{\sigma_\eta} \frac{\partial \overline{\eta}}{\partial x_j}\right) - \sum_k n_k \dot{m}_k$$

Besides, we need to solve the equation of RMS fluctuation of η , the $\overline{\eta'^2}$ equation. When inlet gas fuel and fuel gas originating from the solid or liquid fuel coexist, the local instantaneous mass fraction of any element k , namely b_k is the function of f and η , hence we have

$$b_k = b_{kc}\eta + (1-\eta)\left[f b_{kp} + (1-f) b_{ks}\right]$$

where b_{kc}, b_{kp} and b_{ks} express the mass fractions of element k in the solid or liquid fuel, primary stream and secondary stream respectively. Each dependent variable φ in the gas flow field is also a function of f and η . The time-averaged value of φ is

$$\overline{\varphi} = \alpha_c \varphi_c + \alpha_1\left[\alpha_p \varphi_p + \alpha_s \varphi_s + \int_0^1 \varphi(0, f)\, p(f)\, df\right] + \alpha_p \int_0^1 \varphi(\eta, 1)\, p(\eta)\, d\eta$$

$$+ \alpha_s \int_0^1 \varphi(\eta, 0)\, p(\eta)\, d\eta + \int_0^1 \int_0^1 \varphi(\eta, f)\, p(f)\, p(\eta)\, df d\eta$$

where α_c, α_1 denote the probability of particle-off gases and pure inlet gases (including primary and secondary streams) respectively.

Actually, even if we use the local instantaneous unmixedness model, the mixture fraction f is not a conservative scalar, because we have

$$\frac{\partial}{\partial t}(\rho f) + \frac{\partial}{\partial x_j}(\rho v_j f) = \frac{\partial}{\partial x_j}\left(D\rho \frac{\partial f}{\partial x_j} \right) + S$$

where the defination of f is the same as that defined in Sec. 9. 3. In this case, the k, ε, f, g equations must be solved together with the particle-phase equations.

9. 7. 4 Modeling of Reacting Particle Phase

In order to fully account for the turbulent diffusion of particle mass, momentum and energy and the particle history effect, a continuum-trajectory model (Lagrangian-Eulerian model) of reacting particle phase, and a pure multifluid model accounting for the particle history effect have been proposed by the present author.

The idea of the continuum-trajectory model is that the particle-phase continuity and momentum equations in Eulerian coordinates, that is, Eqs. (8. 52) and (8. 54) or Eqs. (8. 86) and (8. 89) are solved by using the multifluid model to find particle velocity and concentration distributions, whereas the particle mass and temperature change due to reaction and heat transfer is tracked along the trajectories or the streamlines obtained from the Eulerian predictions, by using the ordinary differential equations and algebraic expressions, for example, Eqs. (6. 22) to (6. 30). On the other hand, in order to account for the particle history effect in the pure multifluid model, we need to transform the equations describing the particle total mass change and the daf coal mass change in Lagrangian coordinates into those in Eulerian coordinates. In this case, besides previously derived particle-phase equations for turbulent reacting two-phase flows in the multifluid model, such as Eqs. (8. 51), (8. 52), (8. 54), (8. 56) and other closure equations, we need to add a daf coal transport equation in the form of

$$\frac{\partial}{\partial t}(n_k m_{ck}) + \frac{\partial}{\partial x_j}(n_k v_{kj} m_{ck}) = m_{ck} \frac{\partial}{\partial x_j}\left(\frac{\gamma_k}{\sigma_k} \frac{\partial n_k}{\partial x_j} \right) - n_k m_{ck} \exp(-E_v/RT_k)$$

The solution procedure of the turbulent two-phase reacting flows using the continuum-trajectory model of particle phase can be illustrated by the flow chart shown in Fig. 9. 11. There are three kinds of iterations: iterations inside the gas phase; iterations between the gas-phase and particle-phase Eulerian predictions; iterations between trajectory predictions and the gas-phase flow field predictions. Besides, there are three kinds of coupling: gas flow field-particle flow field; particle flow field-particle history; particle history-gas flow field. The application of this model will be discussed in Chater 10. As for the solution procedure of the multifluid model, it has already been discussed in Chapter 8 (Fig. 8. 6).

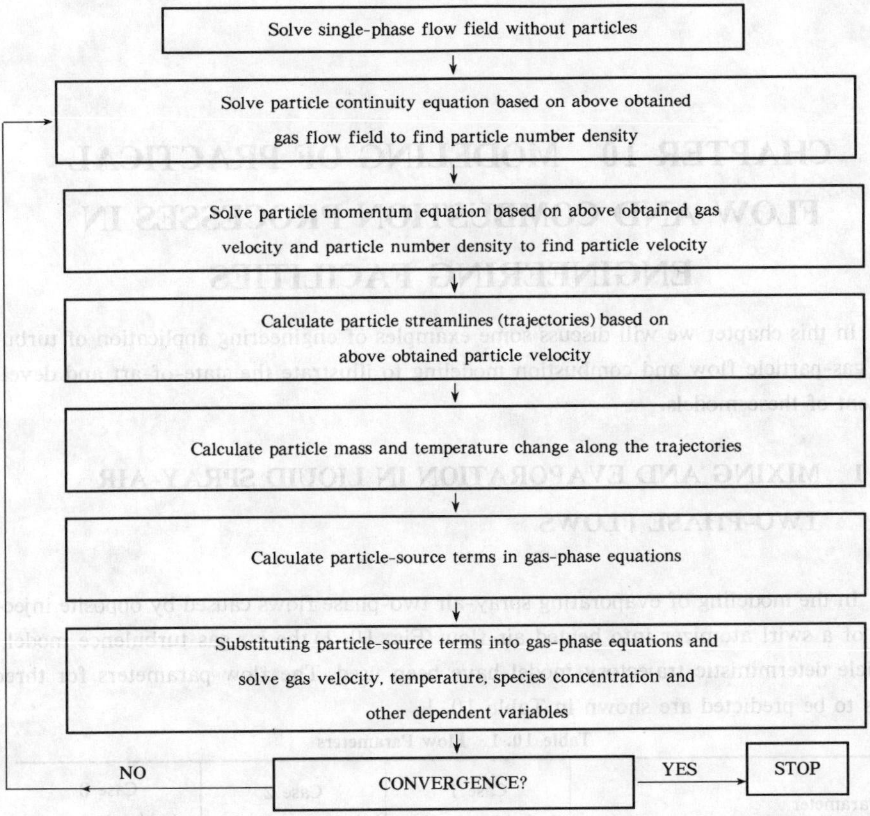

Fig. 9. 11 LEAGAP Flow Chart for Turbulent Reacting Two-phase Flows

CHAPTER 10 MODELING OF PRACTICAL FLOW AND COMBUSTION PROCESSES IN ENGINEERING FACILITIES

In this chapter we will discuss some examples of engineering application of turbulent gas-particle flow and combustion modeling to illustrate the state-of-art and development of these models.

10.1 MIXING AND EVAPORATION IN LIQUID SPRAY-AIR TWO-PHASE FLOWS

In the modeling of evaporating spray-air two-phase flows caused by opposite injection of a swirl atomizer into heated air flow (Fig. 10. 1) the k-ε gas turbulence model, particle deterministic trajectory model have been used. The flow parameters for three cases to be predicted are shown in Table 10. 1.

Table 10.1 Flow Parameters

Parameter	Case 1 (with evap.)	Case 2 (without evap.)	Case 3 (without evap.)
Inlet air velocity (m/s)	86	58	75
Inlet air temperature (K)	473	323	323
Inlet kerosene tempetaure (K)	318	323	323
Kerosene flow rate (kg/s)	8.6×10^{-3}	5.3×10^{-3}	5.3×10^{-3}
Aromization pressure (10 Pa)	2	5	20
Spray angle	72°, 82°, 92°,	50°, 60°, 70°	50°, 60°, 70°
Inlet turbulence intensity	5%	5%	5%
Inlet turbulence scale (m)	0.0027	0.0027	0.0027

The initial droplet size distribution is taken as

$$\exp\left[-\left(\frac{d}{70}\right)^{1.84}\right], \quad \exp\left[-\left(\frac{d}{44}\right)^{1.78}\right]$$

The predicted results are shown in Fig. 10. 2 to 10. 4. Predicted droplet trajectories of medium sizes (41. 2μm to 103μm) for case 2 are in fairly good agreement with experiments (Fig. 10. 2). The predicted gas velocity profiles in two-phase flows for case 3 are in good agreement with experiments (Fig. 10. 3). It can be seen that the dense liquid spray near the nozzle of the atomizer, like a bluff body, causes a wake in the gas-phase

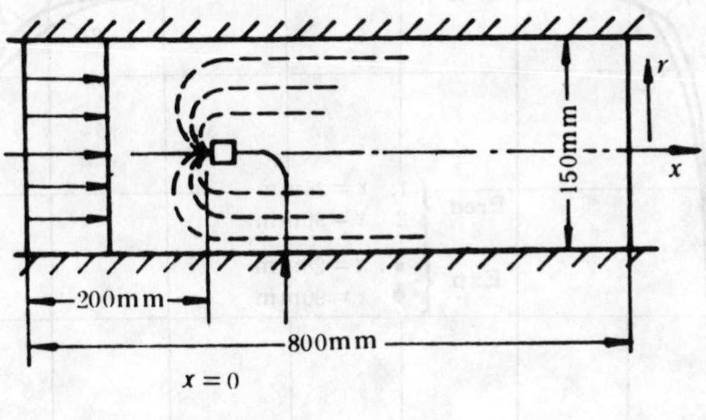

Fig. 10. 1

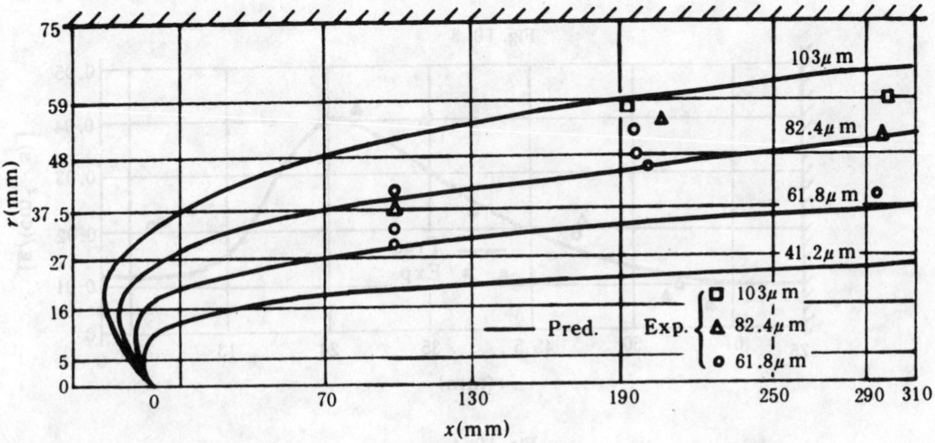

Fig. 10. 2

velocity field. The predicted liquid-phase mass flux at $x = 355$mm behind the atomizer for case 1 is in general agreement with experiments (Fig. 10. 4) except in the near-axis region where the measured mass flux is higher than that predicted, since actually there is turbulent diffusion of small droplets which is neglected in the particle deterministic trajectory model. In a similar way the evaporating spray-air two-phase flows in the diffusor of a turbofan afterburner (Fig. 10. 5) were simulated. Predicted gas velocity (Fig. 10. 6) and temperature (Fig. 10. 7) distributions in the absence of liquid spray are in good agreement with experiments. Predicted droplet trajectories (Fig. 10. 8) and fuel vapor concentration field (Fig. 10. 9) indicate higher fuel concentration in the near-wall low temperature zone which is unfavorable to flame stabilization and combustion.

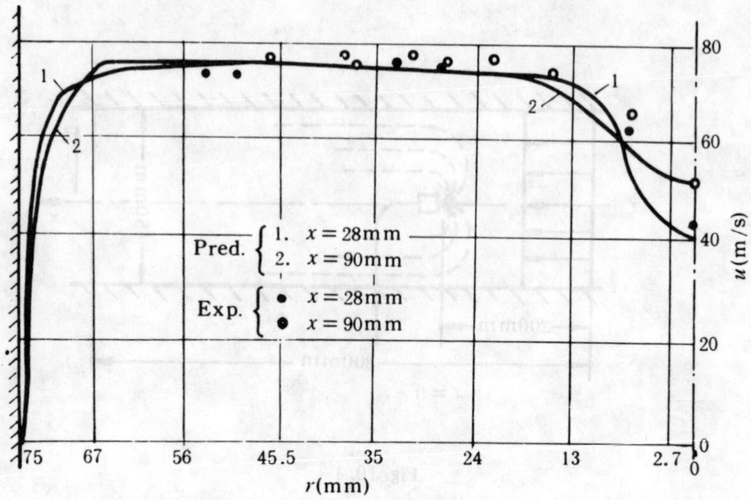

Fig. 10. 3

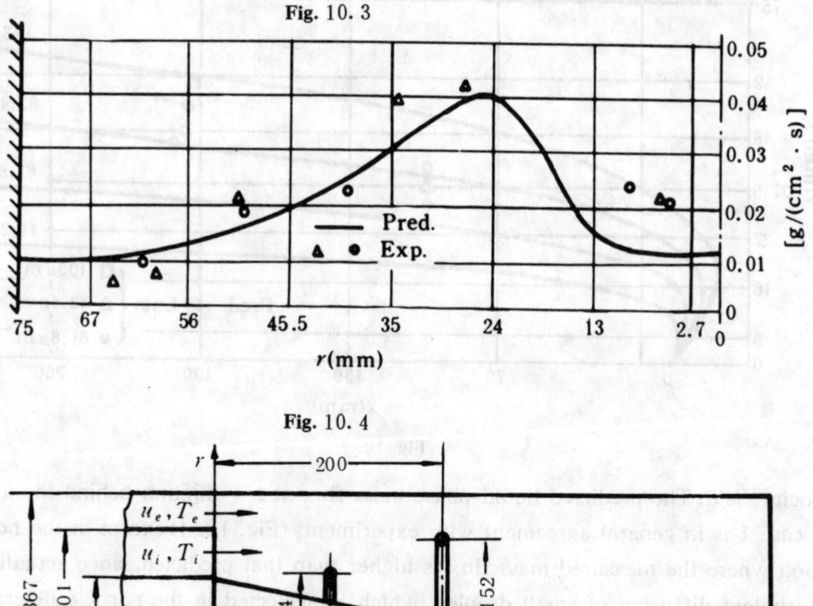

Fig. 10. 4

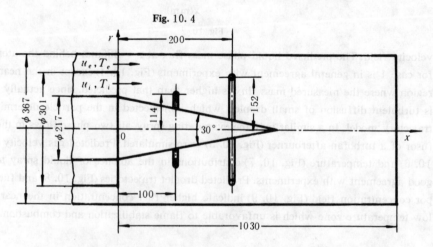

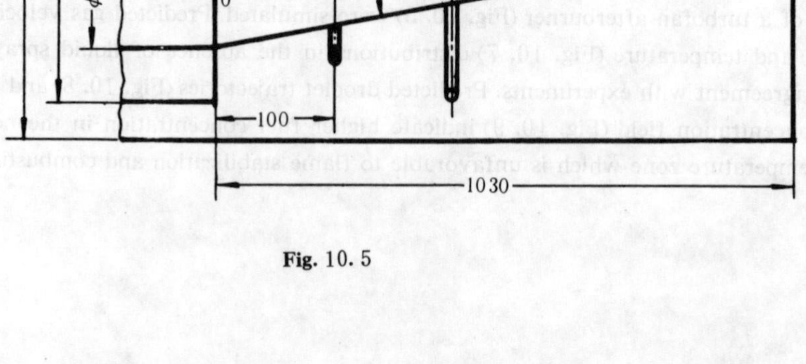

Fig. 10. 5

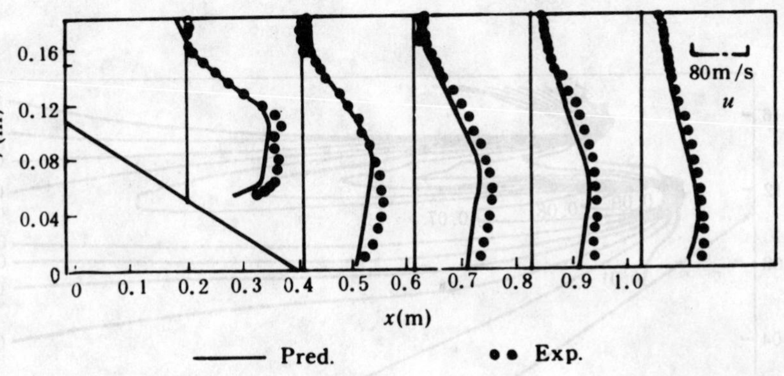

Fig. 10. 6

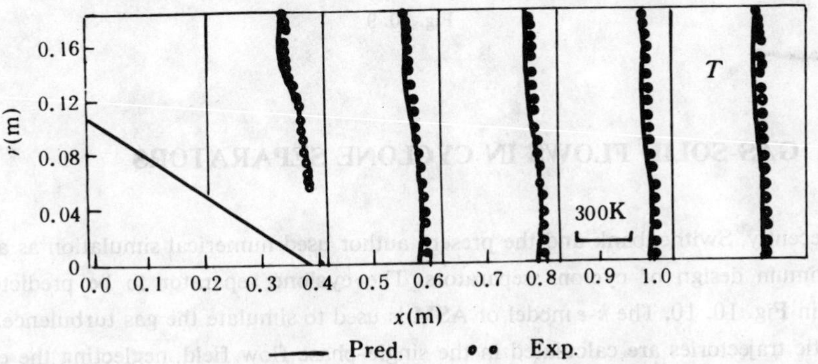

Fig. 10. 7

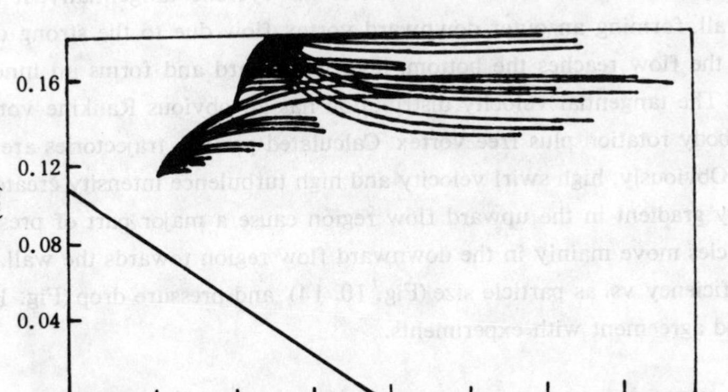

Fig. 10. 8

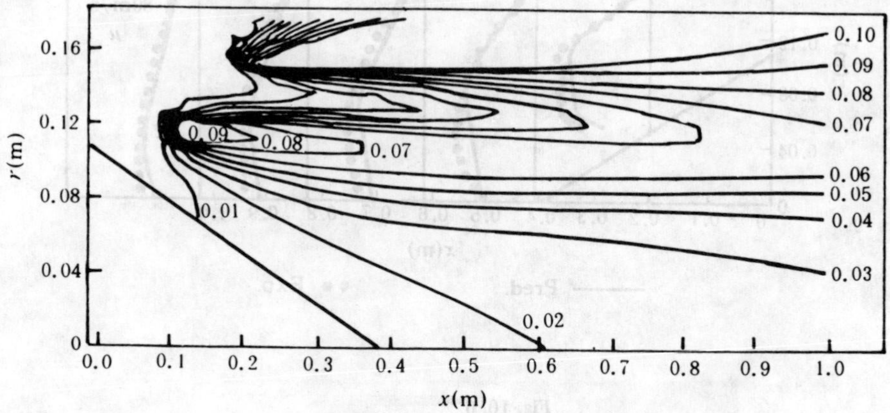

Fig. 10. 9

10. 2 GAS-SOLID FLOWS IN CYCLONE SEPARATORS

Recently, Swithenbank and the present author used numerical simulation as a tool for optimum design of cyclone separators. The cyclone separator to be predicted is shown in Fig. 10. 10. The k-ε model or ASM is used to simulate the gas turbulence. The stochastic trajectories are calculated in the single-phase flow field, neglecting the effect of particles on gas-phase flow field. Predicted axial velocity, tangential velocity and pressure profiles are shown in Figs. 10. 11 and 10. 12. Predictions are in general agreement with experiments. As the air flow enters the cyclone tangentially, it moves towards the wall, forming an outer downward vortex flow due to the strong centrifugal force. After the flow reaches the bottom, it turns upward and forms an inner upward vortex flow. The tangential velocity distribution has an obvious Rankine-vortex structure—solid-body rotation plus free vortex. Calculated particle trajectories are shown in Fig. 10. 13. Obviously, high swirl velocity and high turbulence intensity created by high swirl velocity gradient in the upward flow region cause a major part of pressure loss, but the particles move mainly in the downward flow region towards the wall. Predicted collection efficiency vs. as particle size (Fig. 10. 14) and pressure drop (Fig. 10. 15) are in fairly good agreement with experiments.

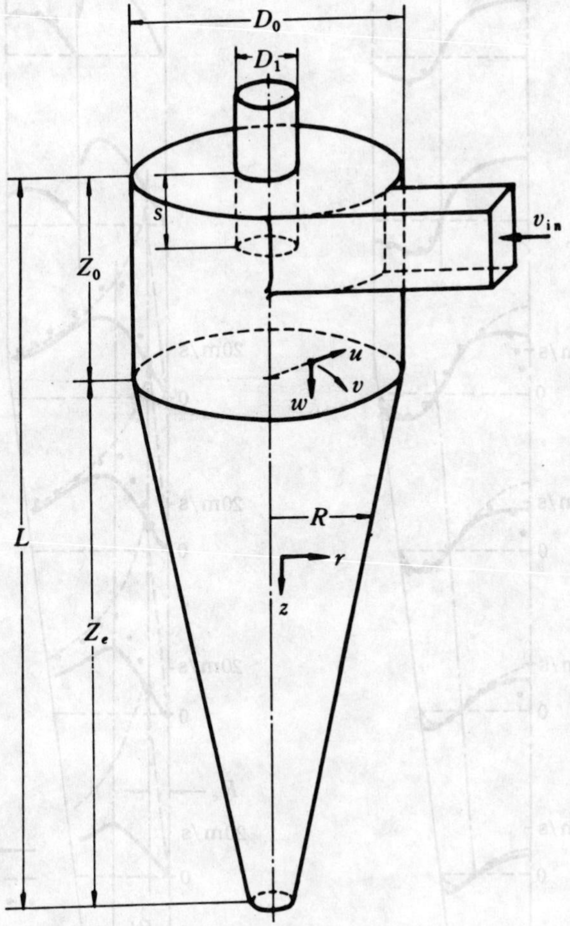

Fig. 10. 10

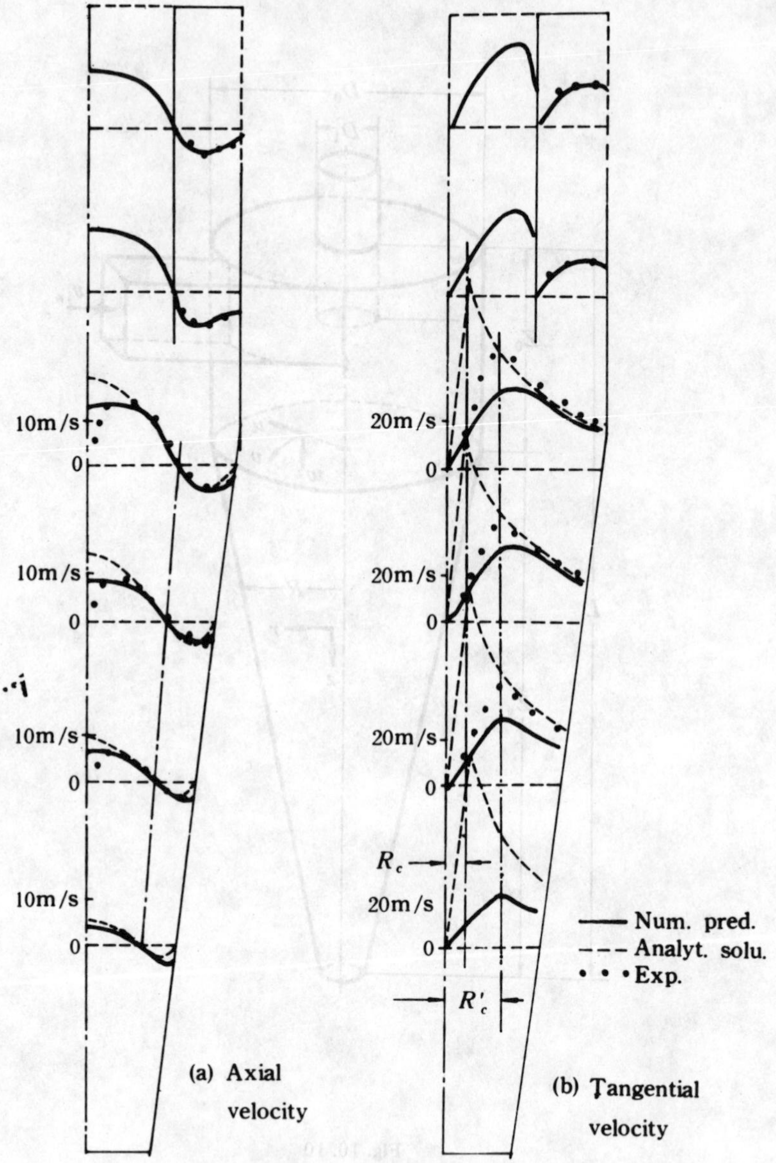
10m/s
0
10m/s
0
10m/s
0
10m/s
0
(a) Axial velocity
20m/s
0
20m/s
0
20m/s
0
R_c
20m/s
0
R'_c
Num. pred.
Analyt. solu.
Exp.
(b) Tangential velocity

Fig. 10. 11

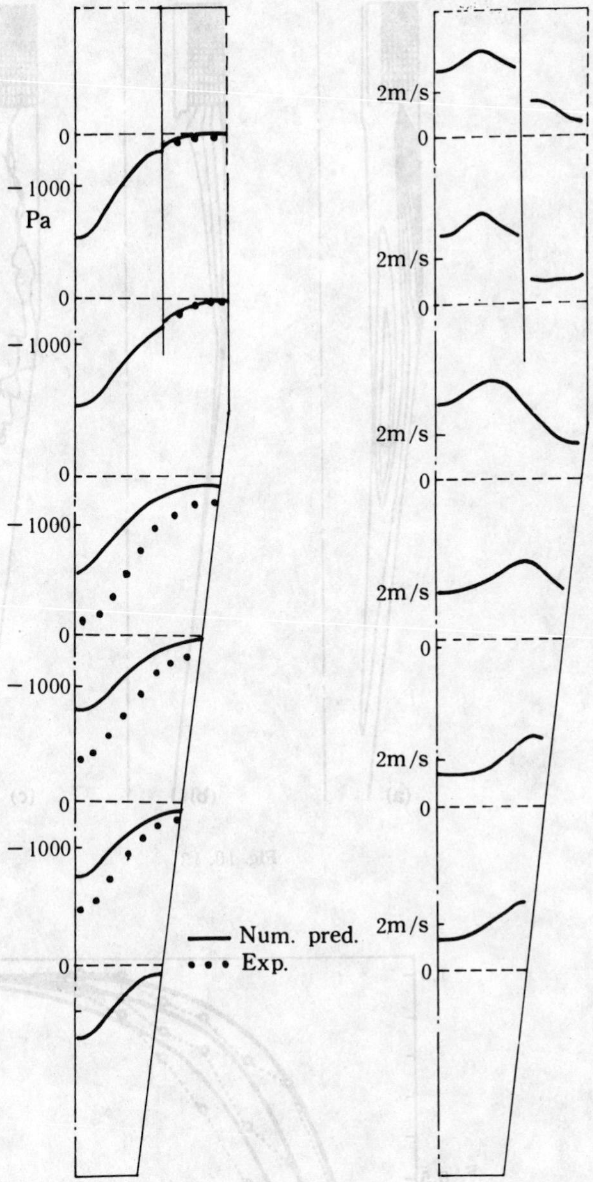
0
− 1000
Pa
0
− 1000
0
− 1000
0
− 1000
0
− 1000
0
Num. pred.
Exp.
2m/s
0
2m/s
0
2m/s
0
2m/s
0
2m/s
0
2m/s
0

Fig. 10. 12

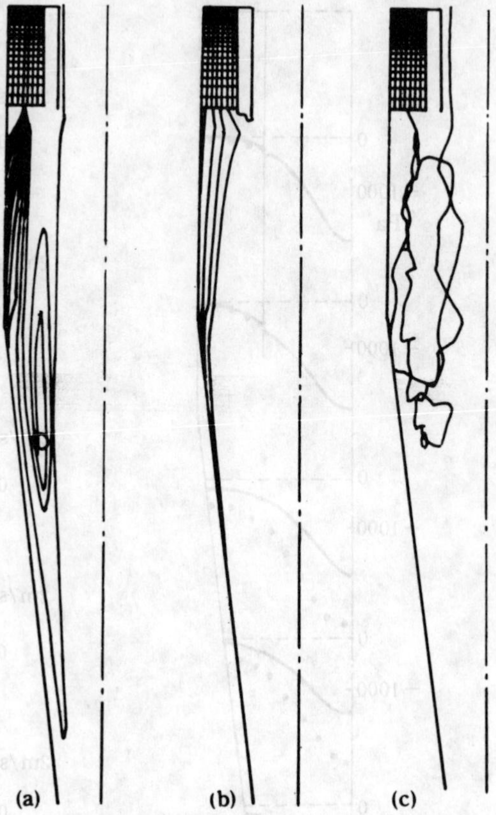

Fig. 10. 13

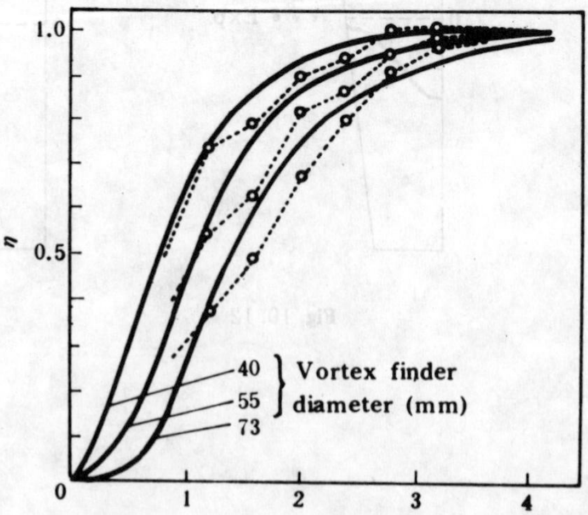

Fig. 10. 14

10.3 LIQUID SPRAY COMBUSTION

The liquid spray combustion in a confined two-phase jet has been simulated by X. Q. Zhou using a multifluid model, in which the particle mass diffusion is taken into account, but the particle momentum and energy diffusion (Reynolds stresses and heat fluxes) are neglected. The k-ε gas turbulence model. algebraic particle turbulence model and EBU-Arrhenius gas combustion model are adopted. Radiation has not been taken into account. Predicted gas velocity, temperature, particle velocity and liquid-phase mass fraction for evaporation without combustion are shown in Fig. 10.16 to 10.18. Predicted gas temperature, species and particle number density for liquid-spray combustion are shown in Figs. 10.19 and 10.20. The results are qualitatively reasonable, although no experimental verification has been reported. Banhawy and Whitelaw have simulated kerosene-spray combustion in a sudden-expansion combustor by using the k-ε gas turbulence model, deterministic particle trajectory model and pdf-fast-reaction gas combustion model. The predicted gas velocity and temperature during spray combustion for different cases and their comparison with experiments are shown in Figs. 10.21. and 10.22. Predicted temperature is much lower than that measured in the near-axis region. This comparison indicates the shortcomings of the deterministic trajectory model which neglects the actually existing particle diffusion.

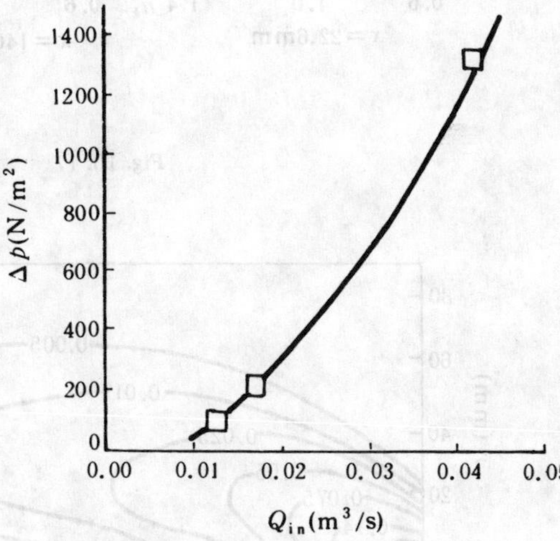

Fig. 10.15

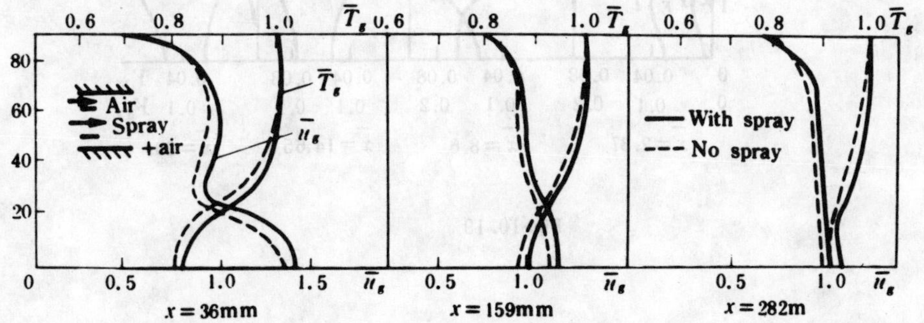

Fig. 10.16

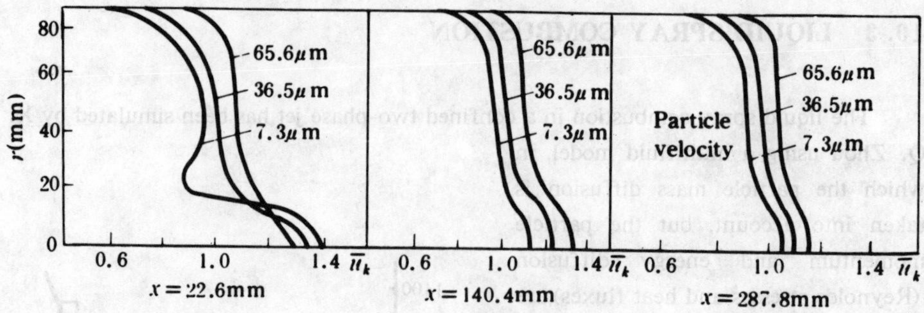

Fig. 10. 17

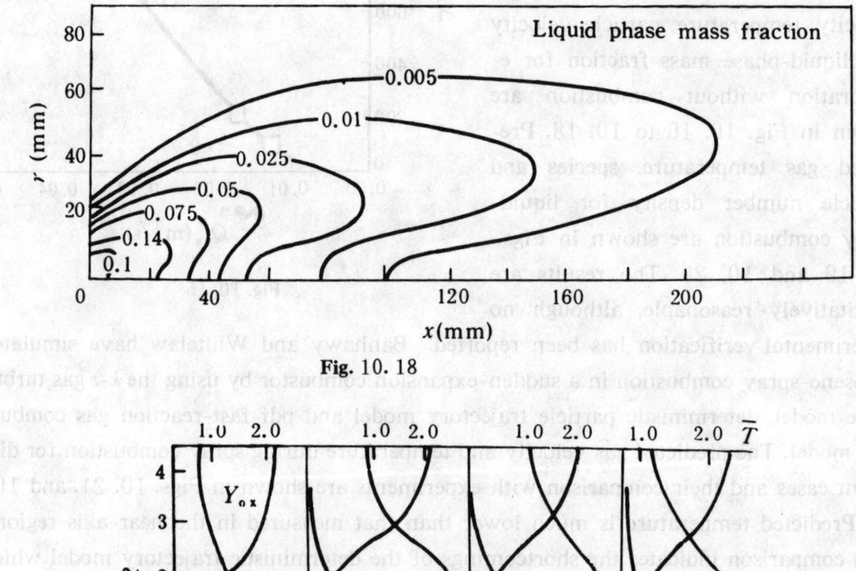

Fig. 10. 18

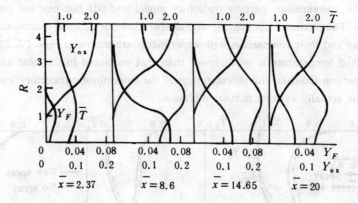

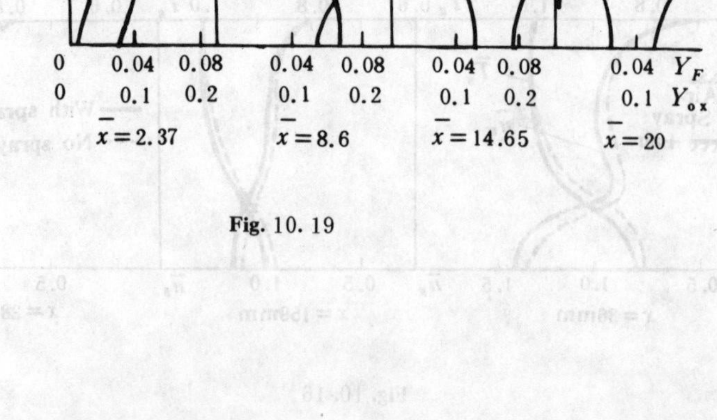

Fig. 10. 19

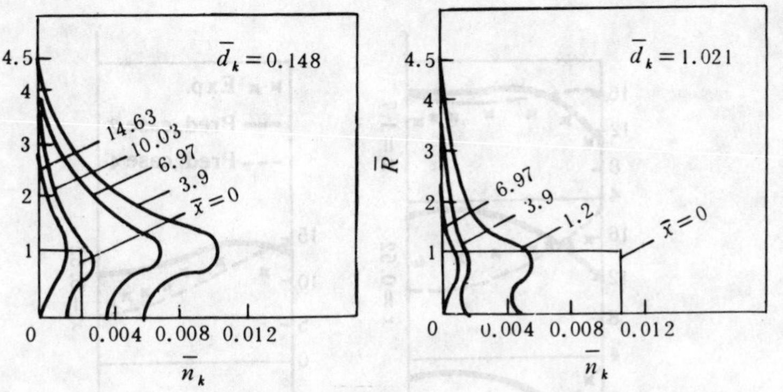

Fig. 10.20

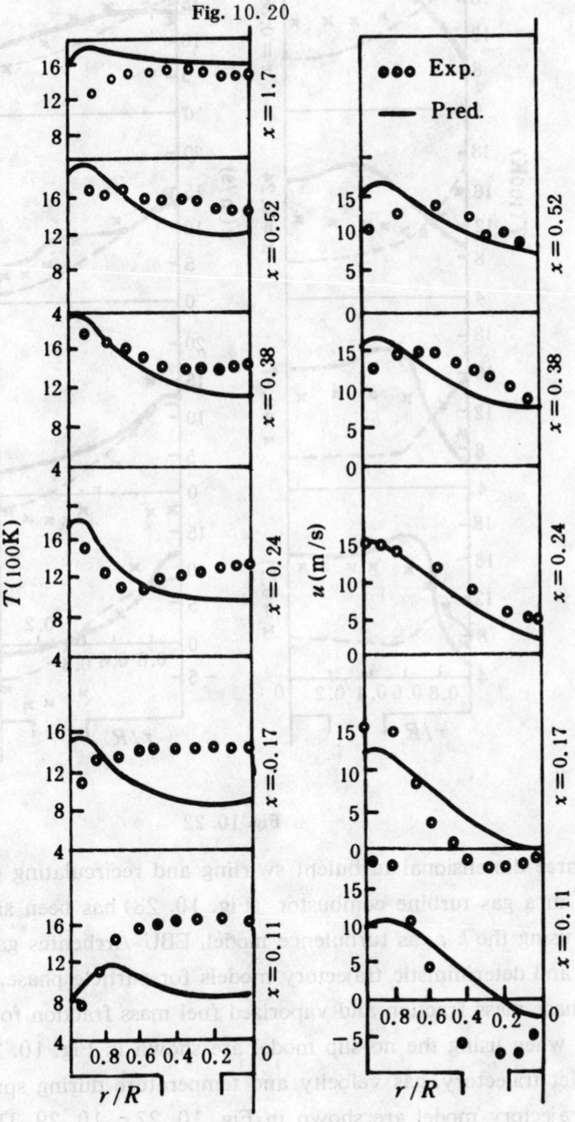

Fig. 10.21

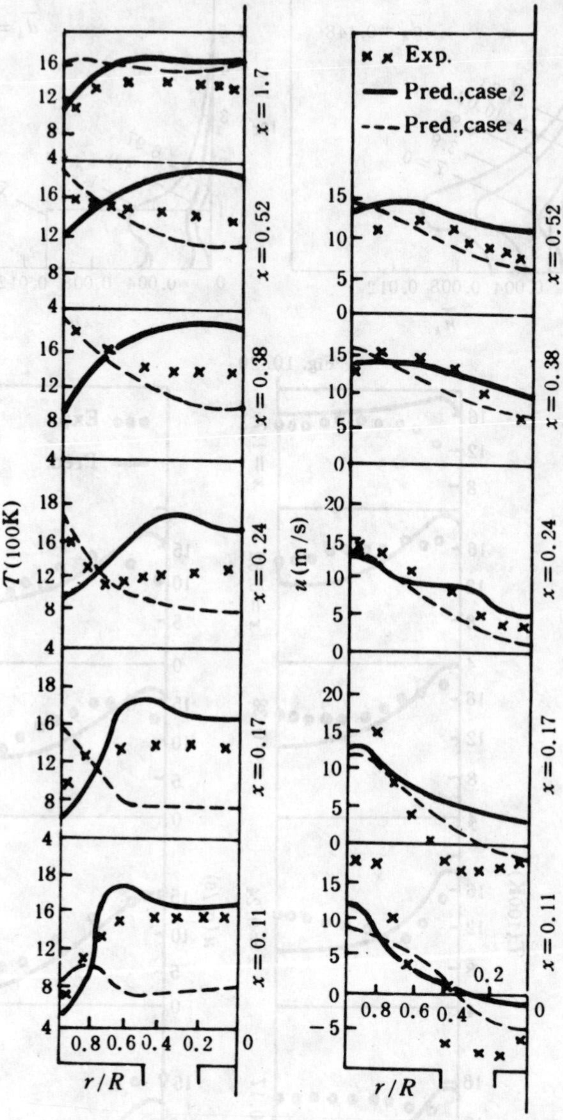

Fig. 10. 22

The three-dimensional turbulent swirling and recirculating gas-droplet flows and combustion in a gas-turbine combustor (Fig. 10. 23) has been simulated by Swithenbank et al. using the k-ε gas turbulence model, EBU-Arrhenius gas combustion model, the no-slip and deterministic trajectory models for particle phase. Predicted gas velocity, liquid-phase mass fraction and vaporized fuel mass fraction for evaporation without combustion when using the no-slip model are shown in Fig. 10. 24—10. 26. The predicted droplet trajectory, gas velocity and temperature during spray combustion when using the trajectory model are shown in Fig. 10. 27—10. 29. The droplet trajectories are concentrated in the head zone (primary zone) where a strong gas recirculation exists. The high temperature is developed in the head zone and near the wall. No comparison with experiments was made for velocity or temperature field inside the combustor, although comparison was made for the exit velocity and temperature profiles.

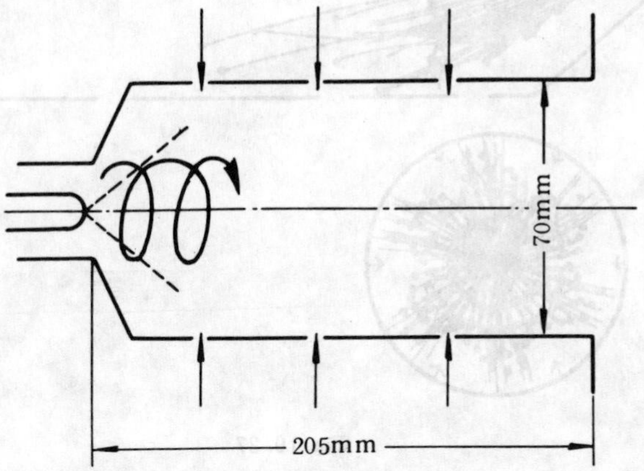

Fig. 10. 23

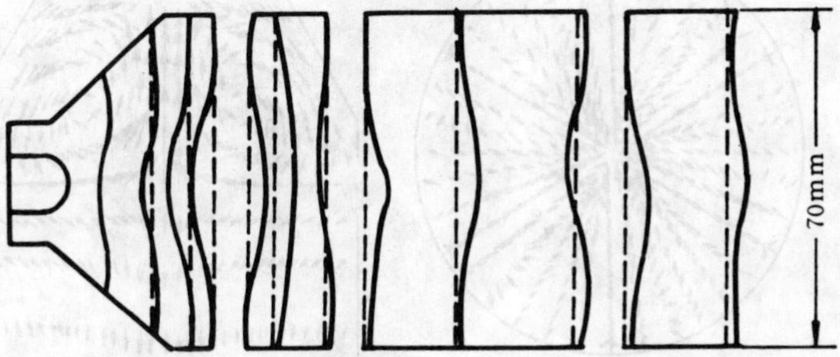

Fig. 10. 24

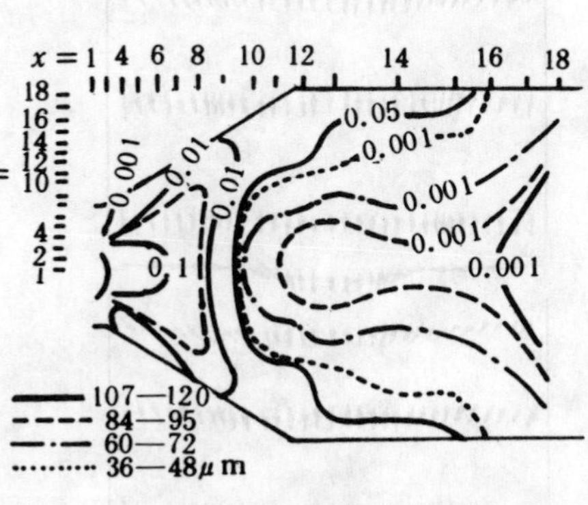

Fig. 10. 25

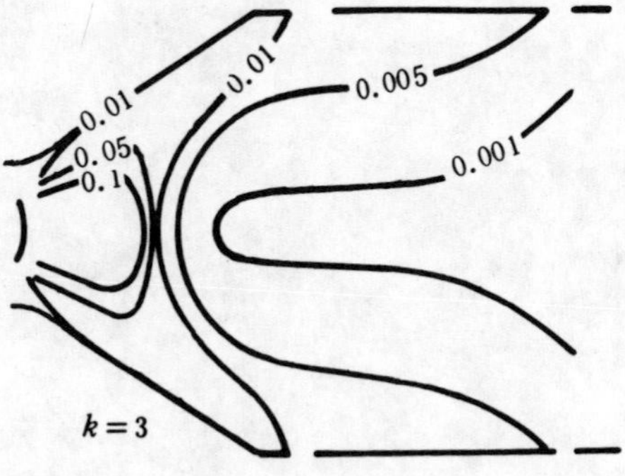

Fig. 10. 26

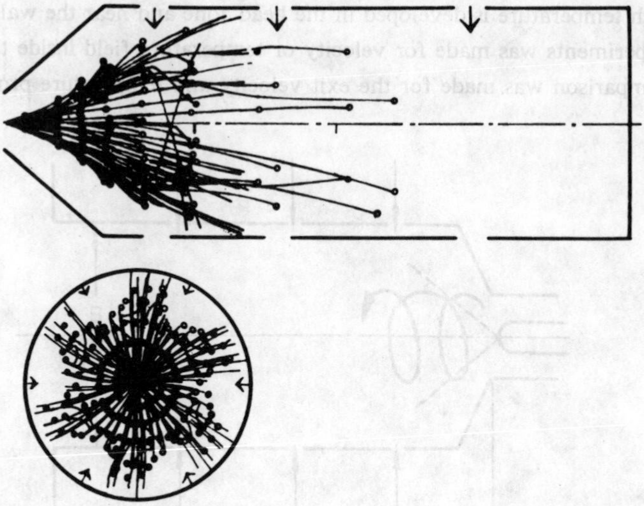

Fig. 10. 27

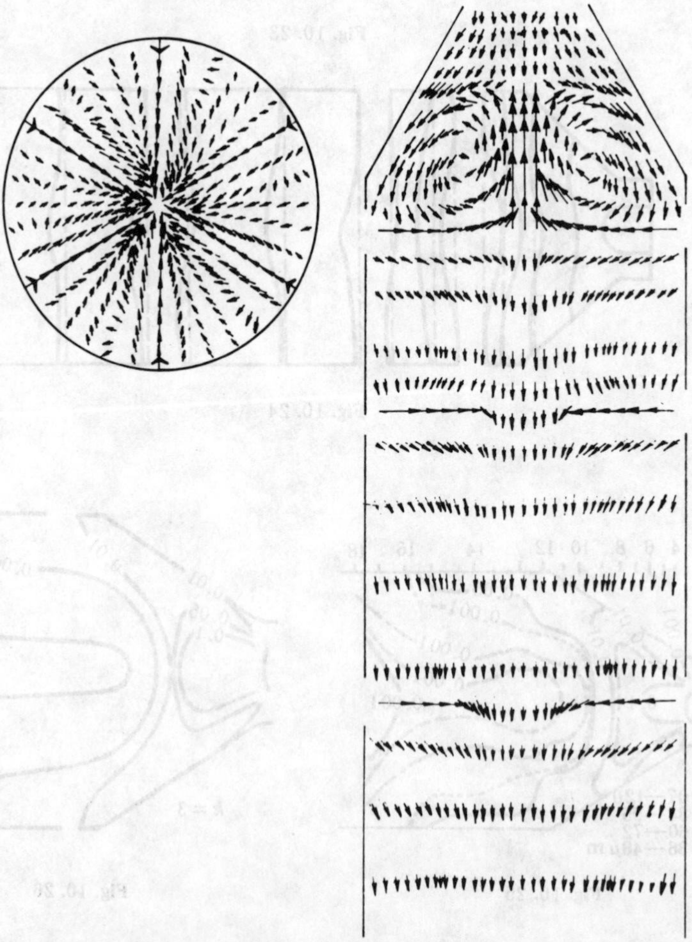

Fig. 10. 28

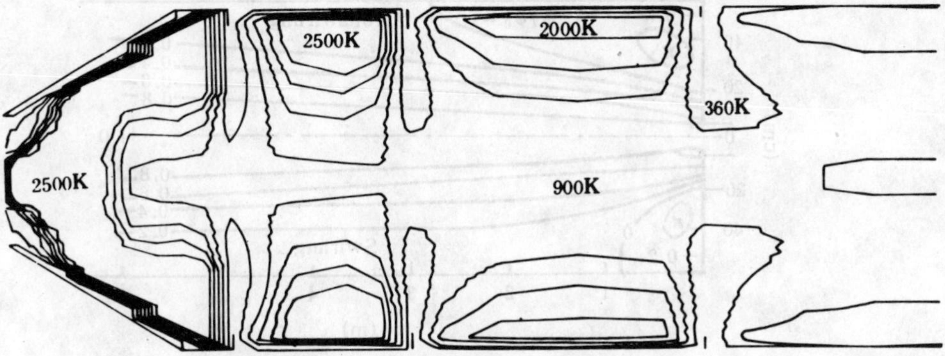

Fig. 10. 29

10. 4 PULVERIZED-COAL-AIR TWO-PHASE FLOWS AND COMBUSTION

The pioneer work on numerical modeling of pulverized-coal-air two-phase flows and combustion in coaxial jets discharged into a sudden-expansion combustor (Fig. 10. 30) was done by Gibson and Morgan. They used an empirical algebraic turbulence model, no-slip particle-phase model, pdf-fast-reaction gas combustion model, two-flux radiative heat transfer model, d^2 -law of char diffusion combustion and an assumption of infinitively fast initial devolatilization. The predicted gas streamlines, gas temperature field, wall heat flux and coal burnout rate are shown in Figs. 10. 31—10. 34. Predictions are in qualitative agreement with experiments, however, quantitatively, the predicted high temperature zone and heat flux peak are much nearer to the entrance than those observed in experiments, or in other words, the process predicted in numerical modeling develops much faster than what observed in experiments. Obviously, this discrepancy is caused by the drawback of the no-slip model which neglects the velocity lag and the preheating of particles, and by the simplification of infinitively fast devolatilization.

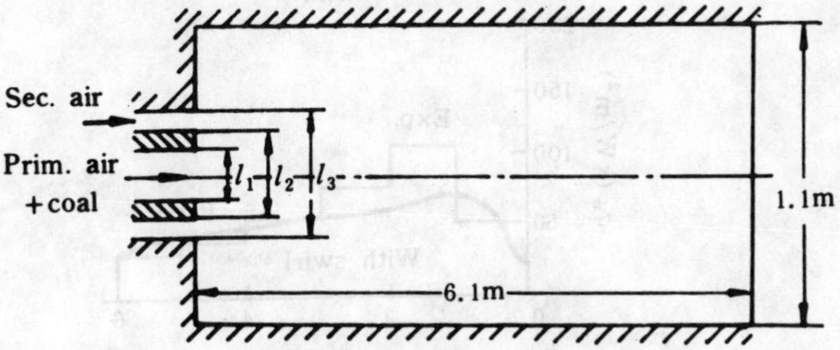

$l_1 = 0.08 - 0.1\text{m}$

$l_2 = 0.15 - 0.3\text{m}$

$l_3 = 0.38 - 0.4\text{m}$

Fig. 10. 30

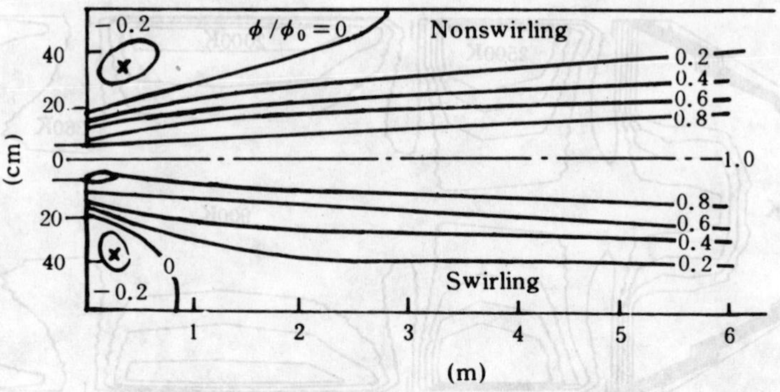

Fig. 10. 31

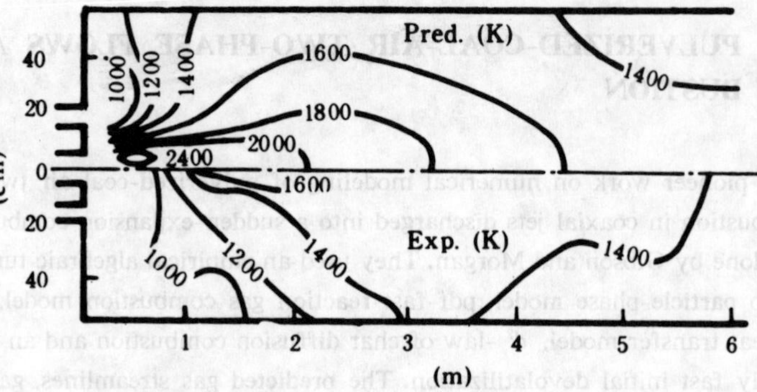

Fig. 10. 32

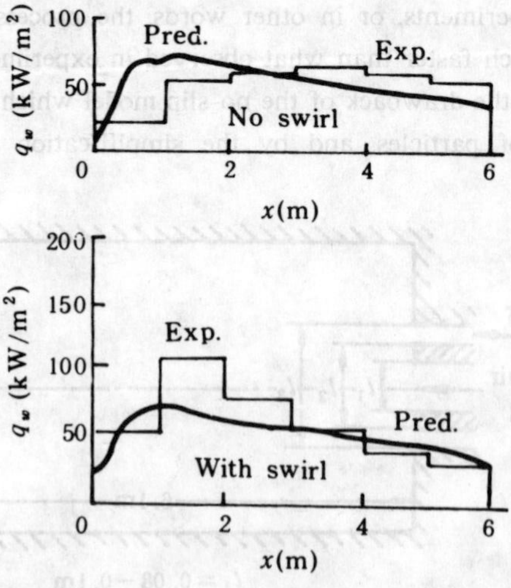

Fig. 10. 33

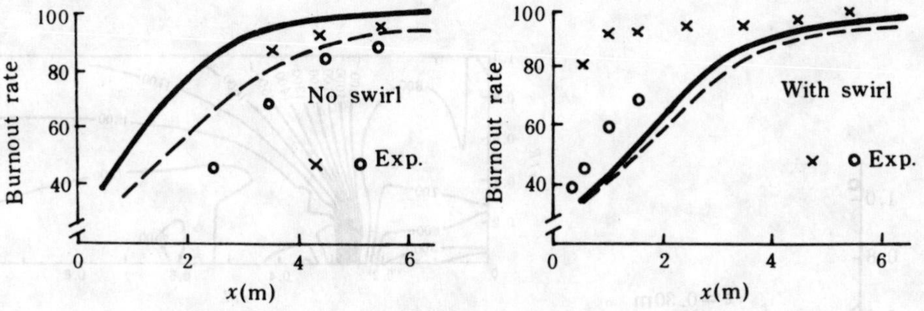

Fig. 10. 34

The same case of pulverized-coal combustion in a sudden-expansion combustor with primary-air and secondary-air coaxial jets was simulated by L. D. Smoot and P. J. Smith in recent years using much more thoroughly elaborated comprehensive models, based on which the computer codes PCGC-2 and PCGC-3 have been developed. In PCGC-2 the k-ε gas turbulence model, deterministic particle trajectory model with a drift-velocity modification, pdf-local-instantaneous-equilibrium model of gas-phase combustion, four-flux radiative heat transfer model, two-equation devolatilization model and a semiempirical diffusion-kinetic char combustion model are used. The predicted particle trajectories (Fig. 10. 35) show the entrainment of some particles into the near-wall recirculation zone. The prediction results of particle mass flux (Fig. 10. 36) are obtained actually by using the no-slip model with a particle Schmidt number of $\sigma_p = 0.35$. Figs. 10. 37 and 10. 38 are predicted gas temperature and NO_x concentration maps for cases of different heat transfer conditions and different treatment of effect of turbulence on reaction. Fig. 10. 39 gives predicted O_2, CO_2 and H_2O concentration profiles and their comparison with experiments. The agreement is fairly good, but there is still discrepancy between predictions and experiments, which shows that there is a need to further improve the submodels.

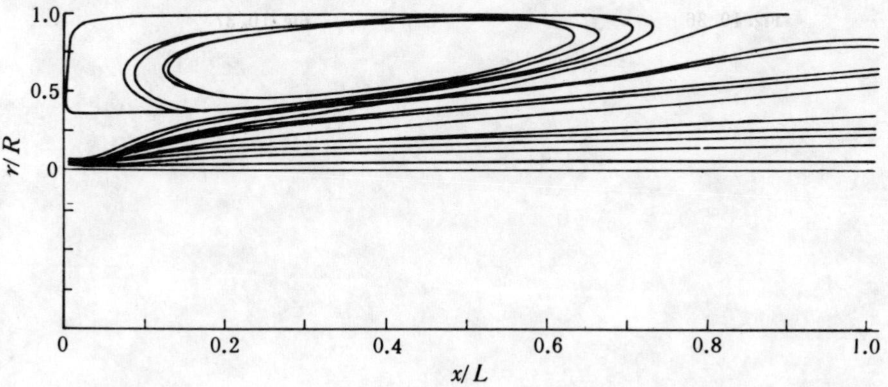

Fig. 10. 35

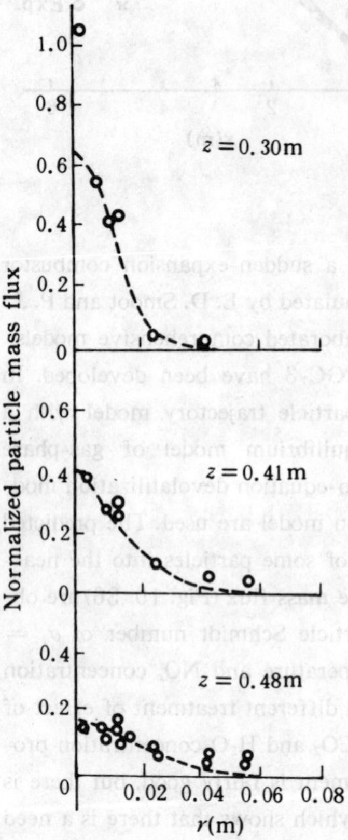

Fig. 10.36

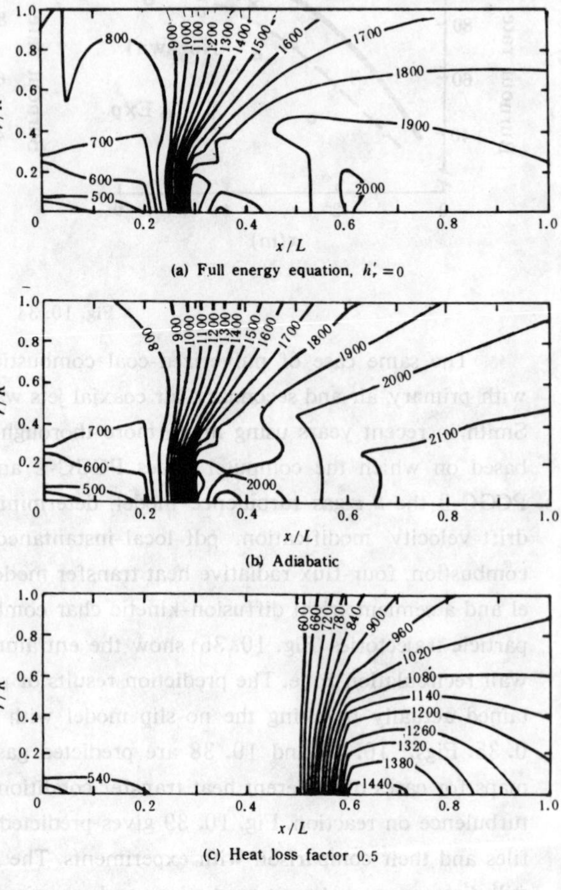

Fig. 10.37

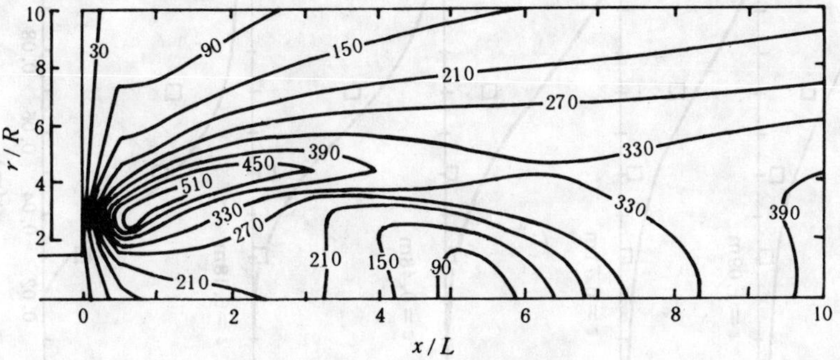

(a) No fluctuation

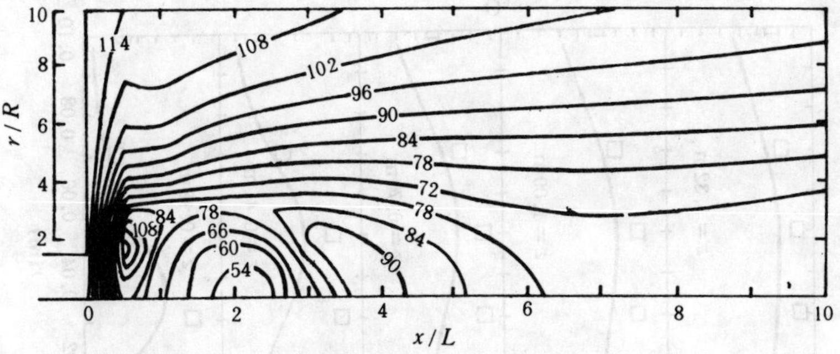

(b) With fluctuation

Fig. 10. 38

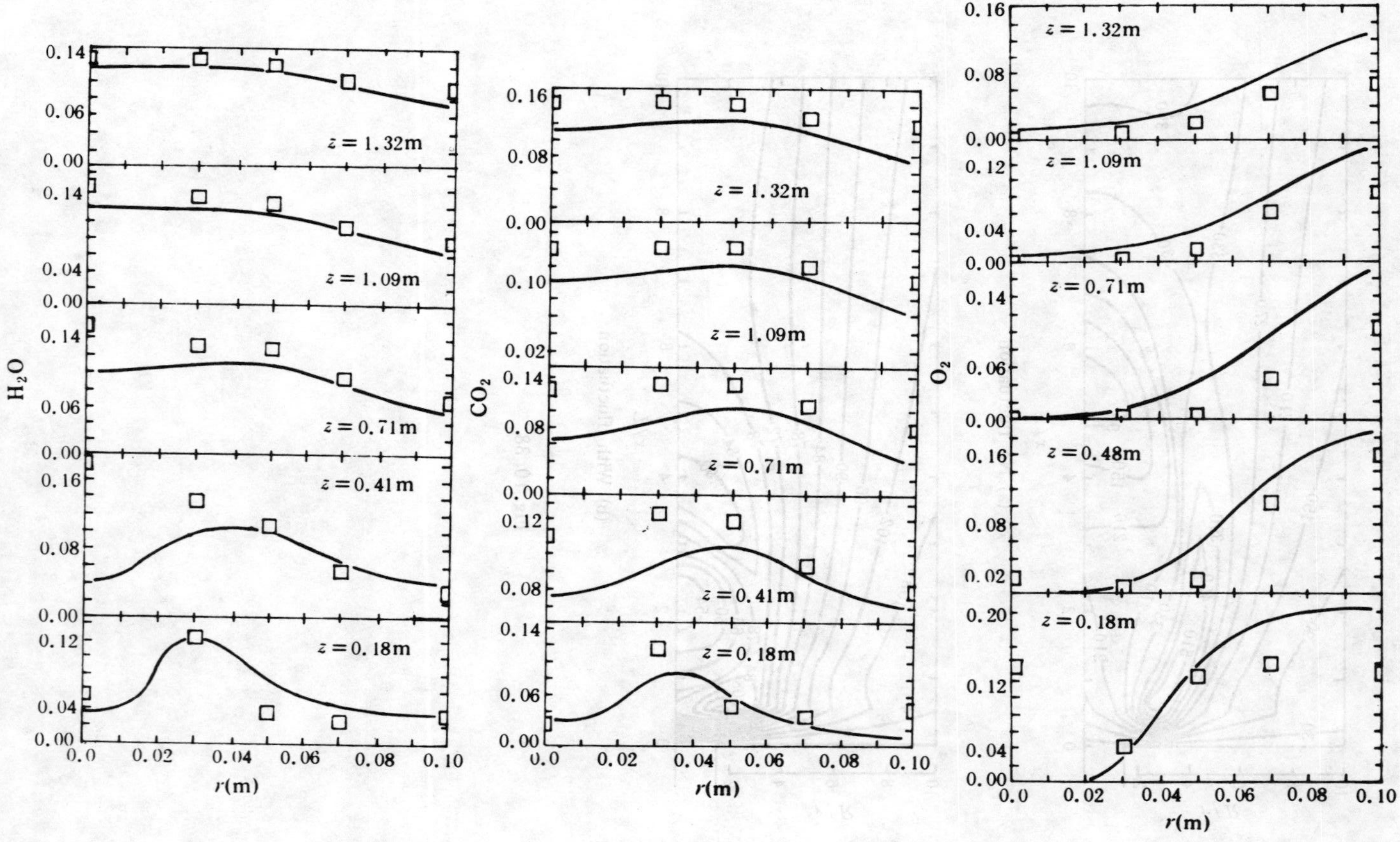

H₂O
CO₂
O₂
z = 1.32m
z = 1.09m
z = 0.71m
z = 0.48m
z = 0.41m
z = 0.18m
r(m)
0.0
0.02
0.04
0.06
0.08
0.10
0.00
0.12
0.14
0.16
0.20

Fig. 10. 39

In recent five years the three-dimensional turbulent recirculating gas-particle flows and pulverized-coal combustion in a sudden-expansion combustor with high-velocity jets discharged from the backward facing step (Fig. 10. 40) (so-called co-flow jets with large velocity difference) have been simulated by the present author and his colleagues. The computer code developed in the Laboratory of Two-phase Flows and Combustion, Tsinghua University, is called LEAGAP. The main feature of the comprehensive model is to use the continuum-trajectory model for the reacting particle phase, and an attempt has been made to use the k_p model for particle turbulence, besides the conventional k-ε model of gas-turbulence and the EBU-Arrhenius model of gas-phase combustion.

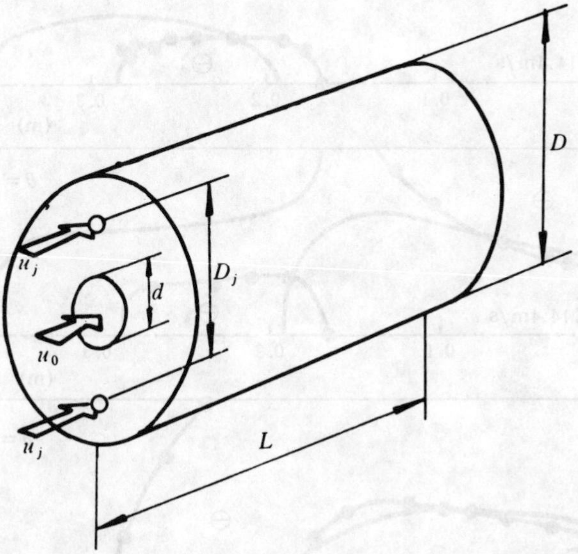

Fig. 10. 40

Predicted isothermal single-phase flow field in two model-combustors of different sizes and its comparison with experiments (Figs. 10. 41 and 10. 42) carried out by using LDV measurements show that in the x-r plane of $\theta = 0°$ (the plane containing two jet holes) a strong reverse flow zone in the near-axis region is induced by the high-velocity jets, while in the plane of $\theta = 90°$ a large size reverse flow zone combining the central and near-wall reverse flows is formed. Predictions and measurements give the same general feature of reverse flow zones, although there is still some quantitative disagreement, perhaps due the shortcomings of the turbulence model and the inaccuracy in measurements. Prediction results for isothermal gas-particle flows are shown in Figs. 10. 43 —10. 48, including the results obtained by using the k-ε-A_p model (A_p denotes the algebraic model of particle turbulence) and the k-ε-k_p model. There is no remarkable difference between the predicted gas-phase and particle axial velocity and particle mass flux given by the k-ε-A_p model (Figs. 10. 43 and 10. 45) and those given by the k-ε-k_p model (Figs. 10. 46 and 10. 48). Predicted gas and particle velocity have the same tendency as those given by LDV measurements. There is very large velocity slip between two phases, in particular in the central reverse flow zone where the particle velocity is much larger than the gas velocity and particles keep their forward motion. Predicted particle mass flux profiles by both two models (Figs. 10. 45 and 10. 48) are in good agreement with experimental results obtained by sampling probe measurements. The results show that a large amount of particles enter the central recirculation zone. All these features of the flow field are favorable to flame stabilization and combustion intensification.

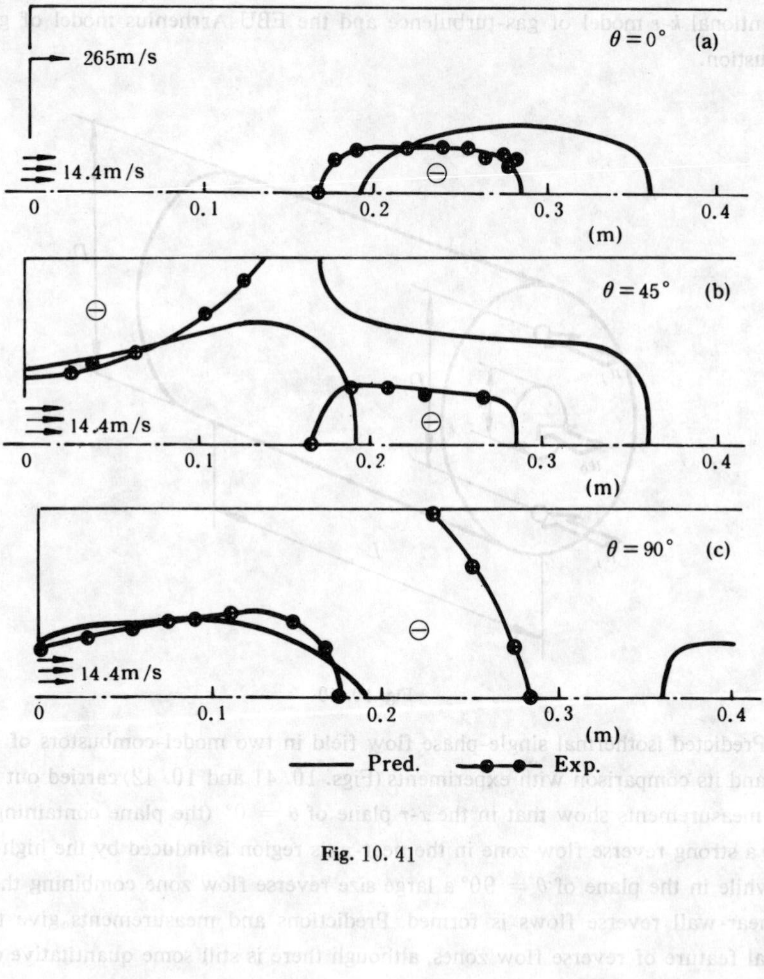

Fig. 10. 41

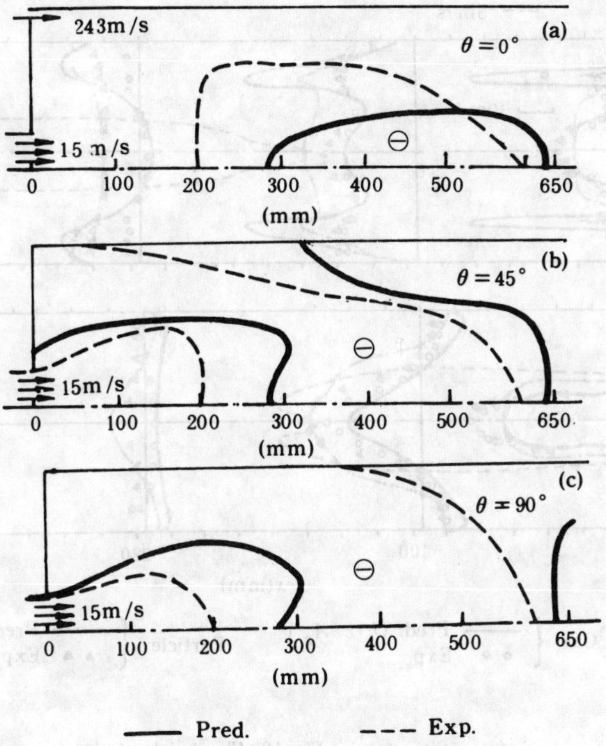
243m/s
(a)
θ = 0°
15 m/s
0
100
200
300
400
500
650
(mm)
(b)
θ = 45°
15m/s
0
100
200
300
400
500
650
(mm)
(c)
θ = 90°
15m/s
0
100
200
300
400
500
650
(mm)
Pred.
Exp.

Fig. 10. 42

Gas $\left\{\begin{array}{l} \text{—— Pred. } (k\text{-}\varepsilon\text{-}A_p) \\ \circ\ \circ\ \text{Exp.} \end{array}\right.$ Particle $\left\{\begin{array}{l} \text{—·— Pred. } (k\text{-}\varepsilon\text{-}A_p) \\ \blacktriangle\ \blacktriangle\ \text{Exp.} \end{array}\right.$

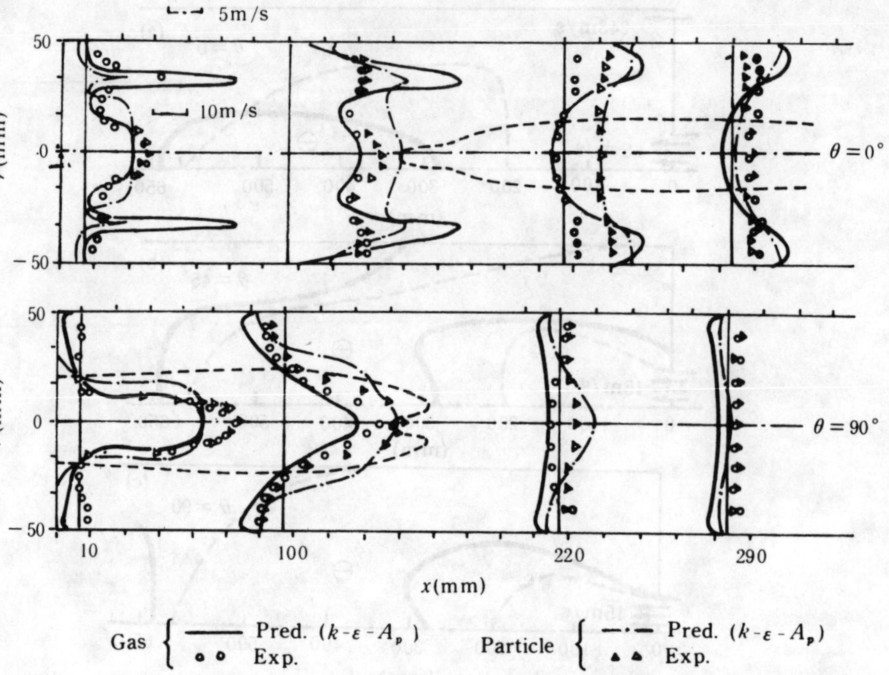

Fig. 10. 43   Axial velocity

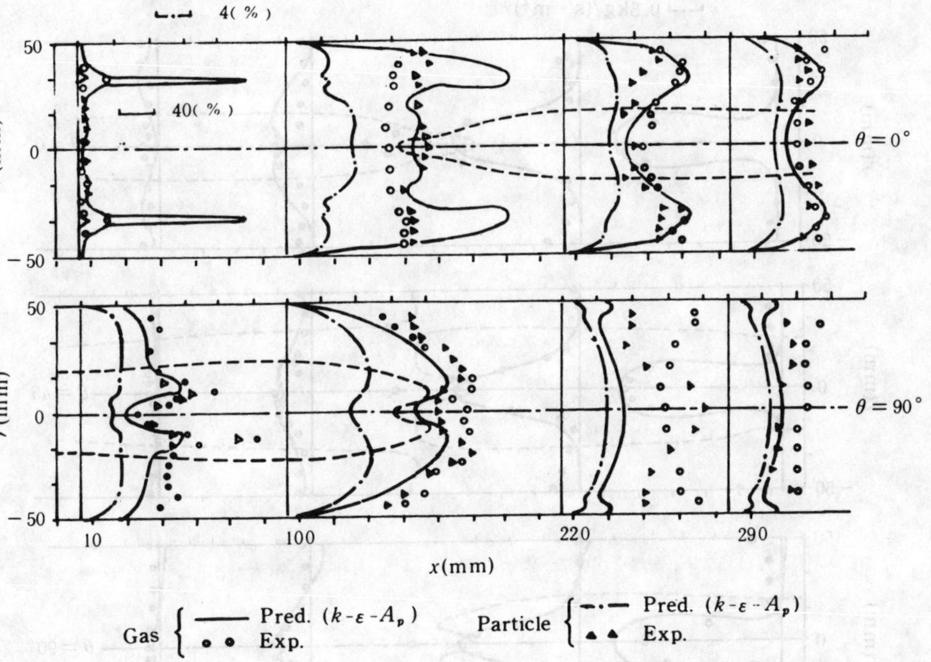

Fig. 10. 44 Turbulence intensity

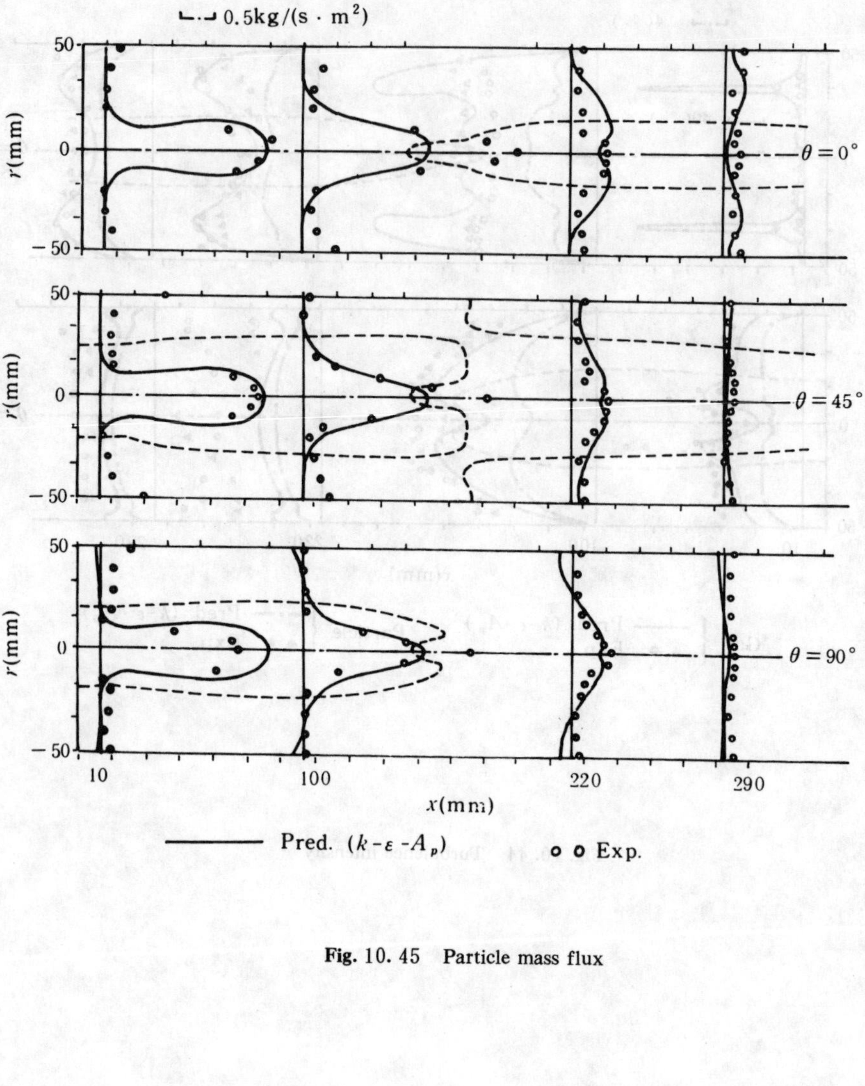

Fig. 10. 45 Particle mass flux

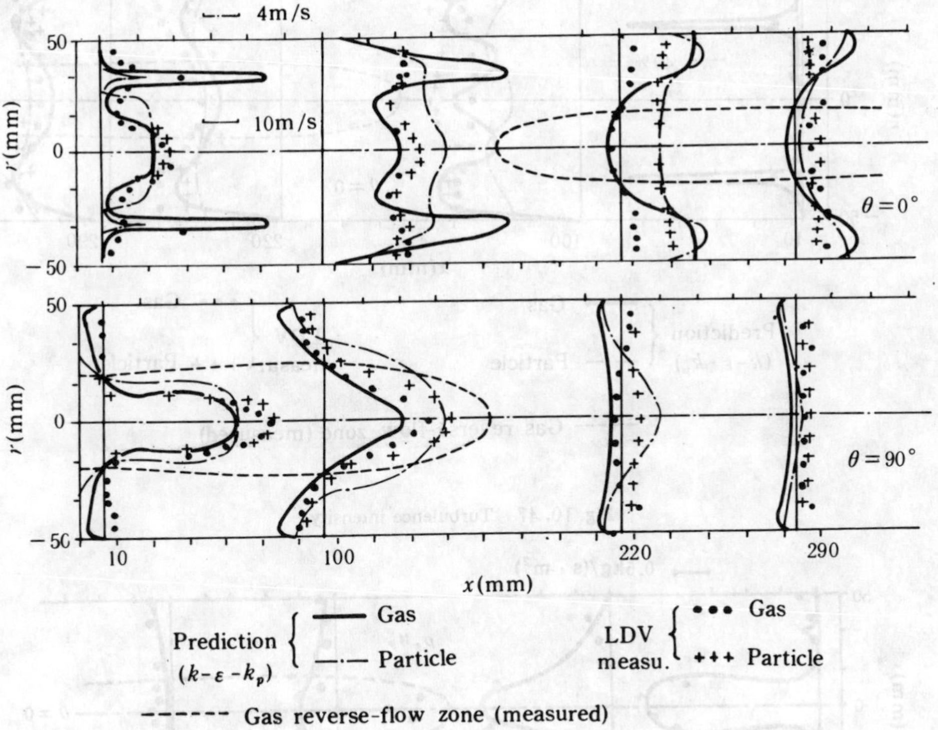

Fig. 10. 46 Axial velocity

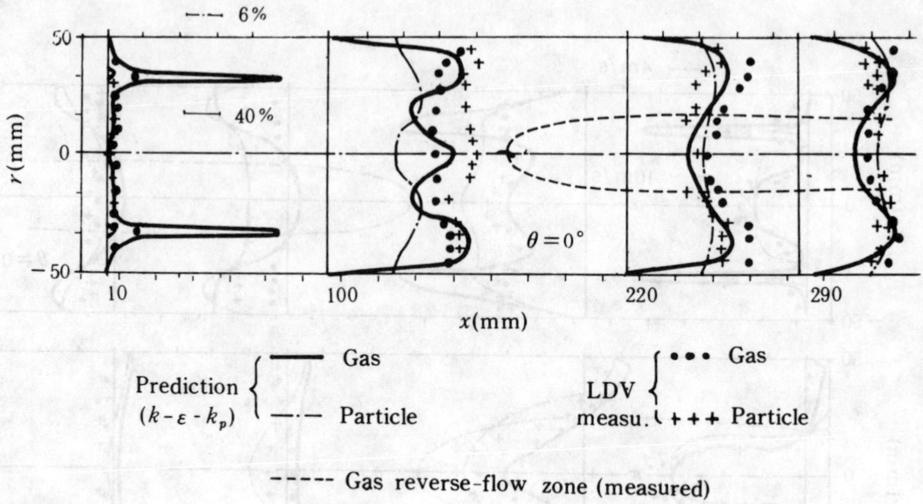

Fig. 10. 47 Turbulence intensity

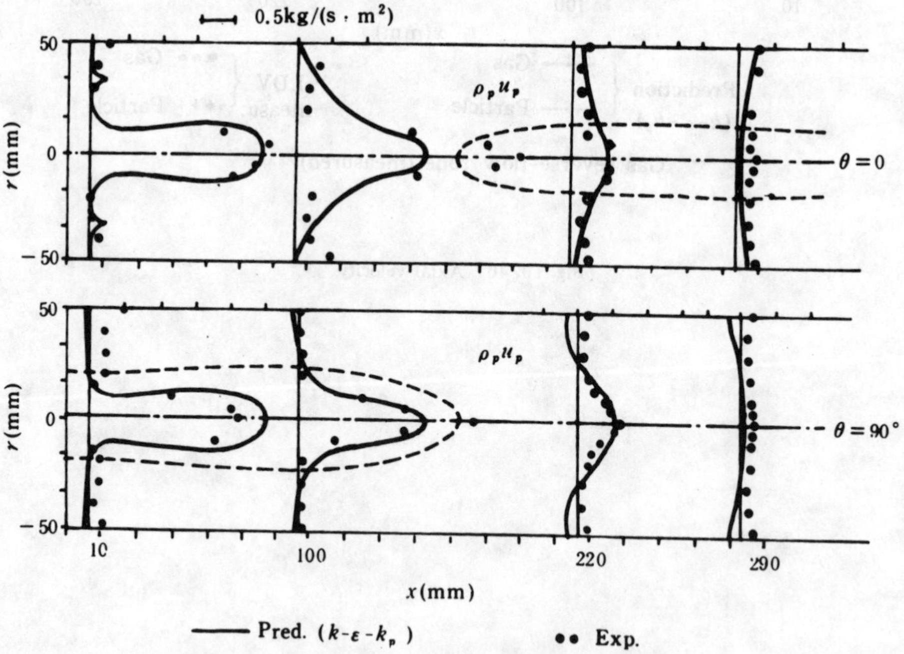

Fig. 10. 48 Particle mass flux

Predicted gas and particle turbulence intensity by using both models (Figs. 10. 44 and 10. 47) are in poor agreement with that measured by LDV. The k-ε-A_p model gives the gas turbulence intensity always larger than the particle turbulence intensity everywhere in the flow field, while experiments indicate that the particle turbulence intensity is larger than gas turbulence intensity in some regions of the flow field. At this point, the k-ε-k_p model gives the tendency nearer to that measured. For example, the particle turbulence intensity given by the k-ε-k_p model is larger than the gas turbulence intensity in the central reverse flow zone at $x = 290$mm, this is in agreement with the experimental results.

In order to further examine two models, the confined gas-particle jet was simulated by using the k-ε-A_p model and the k-ε-k_p model. The prediction results and their comparison with experimental results obtained by Durst et al. and the present author (Figs. 10. 49 to 10. 51) show that the k-ε-k_p model is much better than the k-ε-A_p model. According to the k-ε-k_p model, the turbulent kinetic energy of larger size particles is greater than that of smaller size particles which in turn is greater than that of gas phase (Fig. 10. 49). The k-ε-k_p model gives faster diffusion of 165μm particles than that of 26μm particles which was observed in experiments, while the k-ε-A_p model gives opposite results (Fig. 10. 50). Finally, the k-ε-k_p model gives the particle RMS fluctuation velocity larger than that of gas phase in the flow field of the confined jet everywhere, which is in agreement with the experimental results, but the k-ε-A_p model, once again, gives opposite results (Fig. 10. 51).

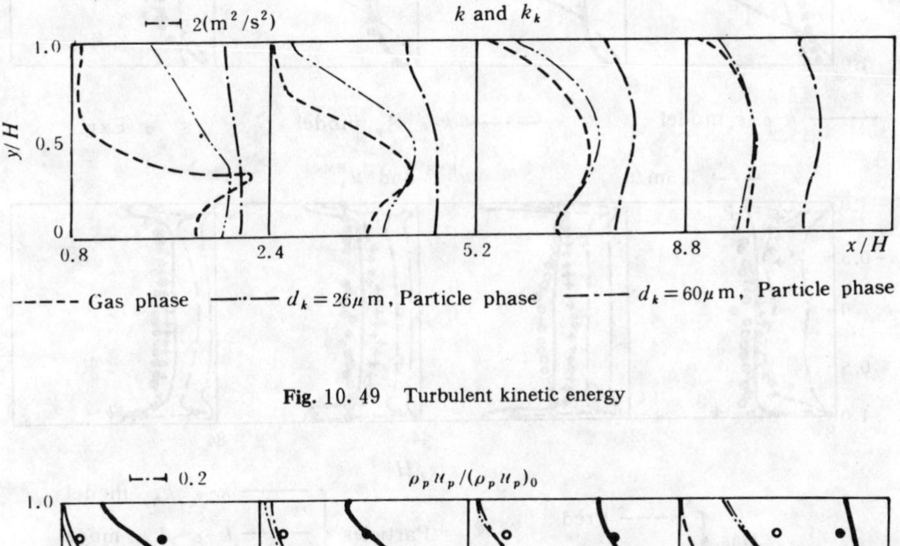

Fig. 10. 49 Turbulent kinetic energy

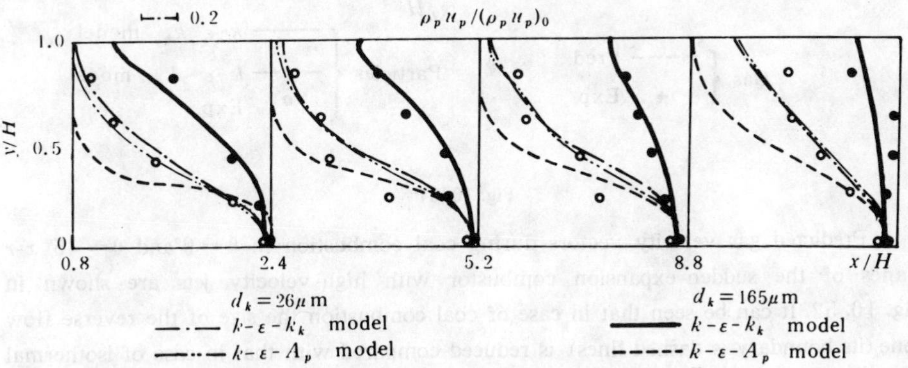

Fig. 10. 50 Particle mass flux

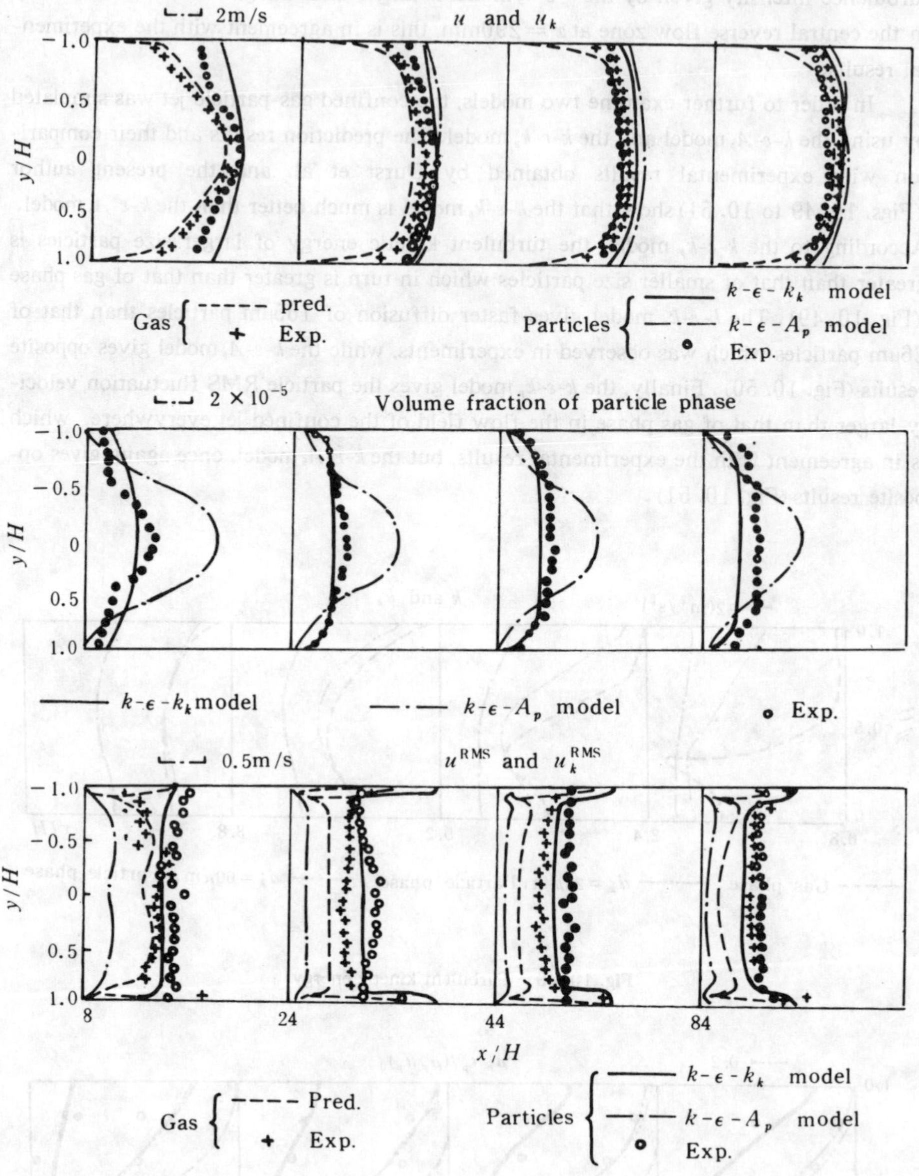

Fig. 10. 51

Predicted gas velocity vectors during coal combustion in θ = 0° and θ = 45° x-r planes of the sudden-expansion combustor with high-velocity jets are shown in Fig. 10. 52. It can be seen that in case of coal combustion the size of the reverse flow zone (its boundary is dashed lines) is reduced compared with that in case of isothermal gas-particle flows (its boundary is dot-dashed lines). Fig. 10. 53 gives the gas temperature field during coal combustion which indicates that ignition takes place right at the stagnation point of the central reverse flow zone. The volatile concentration map (Fig.

10. 54) implies that volatiles complete their combustion nearly at $x = 300$mm where the reverse flow almost disappears. This tells us the important role of the central reverse flow zone in coal ignition and volatile combustion.

Recently many investigators in the United States, England, Australia, Denmark, Germany, Japan and China have been attempting to model the three-dimensional gas-particle flows and coal combustion in boiler furnaces. The present author and his colleagues have simulated the three-dimensional turbulent swirling recirculating gas-particle flows in a tangentially fired boiler furnace by using the algebraic stress model of gas turbulence and continuum model of particle phase with Hinze-Tchen's algebraic model of particle turbulence for the first time. Some of the typical results are shown in Figs. 10. 55—10. 58. Unlike the results obtained by other investigators which gives only the information of particle trajectories with almost no information of particle concentration (by using the stochastic trajectory model) or only the information of particle concentration (by using the no-slip model) with no information of particle velocity, these results give the detailed information of particle velocity and concentration distributions throughout the whole flow field . The gas velocity vectors in vertical planes

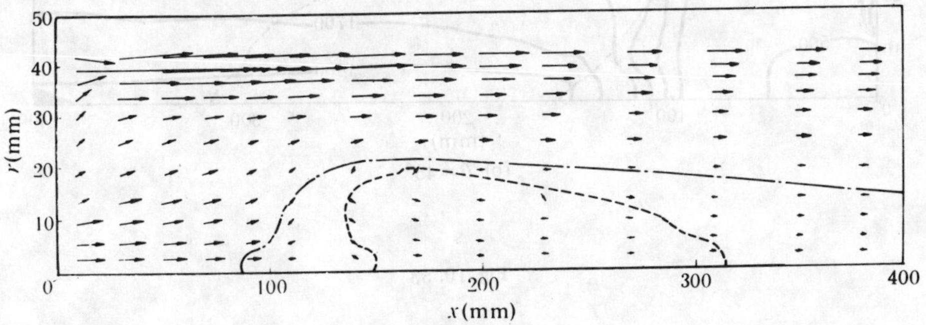

(a) $\theta = 0°$

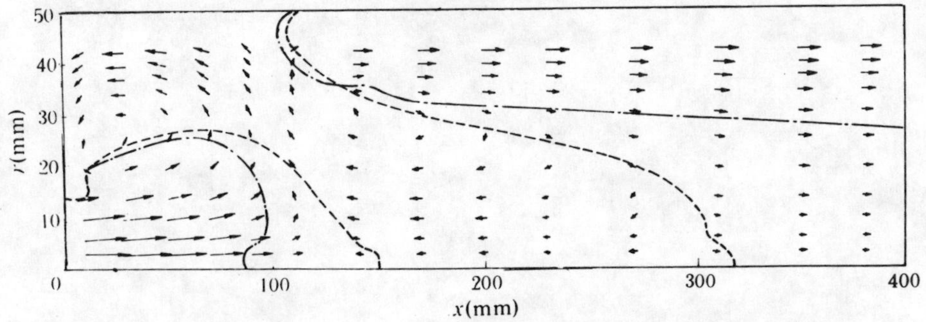

(b) $\theta = 45°$

—·—·— Gas reverse flow boundary (isothermal flow)

- - - - - - Gas reverse flow boundary (coal combustion)

Fig. 10. 52

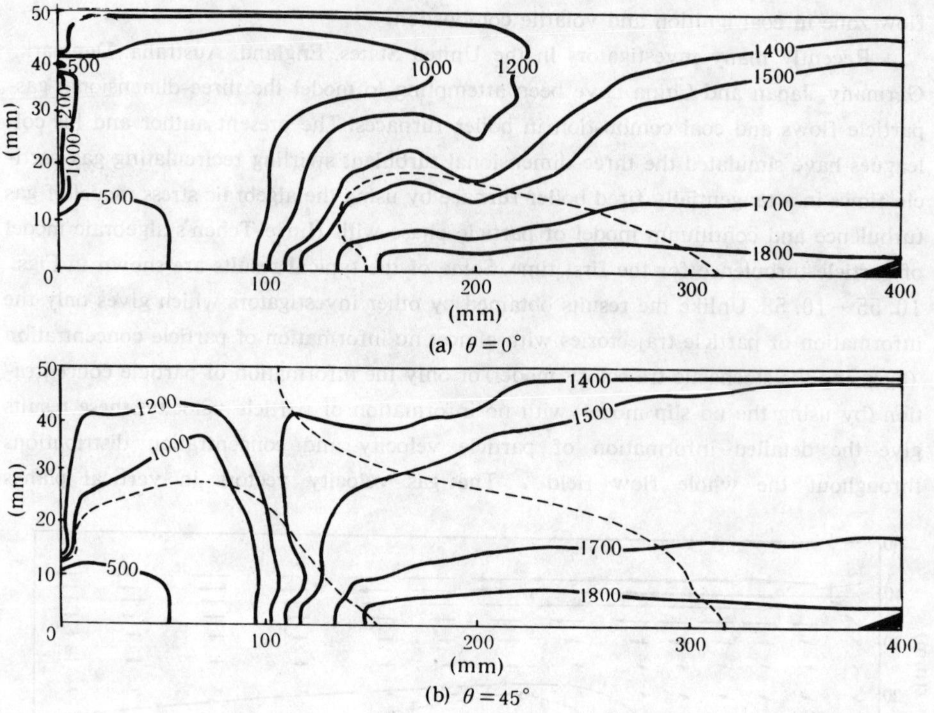

Fig. 10. 53

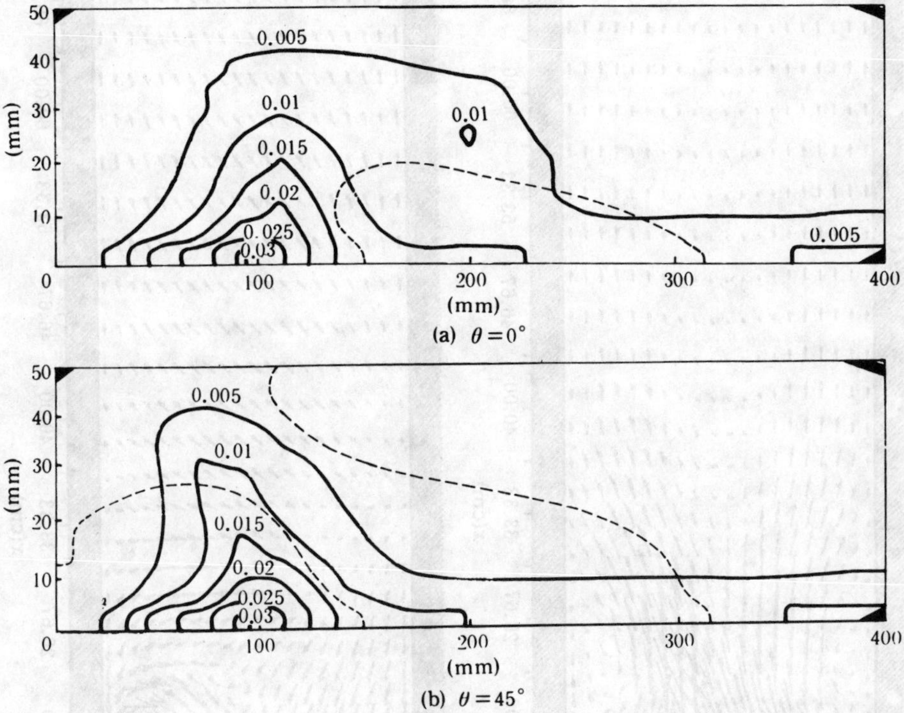

Fig. 10. 54

(Fig. 10. 55) show several near-wall recirculation zones above and below the hight of burners located near four corners and a near-axis central recirculation zone at the hight above burner inlets. The particle velocity is different from the gas velocity both in direction and magnitude due to different inertia of two phases, especially in the near-burner region where the gas flow changes its direction more rapidly. The swirling motion caused by partially tangential inlets is strongest at the hight of burners (Fig. 10. 56) and is much weaker at the hights above burner inlets. The particle concentration distribution (Figs. 10. 57 and 10. 58) shows a ring-shaped concentration peak at the hight of burner inlets caused by impingement of burner jets and this saddle-shaped distribution gradually is deformed into a cup-shaped distribution with slightly higher concentration near the wall due to the effect of centrifugal force and particle turbulent diffusion.

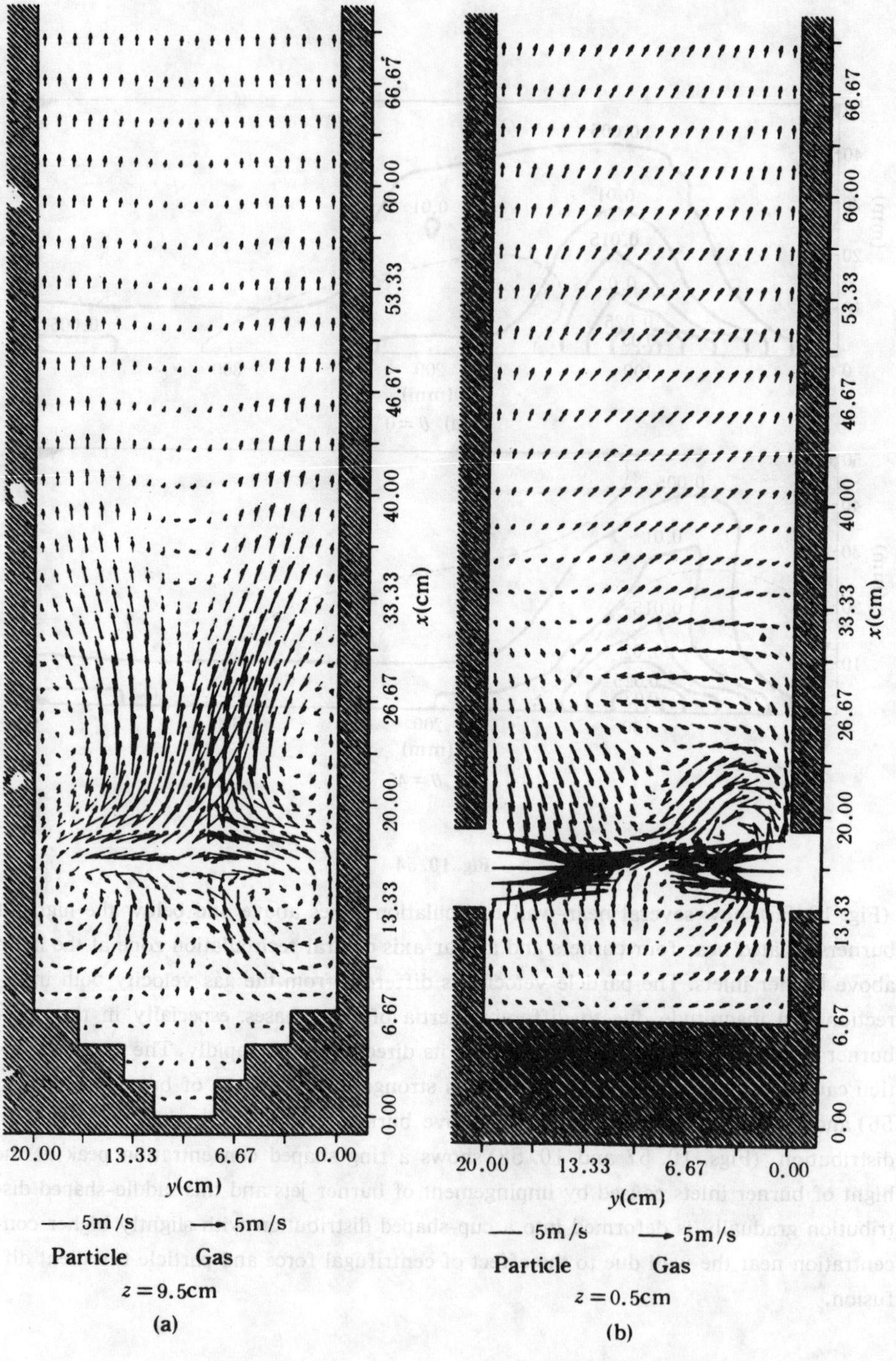

— 5m/s   → 5m/s
Particle   Gas
z = 9.5cm
(a)

— 5m/s   → 5m/s
Particle   Gas
z = 0.5cm
(b)

Fig. 10. 55

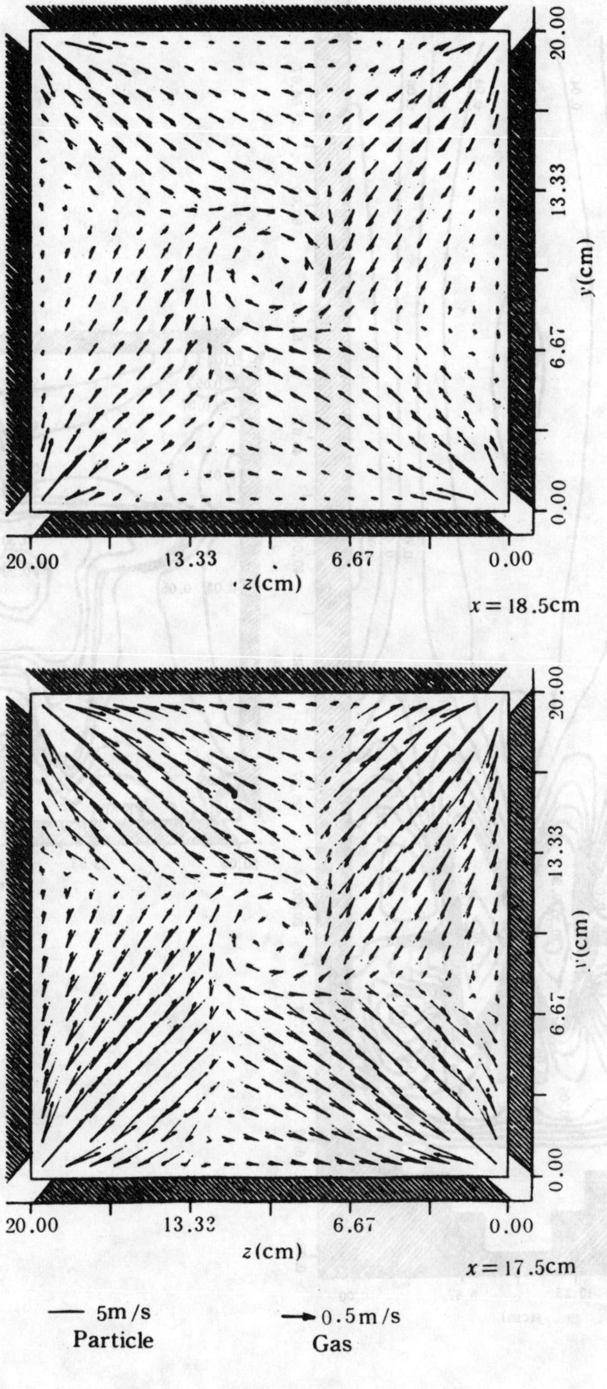
20.00
13.33
6.67
0.00
y(cm)
20.00
13.33
6.67
0.00
z(cm)
x = 18.5cm
20.00
13.33
6.67
0.00
y(cm)
20.00
13.32
6.67
0.00
z(cm)
x = 17.5cm
5m/s
Particle
0.5m/s
Gas

Fig. 10.56

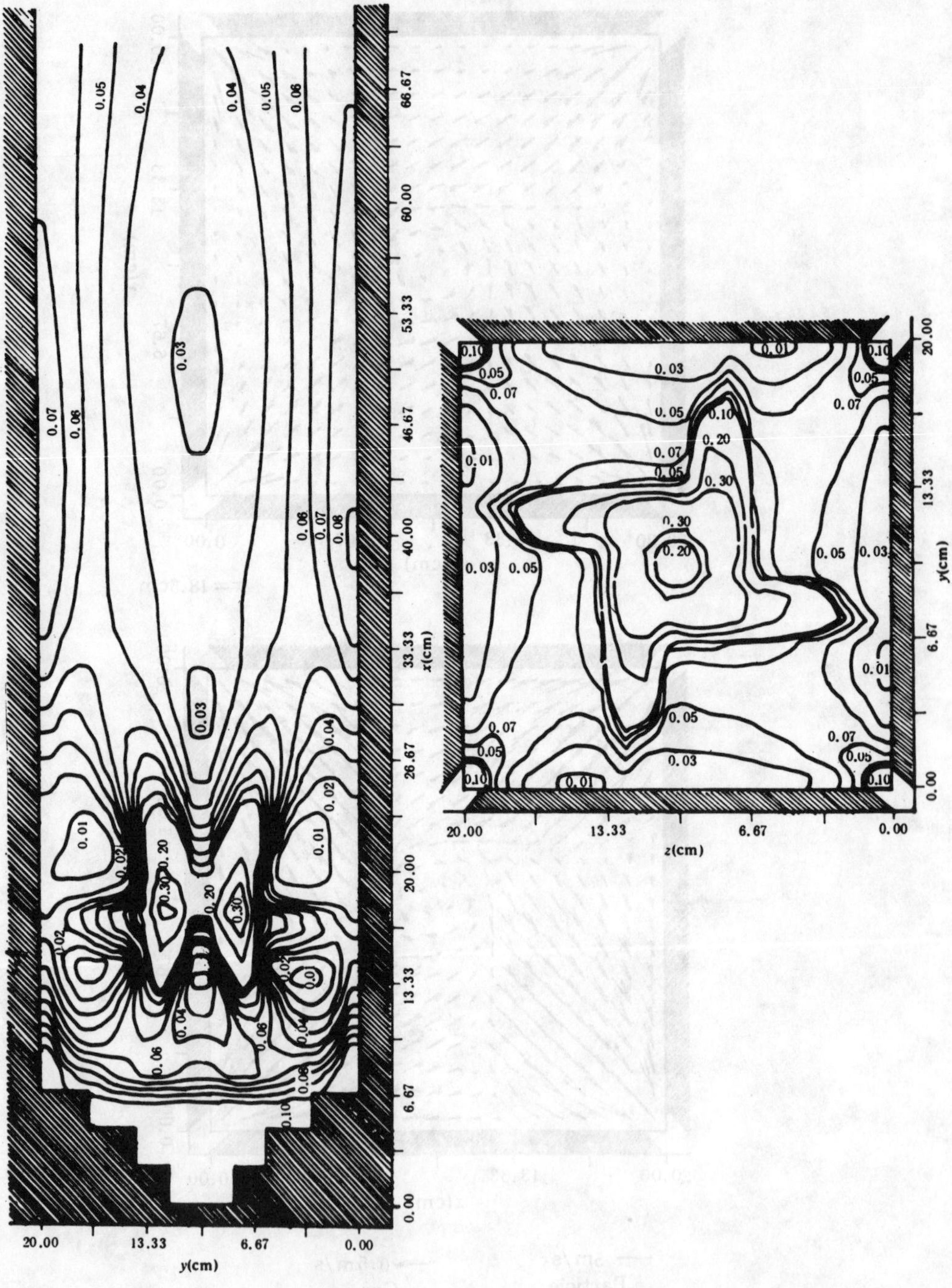

Fig. 10.57

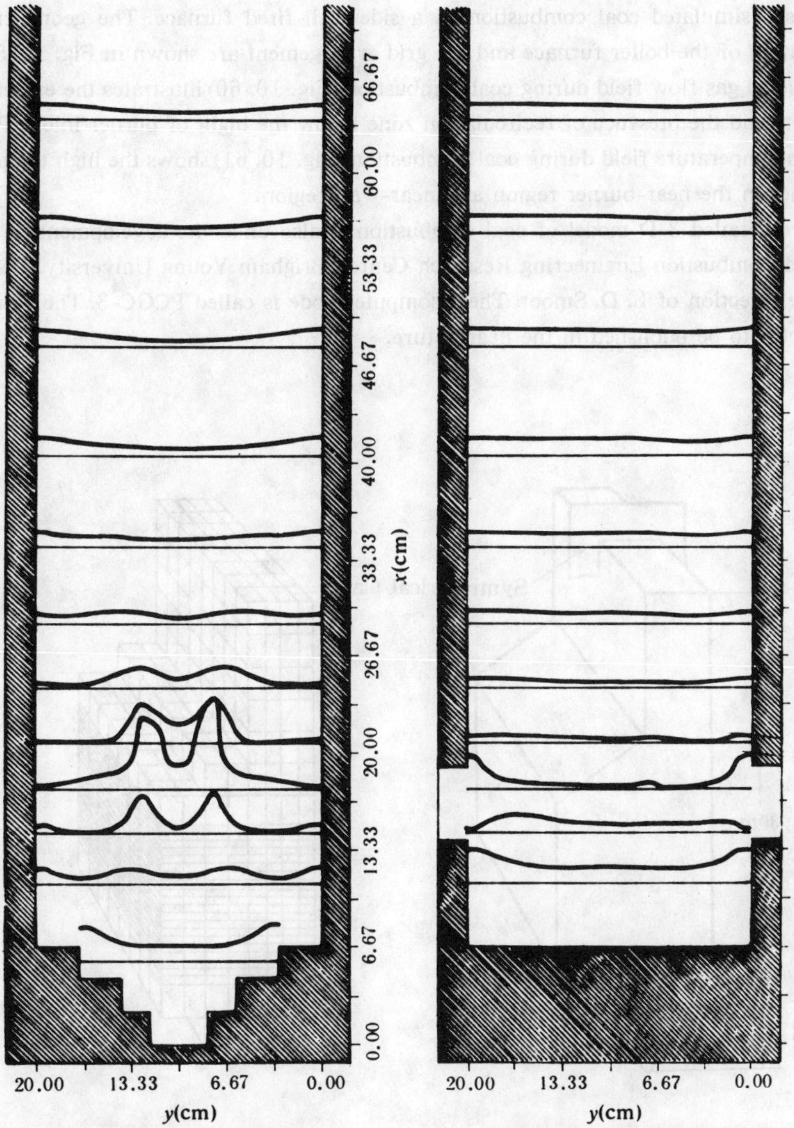

Particle density profiles(%) in x-y plane, $z=0.105$m

(a)

Particle density profiles(%) in x-y plane, $z=0.015$m

(b)

Fig. 10.58

Jensen simulated coal combustion in a side-wall-fired furnace. The geometrical configuration of the boiler furnace and the grid arrangement are shown in Fig. 10. 59. The predicted gas flow field during coal combustion (Fig. 10. 60) illustrates the effect of burner jets and the presence of recirculation zone below the hight of burner inlets. Predicted gas temperature field during coal combustion (Fig. 10. 61) shows the high temperature zones in the near-burner region and near-wall region.

More detailed 3-D model of coal combustion is also under its development in the Advanced Combustion Engineering Research Center, Brigham Young University, USA, under the direction of L. D. Smoot. Their computer code is called PCGC-3. The results are expected to be published in the near future.

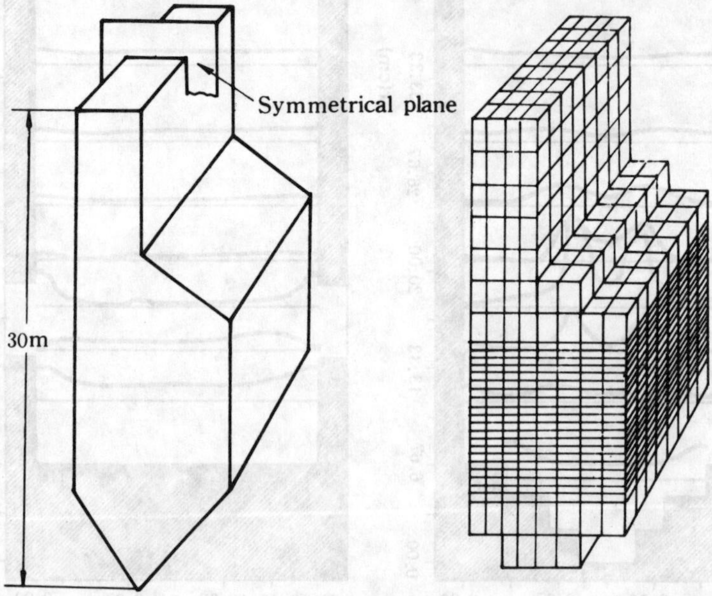

Fig. 10. 59

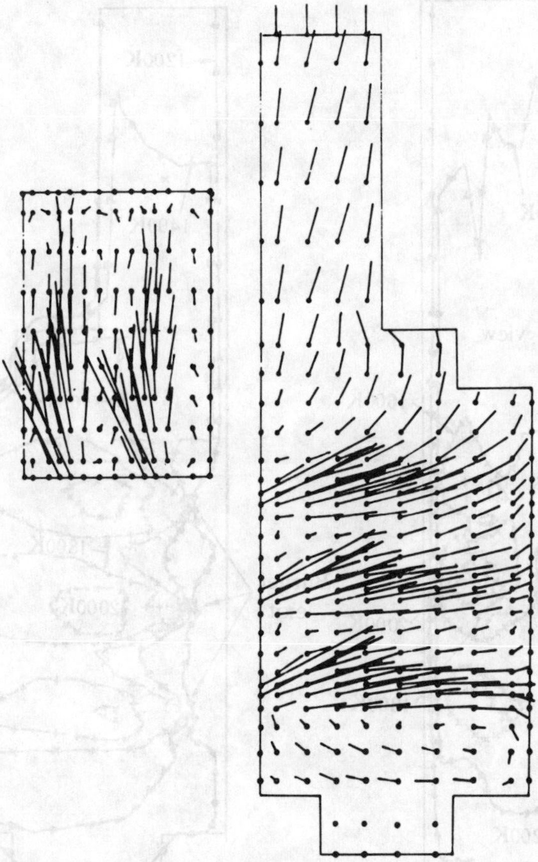

Fig. 10. 60

10. 5 CONCLUDING REMARKS

It can be seen from the discussions in this book that tremendous strides have been made in the theory and modeling of turbulent gas-particle flows and combustion within the last 20 years. Breakthroughs can be found in theory and models, numerical methods and computer codes, and experimental verification of prediction results. Application of modeling results in engineering design has been done on different levels. However, it should be pointed out that the turbulent two-phase flows and combustion are extremely complex processes which involve many physical and chemical processes, such as fluid flow, heat and mass transfer, mass, momentum and energy interaction between phases, gas-phase and particle/droplet-phase turbulence, gas-phase and particle/droplet-phase reactions. Therefore, there are still many problems remaining to be solved. For example, we need to develop more reasonable and economical particle-phase global models, particle turbulence models, gas turbulence models for very strongly swirling and recir-

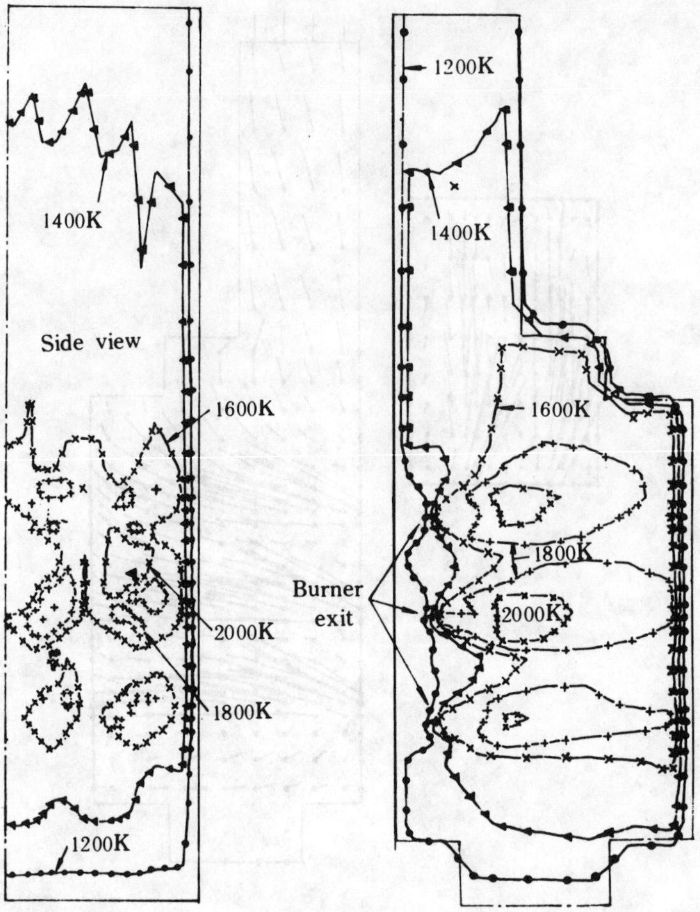

Fig. 10. 61

culating flows, models for reacting particle phases, and gas combustion models in two-phase flows. Besides, it is necessary to further study the reliable and economical numerical methods for solving 3-D two-phase reacting flow problems. Problems, such as boundary conditions of particle phase accounting for deposition/erosion, need to be studied. Finally, it is worthwhile to note that experimental verification of submodels (turbulence model, combustion model, two-phase model, ...) is of vital importance. What we need is the detailed measurements of the two-phase flow field which becomes possible due to the recent development of various laser diagnostics. In this aspect, in the personal opinion of the present author, special attention should be paid to the work done by F. Durst at Nürnberg-Erlangen University on verification of two-phase flow models (special international workshop on two-phase flow predictions is held every two years) and the work done by L. D. Smoot at Brigham Young University on verification of each submodel in the comprehensive model of coal combustion (the Computational Laboratory of Advanced Combustion Engineering Research Center is the largest one working on modeling of gas-particle flows and coal combustion). In these research institutes or laboratories, the theory, numerical modeling, laser measurements and engineering application are successfully combined together, allowing them to systematically obtain fruitful research results.

In spite of many existing difficulties, it is undoubted that the theory and modeling of turbulent gas-particle flows and combustion will certainly be fully applied in engineering design in the not too far future.

NOMENCLATURE

A Area
a Sound velocity; thermodiffusivity
B Preexponential factor; transfer number
c Empirical constants;
c_d Drag coefficient
c specific heat
d Diameter
D Diffusivity; width
E Activation energy
e Internal energy
F Force; area
f Mixture fraction
G Total mass flux; mass flow rate
g Gravitational acceleration; mass flux; mean square value of concentration fluctuation
G_k Mean production
G_b Bouyancy production
H Stagnant enthalpy; hight
h Enthalpy; hight;
J Diffusion flux
k Turbulent kinetic energy; rate coefficient
K Evaporation constant; Equilibrium constant
$\mathscr{L}$ Latent heat
L Length; air/fuel ratio
l Turbulent scale; length
Le Lewis number
M Molecular weight
m Mass
N Total particle number flux
n Particle number density; molar density; reaction order; index in Rosin-Rammler distribution
Nu Nusselt number
p Pressure; probability density distribution function
Pe Peclet number
Pr Prandtl number
Q Heat; heating effect
q Heat flux
R Universal gas constant; weight fraction of particles
r Radius
Re Reynolds number
Ri Gradient Richardson number
R_f Flux Richardson number
S Flame propagation velocity; source term; surface
s Particle diameter square
S Deformation rate tensor
Sc Schmidt number
Sh Sherwood number
T Temperature
t Time
u, v, w Velocity components
V Volume; drift velocity
w Reaction rate
x, y, z Coordinates
X Molar fraction
Y Mass fraction
Z Combined mass fraction

GREEK ALPHABETS

α Coefficient of excess air
β Stoichiometric ratio
γ Specific heat ratio
δ Thickness

Γ Transport coefficient

μ Dynamic viscosity

η, ξ Dimensionless coordinates

η Mixture fraction

θ Dimensionless temperature

θ Dimensionless temperature

λ Thermoconductivity

ε Turbulent kinetic energy dissipation rate; reactiveness; emissivity

ν Kinematic viscosity

ρ Density

σ Stefan-Boltzmann constant; Prandtl number or Schmidt number

φ Generalized dependent variable

Φ Volume fraction; dissipation energy

τ Shear stress; time scale

ψ Stream function

SUBSCRIPTS

A Air; ash

c Raw coal; convection; reaction

ch Reaction; char

cr Critical

d Diffusion

e Effective; exit; evaporation

E Extinction

F, fu Fuel

f Flame; fluid

g Gas

h Char; heterogeneous

hr Heterogeneous reaction

i Ignition

i, in Initial, inlet

iner Inert gas

i, j, k Coordinate directions

k k - th particle phase

l Laminar; liquid

m Mixture; mean value; maximum

n Normal

0 Initial

ox Oxygen; oxidizer

p Particle; primary

pr Combustion products

r Radiation

s s -species; surface; secondary

st Steady

t Tangential

T Turbulent

v Volatile

w Wall; moisture

∞ Infinity; oncoming flow

SUPERSCRIPTS

— Time-averaged

~ Favre-averaged

$'$, $''$ Fluctuation component

$*$ Characteristic value

• Changing rate in time

REFERENCES

1. El Banhawy, H and Whitelaw. J. H., *AIAA J*, 18, 12, 1053 (1980).
2. Boysan, F., Ayers, W. H., Swithenbank, J. and Pan, Z., Model of Spray Combustion in Gas Turbine Combustors, Rep. BIC 354, Univ. of Sheffield, 1980.
3. Boysan, F., Ayers, W. H. and Swithenbank, *J., Trans. Instn. Chem. Engrs.*, **60**, 222—230 (1982).
4. Bradshaw, P., An Introduction to Turbulence and Its Measurements, Pergamon Press, 1971.
5. Chapman, S. and Cowling, T. G., The Mathematical Theory of Non-uniform Gases, Cambridge Univ. Press, 1972.
6. Crowe, C. T., The State-of-the-art in the Development of Numerical Models for Dispersed Two-phase Flows, Proc. Inter. Conf. on Multiphase Flows, '91-Tsukuba, **3**, 49—60, 1991.
7. Crowe, C. T. et al., *J. Fluid Eng.*, **99**, 2, 325 (1977).
8. Dooley, D. A., Proc. Heat Transfer and Fluid Mechanics Institute, Stanford Univ., 321, 1957.
9. Emmons, H. W., *Z. Angew. Math. Mech.*, **36**, 60, 1956.
10. Field, M. A. et al., Combustion of Pulverized Coal, BCURA, 1967.
11. Frank-Kamenetskii, D. A., Diffusion and Heat Exchange in Chemical Kinetics, Princeton Univ. Press, 1955.
12. Fuchs, N. A., Mechanics of Aerosols, Macmillan, 1964.
13. Gibson, M. M. and Morgan, B. B., *J. Inst. Fuel*, **43**, 517 (1970).
14. Glassman, I., Combustion, Academic Press, 1977.
15. Gutheil, E., Bockhorn, H. and Fetting, F., Elements of Modeling of Turbulent Diffusion Flames, Proc. Inter. Symp. on Coal Comb., Hemisphere, 181—188, 1988.
16. Hinze, J. O., Turbulence, McGraw-Hill, 1959.
17. Hong, T. and Zhou, L. X., Numerical Simulation of 3-D Gas-particle Flows in Boiler Furnace by a Continuum Model of Particle Phase, Proc. First Asian-Pacific Symp. on Comb. & Energy Utiliz., Inter. Acad. Publishers, 184—189, 1990.
18. Huang, X. Q., and Zhou, L. X., Simulation of 3-D Turbulent Recirculating Gas-particle Flows by an Energy-equation Model of Particle Turbulence, FED-v. 121, Gas-solid Flows, ASME, 261—265, 1991.
19. Huang, X. Q., Chen, L. H. and Zhou, L. X., Numerical Simulation of 3-D Turbulent Recirculating Gas-particle Flows in a Combustor of Co-flow Jets with Large Velocity Difference, Proc. 3rd Inter. Symp. of Gas-solid Flows, La Jolla, CA, 85—88, 1989.
20. Kanury, A. M., Introduction to Combustion Phenomena, Gordon and Beach, 1977.
21. Khalil, E. E., Modeling of Furnaces and Combustors, Abacus Press, 1982.
22. Khitrin, L. N. and Goldenberg, S. A., Proc. Sixth Inter. Symp. on Combu., 545, 1957.
23. Khitrin, L. N., Physics of Combustion and Explosion, National Science Foundation, 1962.
24. Knorre, G. F., Combustion Processes (in Russian), Gosenergoizdat, Moscow, 1961.
25. Landau, L. and Lifshitz, E., Continuum Mechanics (in Russian), Gostechizdat, Moscow, 1944.
26. Launder, B. E. and Spalding, D. B., Mathematical Models of Turbulence, Academic Press, 1972.
27. Levich, V. G., Physico-chemical Hydrodynamics (in Russian), Gostechizdat, Moscow, 1959.
28. Lewis, B. and Elbe, Von, Combustion, Flames and Explosion of Gases, Academic Press, 1961.
29. Libby, P. A. and Williams, F. A. (Edit.), Turbulent Reacting Flows, Springer-Verlag, 1979.
30. Loitsiangsky, L. G., Laminar Boundary Layers (in Russian), Fizmatgiz, Moscow, 1962.
31. Loitsiangsky, L. G., Mechanics of Fluids and Gases (in Russian), Gostechizdat, Moscow, 1957.
32. Ni, H. Q., Wang, N. J. and Zhou, L. X., Acta Mechanica Sinica (in Chinese), **21**, 1, 26—33 (1989).
33. Paleev, I. I. et al., Theory of Combustion Processes (in Russian), Energia, Leningrad, 1966.
34. Penner, S. S., Introduction to the Study of Chemical Reactions in Flow Systems, Butterworths, 1955.
35. Pomerantsev, V. V., Fundamentals of Applied Combustion Theory (in Russian), Energia, Leningrad, 1973.
36. Predvoditelev, A. S., Carbon Combustion (in Russian), Izd. ANSSSR, 1949.
37. Rodi, W., Turbulence Models in Environmental Problems, "Prediction Methods for Turbulent Flows", ed. by Kollman, W., Hemisphere, 1980.

38. Schlichting, H., Boundary-layer Theory, McGraw-Hill, 1979.
39. Shetinkov, E. S., Physics of Gas Combustion (in Russian), Nauka, Moscow, 1965.
40. Smoot, L. D. and Smith, P. J., Coal Combustion and Gasification, Plenum Press, 1985.
41. Soo, S. L., Multiphase Fluid Dynamics, Science Press (Beijing) and Gower Technical (U. S.), 1990.
42. Spalding, D. B., *J. Combus. Sci. & Tech.*, **13**, 1—6, 3, 1976.
43. Spalding, D. B., HTS/76/6, Imper. College of Sci. & Tech., 1976.
44. Spalding, D. B., Mathematical models of Continuous Combustion, "Emissions from Continuous Combustion Systems". Plenum Press, 1972.
45. Spalding, D. B., Some Fundamentals of Combustion, Butterworths, 1955.
46. Stambuleanu, A., Flame Combustion Processes in Industry, Abacus Press, 1976.
47. Tennekes, H. and Lumley, J. L., A First Course to Turbulence, MIT Press, 1972.
48. Toong, T. Y., Combustion Dynamics: The Dynamics of Chemically Reacting Fluids, McGraw-Hill, 1983.
49. Vulis, L. A., Thermal Regimes of Combustion, McGraw-Hill, 1961.
50. Vulis, L. A., Proc. 3rd All-Union Symp. on Combustion Theory (in Russian), Moscow, 1960.
51. Wang, Y. S., Fan W. C., Zhou, L. X. and Xu, X. C., Numerical Computation of Combustion Processes (in Chinese), Science Press, Beijing, 1986.
52. Williams, F. A., Combustion Theory, Addison Wesley, 1965.
53. Zeldovich, Ya. B., The Mathematical Theory of Combustion and Explosions, McNeill, 1985.
54. Zhou, L. X. and Soo S. L., *J. Powder Tech.*, **64**, 213—220 (1991).
55. Zhou, L. X. and Soo S. L., *Acta Mechanica Sinica*, English Edition, 7, 4, 309—315 (1991).
56. Zhou, L. X., Numerical Modeling of Turbulent Gas-particle Flows and Combustion (in Chinese), Tsinghua Univ. Press, 1991.
57. Zhou, L. X., An Algebraic Stress/Flux/Reaction-rate Model of Gas-particle Turbulence and Turbulent Combustion for Reacting Gas-particle Flows, Proc. 2nd Inter. Symp. on Coal Comb., China Machine Press, 90—97 (1991).
58. Zhou, L. X. and Soo, S. L., *J. Powder Tech.*, **63**, 45—53 (1990).
59. Zhou, L. X. and Huang, X. Q., Science in China, English Edition, **33**, 1, 52—59 (1990).
60. Zhou, L. X., Nieh, S. and Yang, G., *J. Eng. Thermophysics*, English Edition, **2**, 2, 231—241 (1990).
61. Zhou, L. X., *J. Chem. Ind. & Eng.*, English Edition, **4**, 2, 240—247 (1989).
62. Zhou, L. X., A Multi-fluid Model of Two-phase Flows with Pulverized-coal Combustion, Proc. Inter. Symp. on Coal Combus., Hemisphere, 173—179 (1988).
63. Zhou, L. X., Lin, W. Y., Zhang, J. and Wang, Z. L., Numerical Modeling of 3-D Flow Field and 2-D Coal Combustion in a Combustor of Coflow Jets with Large Velocity Difference, Proc. 21st Inter. Symp. on Combu., 257—264 (1987).
64. Zhou, L. X. and Zhang, J., A Lagrangian-Eulerian Particle Model for Turbulent Two-phase Flows with Reacting Particles, Proc. 10th Inter. Conf. on Num. Meth. in Fluid Dyn., Springer-Verlag, 705—709, 1986.
65. Zhou, L. X., Combustion Theory and Dynamics of Reacting Fluids (in Chinese), Science Press, Beijing, 1986.
66. Zhou, L. X., Lin, W. Y. and Jiang, Z., Numerical Modeling of Evaporating Spray Two-phase Flows Caused by Opposite Injection of a Swirl Atomizer in a Duct, *Combus. Sci. & Tech.* (China) (in Chinese), **3**, 56—64 (1984).
67. Zhou, L. X. and Zhang, J., *J. Eng. Thermophysics* (in Chinese), **5**, 4, 396—401 (1984).
68. Zhou, L. X., Advances in Mechanics (China) (in Chinese), **12**, 2, 141—150 (1982).
69. Zhou, L. X., A Multicontinuum Model of Multiphase Flows with Reacting Particles in Eulerian Coordinates, Proc. 2nd Chinese Symp. on Multiphase Fluid Dynamics, Beijing, 1, paper 007, 1982.
70. Zhou, L. X., The Similarity Parameters and Combustion Efficiency Analysis of Liquid Spray Combustion, Proc. Symp. on Experim. Studies of Combu. in Gas-turbine Combustors (in Chinese), China Engineering Press, 1965.
71. Zhou, L. X., Evaporation and Combustion of Single Liquid Droplets and Liquid Spray, Ph. D. dissertation, Department of Physics and Mechanics, Leningrad Polytechnical Institute, USSR, 1961.

INDEX

R

S

T